# Gulliver in the Country of Lilliput

# Gulliver in the Country of Lilliput

## An Interplay of Noncovalent Interactions

Editor

**Ilya G. Shenderovich**

MDPI • Basel • Beijing • Wuhan • Barcelona • Belgrade • Manchester • Tokyo • Cluj • Tianjin

*Editor*
Ilya G. Shenderovich
University of Regensburg
Germany

*Editorial Office*
MDPI
St. Alban-Anlage 66
4052 Basel, Switzerland

This is a reprint of articles from the Special Issue published online in the open access journal *Molecules* (ISSN 1420-3049) (available at: https://www.mdpi.com/journal/molecules/special_issues/Interplay_Noncovalent_Interactions).

For citation purposes, cite each article independently as indicated on the article page online and as indicated below:

LastName, A.A.; LastName, B.B.; LastName, C.C. Article Title. *Journal Name* **Year**, *Volume Number*, Page Range.

**ISBN 978-3-0365-0430-8 (Hbk)**
**ISBN 978-3-0365-0431-5 (PDF)**

Cover image courtesy of Ilya G. Shenderovich.

© 2021 by the authors. Articles in this book are Open Access and distributed under the Creative Commons Attribution (CC BY) license, which allows users to download, copy and build upon published articles, as long as the author and publisher are properly credited, which ensures maximum dissemination and a wider impact of our publications.

The book as a whole is distributed by MDPI under the terms and conditions of the Creative Commons license CC BY-NC-ND.

# Contents

About the Editor . . . . . . . . . . . . . . . . . . . . . . . . . . . . . . . . . . . . . . . . . . . . . . . . . . . . . . . . . . . . vii

**Editorial to the Special Issue "Gulliver in the Country of Lilliput: An Interplay of Noncovalent Interactions"**
Editorial to the Special Issue "Gulliver in the Country of Lilliput: An Interplay of Noncovalent Interactions"
Reprinted from: *Molecules* 2021, 26, 158, doi:10.3390/molecules26010158 . . . . . . . . . . . . . . . 1

**Lucija Hok, Janez Mavri and Robert Vianello**
The Effect of Deuteration on the $H_2$ Receptor Histamine Binding Profile: A Computational Insight into Modified Hydrogen Bonding Interactions
Reprinted from: *Molecules* 2020, 25, 6017, doi:10.3390/molecules25246017 . . . . . . . . . . . . . . . 5

**Charlotte Zimmermann, Taija L. Fischer and Martin A. Suhm**
Pinacolone-Alcohol Gas-Phase Solvation Balances as Experimental Dispersion Benchmarks
Reprinted from: *Molecules* 2020, 25, 5095, doi:10.3390/molecules25215095 . . . . . . . . . . . . . . . 23

**Valentine G. Nenajdenko, Namiq G. Shikhaliyev, Abel M. Maharramov, Khanim N. Bagirova, Gulnar T. Suleymanova, Alexander S. Novikov, Victor N. Khrustalev and Alexander G. Tskhovrebov**
Halogenated Diazabutadiene Dyes: Synthesis, Structures, Supramolecular Features, and Theoretical Studies
Reprinted from: *Molecules* 2020, 25, 5013, doi:10.3390/molecules25215013 . . . . . . . . . . . . . . . 37

**Kinga Jóźwiak, Aneta Jezierska, Jarosław J. Panek, Eugene A. Goremychkin, Peter M. Tolstoy, Ilya G. Shenderovich and Aleksander Filarowski**
Inter- vs. Intramolecular Hydrogen Bond Patterns and Proton Dynamics in Nitrophthalic Acid Associates
Reprinted from: *Molecules* 2020, 25, 4720, doi:10.3390/molecules25204720 . . . . . . . . . . . . . . . 51

**Paweł A. Wieczorkiewicz, Halina Szatylowicz and Tadeusz M. Krygowski**
Mutual Relations between Substituent Effect, Hydrogen Bonding, and Aromaticity in Adenine-Uracil and Adenine-Adenine Base Pairs
Reprinted from: *Molecules* 2020, 25, 3688, doi:10.3390/molecules25163688 . . . . . . . . . . . . . . . 73

**Alexander P. Voronin, Artem O. Surov, Andrei V. Churakov, Olga D. Parashchuk, Alexey A. Rykounov and Mikhail V. Vener**
Combined X-ray Crystallographic, IR/Raman Spectroscopic, and Periodic DFT Investigations of New Multicomponent Crystalline Forms of Anthelmintic Drugs: A Case Study of Carbendazim Maleate
Reprinted from: *Molecules* 2020, 25, 2386, doi:10.3390/molecules25102386 . . . . . . . . . . . . . . . 93

**Changming Ke and Zijing Lin**
Catalytic Effect of Hydrogen Bond on Oxyhydryl Dehydrogenation in Methanol Steam Reforming on Ni(111)
Reprinted from: *Molecules* 2020, 25, 1531, doi:10.3390/molecules25071531 . . . . . . . . . . . . . . . 113

**Alexei S. Ostras', Daniil M. Ivanov, Alexander S. Novikov and Peter M. Tolstoy**
Phosphine Oxides as Spectroscopic Halogen Bond Descriptors: IR and NMR Correlations with Interatomic Distances and Complexation Energy
Reprinted from: *Molecules* 2020, 25, 1406, doi:10.3390/molecules25061406 . . . . . . . . . . . . . . . 123

**Ilya G. Shenderovich and Gleb S. Denisov**
Adduct under Field—A Qualitative Approach to Account for Solvent Effect on
Hydrogen Bonding
Reprinted from: *Molecules* **2020**, *25*, 436, doi:10.3390/molecules25030436 . . . . . . . . . . . . . . **141**

**Sławomir J. Grabowski**
Hydrogen Bond and Other Lewis Acid–Lewis Base Interactions as Preliminary Stages of
Chemical Reactions
Reprinted from: *Molecules* **2020**, *25*, 4668, doi:10.3390/molecules25204668 . . . . . . . . . . . . . . **157**

**Gerd Buntkowsky and Michael Vogel**
Small Molecules, Non-Covalent Interactions, and Confinement
Reprinted from: *Molecules* **2020**, *25*, 3311, doi:10.3390/molecules25143311 . . . . . . . . . . . . . . **175**

# About the Editor

**Ilya G. Shenderovich** (Dr.Nat.Sci.) received a BSc degree in Physics (1993, Chemical Physics) and an MSc degree in Physics (1995, Physics of Condensed Matter) from Sankt-Petersburg State University, Russia (Mentor: Dr. G. N. Kuz'min). He became a Candidate of Science in Physics and Mathematics (PhD) in 1999 (Topic: Manifestation of Covalency, Cooperativity, and Symmetry of Strong Hydrogen Bonds in NMR Spectra; Mentor: Prof. Dr. G.S. Denisov) and a Doctor of Science in Physics and Mathematics in 2011 (Topic: Study of Hydrogen Bonds in Amorphous Materials and at Interfaces by NMR). He progressed in his research career under the guidance of Prof. Dr. H.-H. Limbach at the Freie Universität Berlin. He runs the NMR department of the Faculty of Chemistry and Pharmacy at the Universität Regensburg. His main research interests focus on noncovalent interactions in condensed matter. His main research methods are NMR spectroscopy and model DFT calculations. The list of his publications is available at publons.com/researcher/636272/ilya-g-shenderovich and https://www.scopus.com/authid/detail.uri?authorId=6701593020.

*Editorial*

# Editorial to the Special Issue "Gulliver in the Country of Lilliput: An Interplay of Noncovalent Interactions"

Ilya G. Shenderovich

Institute of Organic Chemistry, Faculty of Chemistry and Pharmacy, University of Regensburg, Universitaetstrasse 31, 93053 Regensburg, Germany; Ilya.Shenderovich@ur.de

**Citation:** Shenderovich, I.G. Editorial to the Special Issue "Gulliver in the Country of Lilliput: An Interplay of Noncovalent Interactions". *Molecules* **2021**, *26*, 158. https://doi.org/10.3390/molecules26010158

Received: 22 December 2020
Accepted: 30 December 2020
Published: 31 December 2020

**Publisher's Note:** MDPI stays neutral with regard to jurisdictional claims in published maps and institutional affiliations.

**Copyright:** © 2020 by the author. Licensee MDPI, Basel, Switzerland. This article is an open access article distributed under the terms and conditions of the Creative Commons Attribution (CC BY) license (https://creativecommons.org/licenses/by/4.0/).

Noncovalent interactions allow our world to exist. Their study remains vital to the progress of chemistry and chemical physics. This topic has been specifically addressed in a number of the past and present Special Issues of *Molecules*, and almost every publication touches on the subject of noncovalent interactions in one way or another.

The overarching goal of this Special Issue was to bring together publications that consider effects caused by an interplay of noncovalent interactions. A common case is the situation when there is one dominant interaction that determines the structure of the molecular system, and a large number of weaker interactions that are forced to adapt to this structure. Although it is clear that this "Gulliver in the Country of Lilliput" model is only a rough approximation, this view implicitly prompts the assumption that the net effect of weak interactions is negligible, at least on the structure of the system, since multiple small contributions can cancel each other out. In some cases, this may turn out to be true. However, since the total strength of the "Lilliputians" can exceed that of the "Gulliver", a priori conclusions can lose their predictive power. The dominant interaction can be strong and protected from any direct competition, as in the case of the proton-bound homodimer of pyridine, but the geometry of this complex still differs in the gas, solution, and solid phases [1]. Similarly, the result of competition between two strong interactions can be determined by weak interactions [2]. The current challenge is to learn to incorporate multiple competing interactions into effective working models.

The contributions in this Special Issue can be grouped into three thematic areas: (i) specific properties of selected interactions evaluated for bi- or trimolecular complexes, [3–6] (ii) their role in the crystal packing [7,8], chemical [9,10] and enzymatic [11] reactions as well as (iii) manifestations of competing noncovalent interactions in solution [12] and into porous materials [13].

The experimentally challenging study of Suhm et al. [3] analyses the docking preference of alcohols between the two nonequivalent lone electron pairs of the carbonyl group in pinacolone using supersonic jet expansions of 1:1 solvate complexes. The result of an interplay between the nonequivalence of the lone electron pairs and distant London dispersion and Pauli repulsion was modulated by the size of the alkyl group of the alcohol. The obtained experimental results serve as extremely high level benchmarks for verifying the accuracy of theoretical methods. Some of these methods were tested. It is suggested to note the importance of London dispersion for structure and stability of molecular aggregates [14].

The paper by Szatylowicz et al. [4] describes the effect of substituents on the energy of specific noncovalent interactions in adenine-based bimolecular complexes and the aromaticity of the partners. Understanding these effects is essential for effective control of acid-base interactions in biochemistry, where stronger does not always mean better. Special attention was payed to the comparison of the energies obtained using different models. The reader may be interested in further reading on this subject [15].

The contribution by Filarowski et al. [5] reports on the balance of repulsive and attractive intramolecular interactions between adjacent carboxyl groups in selectively substituted phthalic acids, the dependence of this balance on intramolecular steric crowding, and the effect of these intramolecular properties on intermolecular interactions of these carboxyl groups. This study represents a combination of the Infrared, Raman, Nuclear Magnetic Resonance (NMR), and Incoherent Inelastic Neutron Scattering spectroscopies and the Car–Parrinello Molecular Dynamics and Density Functional Theory calculations. The structural and energetic parameters of the intra- and intermolecular interactions were estimated for the gas, liquid, and solid phases. Note that despite the originality of the outcomes arising from neutron scattering, the number of molecular systems studied by these methods is limited due to the complexity of the experimental equipment required [16,17].

The paper by Tolstoy et al. [6] describes halogen-bonded complexes formed by trimethylphosphine oxide with 128 different halogen donors of various classes. Correlations between the energetic, geometric and spectral properties of these complexes were established and summarized. These correlations make it possible to estimate the energy and geometry of a given halogen bond from the corresponding experimentally measured spectral parameter. Both the halogen bonding and the acceptor properties of the P=O moiety [18] are of considerable current interest. Therefore, the reported correlations can be very useful.

Vener et al. [7] describe how the calculated structural and spectral parameters of two component crystals of organic salts depend on various parameters of the theoretical approximation used. The experimental parameters of such systems cannot be correctly reproduced using the approximation of small molecular clusters. Using the periodic density functional theory is a reasonable compromise that allows the parameters of complex multicomponent pharmaceuticals to be accurately predicted [19].

The paper by Nenajdenko, Tskhovrebov et al. [8] demonstrates that halogen-halogen interactions play a critical role in self-assembly of highly polarizable molecules in crystals. A series of novel halogenated aromatic dichlorodiazadienes were prepared and characterized using X-ray diffraction and Bader's Theory of Atoms in Molecules. Although halogen-halogen interactions are not strong, they can be a tool for fine-tuning the crystal structure [20].

The review by Grabowski [9] highlights the role of noncovalent interactions as a preliminary stage of chemical reactions. Hydrogen bonding assisted proton transfer, halogen bonding in solution, molecular hydrogen elimination via a dihydrogen bond, the intramolecular conformational effect of triel bonds, and tetrel bonds involving $S_N2$ reactions were considered. Note that these short-lived interactions can both facilitate and hinder other steps [21,22]. For example, Ke and Lin [10] report on the catalytic effect of hydrogen bonding on the methanol steam reforming reaction on a metal surface. This and other publications on this topic may be directly in demand for practical use [23].

Vianello et al. [11] demonstrate how small alterations in weak interactions can cause significant changes in biological activity. Their molecular dynamic simulations correctly reproduce experimental data on the binding energy of histamine within the $H_2$ receptor and its change caused by deuteration. The fact that deuteration can affect the kinetics of a chemical reaction [24] and result in measurable structural and spectral changes [25] is well known. However, in the paper at hand, the authors were able to identify the mechanism responsible for these changes.

The paper by Shenderovich and Denisov [12] presents an advanced approach to implicitly accounting for the solvent effect. In this adduct-under-field approach, the solvent effect is simulated using an external electric field. It was shown that solute–solvent interactions remarkably affect the geometry of acid-base complexes even if the active sites of these complexes are not accessible for solvent molecules. Note that this approach is applicable to many other molecular systems in solution and in crystal form [26,27].

The review of Buntkowsky and Vogel [13] describes current trends and perspectives in the study of guest molecules in porous silica materials employing solid-state NMR techniques with particular attention to the effect of an interplay between guest–guest and guest–host interactions. It is shown that such interactions can radically change the physicochemical properties of these systems. Solid-state NMR and relaxometry are among the most effective analytical tools in this area of materials chemistry. They can be applied for probing structure or dynamics of materials themselves as well as the behavior of incorporated guests [28,29].

**Funding:** This research received no external funding. The APC was funded by MDPI.

**Acknowledgments:** I want to sincerely thank everyone that contributed to this Special Issue. Special thanks to the assistant editor Lola Huo and the entire team of *Molecules* for their motivation, professional expertise, and support.

**Conflicts of Interest:** The authors declare no conflict of interest.

## References

1. Kong, S.; Borissova, A.O.; Lesnichin, S.B.; Hartl, M.; Daemen, L.L.; Eckert, J.; Yu Antipin, M.; Shenderovich, I.G. Geometry and Spectral Properties of the Protonated Homodimer of Pyridine in the Liquid and Solid States. A Combined NMR, X-ray Diffraction and Inelastic Neutron Scattering Study. *J. Phys. Chem. A* **2011**, *115*, 8041–8048. [CrossRef] [PubMed]
2. Lesnichin, S.B.; Tolstoy, P.M.; Limbach, H.-H.; Shenderovich, I.G. Counteranion-Dependent Mechanisms of Intramolecular Proton Transfer in Aprotic Solution. *Phys. Chem. Chem. Phys.* **2010**, *12*, 10373–10379. [CrossRef] [PubMed]
3. Zimmermann, C.; Fischer, T.L.; Suhm, M.A. Pinacolone-Alcohol Gas-Phase Solvation Balances as Experimental Dispersion Benchmarks. *Molecules* **2020**, *25*, 5095. [CrossRef] [PubMed]
4. Wieczorkiewicz, P.A.; Szatylowicz, H.; Krygowski, T.M. Mutual Relations between Substituent Effect, Hydrogen Bonding, and Aromaticity in Adenine-Uracil and Adenine-Adenine Base Pairs. *Molecules* **2020**, *25*, 3688. [CrossRef]
5. Jóźwiak, K.; Jezierska, A.; Panek, J.J.; Goremychkin, E.A.; Tolstoy, P.M.; Shenderovich, I.G.; Filarowski, A. Inter-vs. Intramolecular Hydrogen Bond Patterns and Proton Dynamics in Nitrophthalic Acid Associates. *Molecules* **2020**, *25*, 4720. [CrossRef]
6. Ostras', A.S.; Ivanov, D.M.; Novikov, A.S.; Tolstoy, P.M. Phosphine Oxides as Spectroscopic Halogen Bond Descriptors: IR and NMR Correlations with Interatomic Distances and Complexation Energy. *Molecules* **2020**, *25*, 1406. [CrossRef]
7. Voronin, A.P.; Surov, A.O.; Churakov, A.V.; Parashchuk, O.D.; Rykounov, A.A.; Vener, M.V. Combined X-ray Crystallographic, IR/Raman Spectroscopic, and Periodic DFT Investigations of New Multicomponent Crystalline Forms of Anthelmintic Drugs: A Case Study of Carbendazim Maleate. *Molecules* **2020**, *25*, 2386. [CrossRef]
8. Nenajdenko, V.G.; Shikhaliyev, N.G.; Maharramov, A.M.; Bagirova, K.N.; Suleymanova, G.T.; Novikov, A.S.; Khrustalev, V.N.; Tskhovrebov, A.G. Halogenated Diazabutadiene Dyes: Synthesis, Structures, Supramolecular Features, and Theoretical Studies. *Molecules* **2020**, *25*, 5013. [CrossRef]
9. Grabowski, S.J. Hydrogen Bond and Other Lewis Acid–Lewis Base Interactions as Preliminary Stages of Chemical Reactions. *Molecules* **2020**, *25*, 4668. [CrossRef]
10. Ke, C.; Lin, Z. Catalytic Effect of Hydrogen Bond on Oxhydryl Dehydrogenation in Methanol Steam Reforming on Ni(111). *Molecules* **2020**, *25*, 1531. [CrossRef]
11. Hok, L.; Mavri, J.; Vianello, R. The Effect of Deuteration on the $H_2$ Receptor Histamine Binding Profile: A Computational Insight into Modified Hydrogen Bonding Interactions. *Molecules* **2020**, *25*, 6017. [CrossRef] [PubMed]
12. Shenderovich, I.G.; Denisov, G.S. Adduct under Field—A Qualitative Approach to Account for Solvent Effect on Hydrogen Bonding. *Molecules* **2020**, *25*, 436. [CrossRef] [PubMed]
13. Buntkowsky, G.; Vogel, M. Small Molecules, Non-Covalent Interactions, and Confinement. *Molecules* **2020**, *25*, 3311. [CrossRef] [PubMed]
14. Altnoder, J.; Bouchet, A.; Lee, J.J.; Otto, K.E.; Suhm, M.A.; Zehnacker-Rentien, A. Chirality-dependent balance between hydrogen bonding and London dispersion in isolated (+/-)-1-indanol clusters. *Phys. Chem. Chem. Phys.* **2013**, *15*, 10167–10180. [CrossRef] [PubMed]
15. Jezuita, A.; Ejsmont, K.; Szatylowicz, H. Substituent effects of nitro group in cyclic compounds. *Struct. Chem.* **2020**. [CrossRef]
16. Bordallo, H.N.; Boldyreva, E.V.; Buchsteiner, A.; Koza, M.M.; Landsgesell, S. Structure-property relationships in the crystals of the smallest amino acid: An incoherent inelastic neutron scattering study of the glycine polymorphs. *J. Phys. Chem. B* **2008**, *112*, 8748–8759. [CrossRef]
17. Piękoś, P.; Jezierska, A.; Panek, J.J.; Goremychkin, E.A.; Pozharskii, A.F.; Antonov, A.S.; Tolstoy, P.M.; Filarowski, A. Symmetry/Asymmetry of the NHN Hydrogen Bond in Protonated 1,8-Bis(dimethylamino)naphthalene. *Symmetry* **2020**, *12*, 1924. [CrossRef]
18. Yu Tupikina, E.; Bodensteiner, M.; Tolstoy, P.M.; Denisov, G.S.; Shenderovich, I.G. P = O Moiety as an Ambidextrous Hydrogen Bond Acceptor. *J. Phys. Chem. C* **2018**, *122*, 1711–1720. [CrossRef]
19. Melikova, S.M.; Voronin, A.P.; Panek, J.; Frolov, N.E.; Shishkina, A.V.; Rykounov, A.A.; Tretyakov, P.Y.; Vener, M.V. Interplay of pi-stacking and inter-stacking interactions in two-component crystals of neutral closed-shell aromatic compounds: Periodic DFT study. *RSC Adv.* **2020**, *10*, 27899–27910. [CrossRef]
20. Tskhovrebov, A.G.; Novikov, A.S.; Kritchenkov, A.S.; Khrustalev, V.N.; Haukka, M. Attractive halogen···halogen interactions in crystal structure of trans-dibromogold(III) complex. *Z. Kristallogr. Cryst. Mater.* **2020**, *235*, 477–480. [CrossRef]

21. Grabowski, S.J.; Ruiperez, F. Dihydrogen bond interactions as a result of $H_2$ cleavage at Cu, Ag and Au centres. *Phys. Chem. Chem. Phys.* **2016**, *18*, 12810–12818. [CrossRef] [PubMed]
22. Gurinov, A.A.; Rozhkova, Y.A.; Zukal, A.; Čejka, J.; Shenderovich, I.G. Mutable Lewis and Brønsted Acidity of Aluminated SBA-15 as Revealed by NMR of Adsorbed Pyridine-$^{15}$N. *Langmuir* **2011**, *27*, 12115–12123. [CrossRef] [PubMed]
23. Ke, C.; Lin, Z. Elementary reaction pathway study and a deduced macrokinetic model for the unified understanding of Ni-catalyzed steam methane reforming. *React. Chem. Eng.* **2020**, *5*, 873–885. [CrossRef]
24. Mavri, J.; Matute, R.A.; Chu, Z.T.; Vianello, R. Path Integral Simulation of the H/D Kinetic Isotope Effect in Monoamine Oxidase B Catalyzed Decomposition of Dopamine. *J. Phys. Chem. B* **2016**, *120*, 3488–3492. [CrossRef]
25. Golubev, N.S.; Melikova, S.M.; Shchepkin, D.N.; Shenderovich, I.G.; Tolstoy, P.M.; Denisov, G.S. Interpretation of H/D Isotope Effects on NMR Chemical Shifts of [FHF]$^-$ Ion Based on Calculations of Nuclear Magnetic Shielding Tensor Surface. *Z. Phys. Chem.* **2003**, *217*, 1549–1563. [CrossRef]
26. Shenderovich, I.G.; Denisov, G.S. Solvent effects on acid-base complexes. What is more important: A macroscopic reaction field or solute-solvent interactions? *J. Chem. Phys.* **2019**, *150*, 204505. [CrossRef]
27. Shenderovich, I.G. Electric field effect on $^{31}$P NMR magnetic shielding. *J. Chem. Phys.* **2020**, *153*, 184501. [CrossRef]
28. Weigler, M.; Winter, E.; Kresse, B.; Brodrecht, M.; Buntkowsky, G.; Vogel, M. Static field gradient NMR studies of water diffusion in mesoporous silica. *Phys. Chem. Chem. Phys.* **2020**, *22*, 13989–13998. [CrossRef]
29. Gedat, E.; Schreiber, A.; Findenegg, G.H.; Shenderovich, I.; Limbach, H.-H.; Buntkowsky, G. Stray Field Gradient NMR Reveals Effects of Hydrogen Bonding on Diffusion Coefficients of Pyridine in Mesoporous Silica. *Magn. Reson. Chem.* **2001**, *39*, S149–S157. [CrossRef]

Article

# The Effect of Deuteration on the H₂ Receptor Histamine Binding Profile: A Computational Insight into Modified Hydrogen Bonding Interactions

Lucija Hok [1], Janez Mavri [2] and Robert Vianello [1,*]

[1] Division of Organic Chemistry and Biochemistry, Ruđer Bošković Institute, HR-10000 Zagreb, Croatia; lucija.hok@irb.hr
[2] Laboratory for Computational Biochemistry and Drug Design, National Institute of Chemistry, SI-1001 Ljubljana, Slovenia; janez.mavri@ki.si
* Correspondence: robert.vianello@irb.hr

Academic Editor: Ilya G. Shenderovich
Received: 4 December 2020; Accepted: 17 December 2020; Published: 18 December 2020

**Abstract:** We used a range of computational techniques to reveal an increased histamine affinity for its H$_2$ receptor upon deuteration, which was interpreted through altered hydrogen bonding interactions within the receptor and the aqueous environment preceding the binding. Molecular docking identified the area between third and fifth transmembrane α-helices as the likely binding pocket for several histamine poses, with the most favorable binding energy of −7.4 kcal mol$^{-1}$ closely matching the experimental value of −5.9 kcal mol$^{-1}$. The subsequent molecular dynamics simulation and MM-GBSA analysis recognized Asp98 as the most dominant residue, accounting for 40% of the total binding energy, established through a persistent hydrogen bonding with the histamine −NH$_3^+$ group, the latter further held in place through the N–H···O hydrogen bonding with Tyr250. Unlike earlier literature proposals, the important role of Thr190 is not evident in hydrogen bonds through its −OH group, but rather in the C–H···π contacts with the imidazole ring, while its former moiety is constantly engaged in the hydrogen bonding with Asp186. Lastly, quantum-chemical calculations within the receptor cluster model and utilizing the empirical quantization of the ionizable X–H bonds (X = N, O, S), supported the deuteration-induced affinity increase, with the calculated difference in the binding free energy of −0.85 kcal mol$^{-1}$, being in excellent agreement with an experimental value of −0.75 kcal mol$^{-1}$, thus confirming the relevance of hydrogen bonding for the H$_2$ receptor activation.

**Keywords:** deuteration; heavy drugs; histamine receptor; hydrogen bonding; receptor activation

---

## 1. Introduction

Histamine is an important mediator and neurotransmitter that is involved in a broad spectrum of central and peripheral physiological as well as pathophysiological processes, such as allergies and inflammation. It exerts its specific effects by the activation of four receptor subtypes (H$_1$R–H$_4$R) [1]. Histamine receptors are 7-transmembrane receptors, which belong to the family of G-protein coupled receptors (GPCR), a very common target for a wide range of therapeutics used in modern pharmacotherapy, and differ in receptor distribution, ligand binding properties, signaling pathways and functions. Some estimates suggest that GPCRs encompass around 30% of the existing drug targets, while their therapeutic potential might be even larger [2,3].

The literature contains many studies on how GPCRs are activated and transmit their signals from the extracellular side to the G-protein coupling domain located on the intracellular side [4–6]. Instead, we have been interested in how different agonists and antagonists bind to the receptor binding site, and whether these processes are modulated upon non-selective deuteration, which would confirm the assumption that ligand binding is governed by hydrogen bonding interactions. Specifically, the topic of deuterium isotope effects is usually concerned with its impact on chemical reactions that are caused by substituting protium hydrogen (H) atoms with deuterium (D) in a molecule. These effects include changes in the rate of cleavage of covalent bonds to deuterium, or to an atom located adjacent to deuterium, in a reactant molecule. Alternatively, deuterium isotope effects on other, for example, noncovalent interactions between molecules are known to occur, but are generally considered to be insignificant, especially in biological experiments where deuterium substituted molecules are used as tracers. Nevertheless, replacing light hydrogen atoms with their heavier deuterium analogues, typically shortens the donor X–D bonds relative to the X–H bonds (X = heteroatom), as the X–D bonds are stronger, more compact and more stable to oxidative processes. Ultimately, this results in the elongation of the corresponding donor⋯acceptor distance among heteroatoms, also known as the Ubbelohde effect [7], which affects the strength of the involved hydrogen bonds and can, therefore, produce modified affinities during the ligand–target recognition. Indeed, D has a 2-fold higher mass than H, leading to a reduced vibrational stretching frequency of the X–D bond compared to the X–H bond and, consequently, lower ground state energy. To further confirm that, Bordallo and co-workers recently performed a very accurate neutron diffraction study of the alanine zwitterion to show that deuteration reduces the electrostatic attraction in the acidic N–D bonds by 2.3% relative to the corresponding N–H bonds [8]. This results in the shortening of the N–D distances, as already noticed in various papers [9–13].

For many years, researchers have sought ways to incorporate deuterium into drug molecules in order to inhibit metabolic conversion into less active or inactive molecules [14,15], with the first such attempts being made nearly 60 years ago [16]. Because bonds to deuterium are stronger than those to hydrogen, early adopters tweaked molecules to better withstand the ravages of drug-metabolizing enzymes like cytochrome P450s. Deuterated drugs, they hoped, would have longer half-lives than their non-deuterated counterparts that would allow perhaps less frequent dosing and produce different metabolites. With this in mind, the focus was placed on drug fragments that were expected to be sites amenable to metabolic transformations, and typically involved chemical deuteration pertaining to heteroatom–$CH_3$ groups (or other alkyl units) that were converted into heteroatom–$CD_3$ alternatives. As an illustrative example, Falconnet [17], Brazier [18], and Cherrah [19,20] studied the binding of caffeine (Figure 1) to human serum albumin (HSA) by the equilibrium dialysis and demonstrated that the corresponding $K_a$ values for caffeine, caffeine-1-$CD_3$ and caffeine-1,3,7-$(CD_3)_3$ were not significantly different, while those for caffeine-3-$CD_3$, caffeine-1,7-$(CD_3)_2$, and caffeine-3,7-$(CD_3)_2$ were considerably lower than that for caffeine, indicating that HSA had a reduced affinity for the deuterated compounds. On the other hand, very recently, the U.S. Food and Drug Administration granted market approval for the first deuterated drug molecule, deutetrabenazine (Figure 1), which is useful in treating chorea associated with Huntington's disease [21]. Deutetrabenazine is a heavier analogue of the existing drug tetrabenazine, with two $-OCH_3$ groups in the latter being replaced by a pair of $-OCD_3$ groups, thereby altering the rate of metabolism to afford greater tolerability and an improved dosing regimen, thus an enhanced therapeutic potential was achieved. Still, in both of these instances, one can hardly argue that modified affinities came as a result of changed hydrogen bonding strengths, since methyl groups and their deuterated versions show poor hydrogen bonding abilities. Therefore, the observed effects likely originate in the modified dipole-dipole or dipole-charge interactions, which are generally weak. Knowing that hydrogen bonding interactions are significantly stronger than those mentioned, and that they typically dominate the ligand–target recognition [22], led us to offer some insight into the effect of deuteration on the ability of the $H_2$ receptor to accommodate its endogenous agonist histamine, the latter particularly suitable to inspect alternations in the hydrogen bonding patterns and the accompanying affinities. Namely, histamine is a biogenic diamine (Figure 1), consisting of a free ethylamino group and an imidazole ring, thus involving three distinct sites able to either donate or accept hydrogen bonds, which makes it reasonable to expect that these particular interactions will predominantly govern its binding to the $H_2$ receptor, as it was clearly demonstrated in the case of its hydration [23,24].

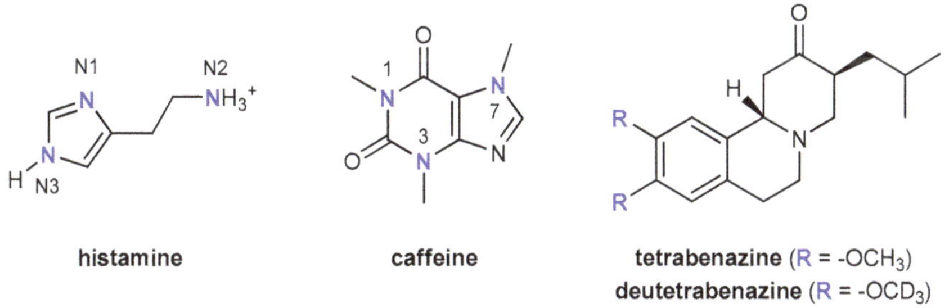

**histamine**

**caffeine**

**tetrabenazine (R = -OCH$_3$)**
**deutetrabenazine (R = -OCD$_3$)**

Figure 1. Chemical structures and atom labeling for the systems relevant to the discussion.

With this in mind, instead of utilizing chemical deuteration described earlier, in our preceding work [25], we have taken a different approach of introducing deuteration through the exchange mechanism by performing binding studies in pure $D_2O$. In this way, we assured that all exchangeable hydrogen atoms, in both aqueous solution and within the $H_2$ receptor will be replaced by deuterium, and that this will allow us to monitor how the hydrogen bonding interactions responsible for both the histamine hydration and its inclusion into the receptor binding site will be affected. Experiments were carried out on the $H_2$ receptor present in cell membranes of cultured neonatal rat astrocytes, where we conducted the saturation and inhibition binding experiments using the antagonist $^3$H-tiotidine as a radiolabel, and histamine as a displacer of a bound radioligand. The results revealed a significant increase in the histamine affinity, as its pIC$_{50}$ values ($p < 0.05$) changed from 7.25 ± 0.11 (control) to 7.80 ± 0.16 ($D_2O$). Building on that, our subsequent work undertook the same approach for the binding of two agonists, 2-methylhistamine and 4-methylhistamine, and two antagonists, cimetidine and famotidine, and showed a notable affinity increase for 4-methylhistamine and a reduced one for 2-methylhistamine, while no change was observed for both antagonists [26]. This was interpreted in the context of the altered hydrogen bonding strength upon deuteration, which impacts ligand interactions with binding sites residues and solvent molecules preceding the binding. Our present work builds on the mentioned results [25,26], and considers the parent agonist histamine through a range of computational techniques, involving docking studies, classical molecular dynamics simulation, and quantum-chemical calculations within a large cluster model of the $H_2$ receptor, in order to offer a more precise insight into the structural and electronic features of the studied ligand with the aim to provide the molecular interpretation to the observed binding differences. The outlined analysis is likely to contribute towards understanding the receptor activation, while the in silico discrimination between agonists and antagonists, based on the receptor structure, remains a distant ultimate goal.

## 2. Results and Discussion

As already mentioned, in our preceding work [25], we used $^3$H-tiotidine as a marker to label histamine $H_2$ receptor binding sites on the cultured neonatal rat astrocytes, and histamine as an agonist to displace it, both in the control system and in deuterated environment. This resulted in a considerable deuteration-induced increase in the histamine affinity, as the measured pIC$_{50}$ values ($p < 0.05$) went from 7.25 ± 0.11 (control) to 7.80 ± 0.16 ($D_2O$). Although the relationship between IC$_{50}$ and $\Delta G_{BIND}$ values is not so straightforward in absolute terms, their relative ratio is connected through the Cheng-Prusoff equation [27] and roughly translates to a difference of $\Delta\Delta G_{BIND} = -0.75$ kcal mol$^{-1}$, which will be used in the rest of the text to evaluate the quality of computational results.

## 2.1. Docking Simulation

To offer some initial insight into the binding of histamine into the $H_2$ receptor, we employed several docking simulations with the aim of obtaining the relevant binding poses and the accompanying binding free energies, and use these as starting points for the subsequent molecular dynamics (MD) simulations. In doing so, we focused on the more stable N3–H ($N^\tau$) tautomer, which was docked into the homology structure of the $H_2$ receptor. Interestingly, although the entire receptor surface was considered equally during the docking procedure, the obtained results reveal that the first four most favorable binding poses correspond to the identical position within the $H_2$ receptor, and only differ in the conformation of the histamine ligand (Figure 2).

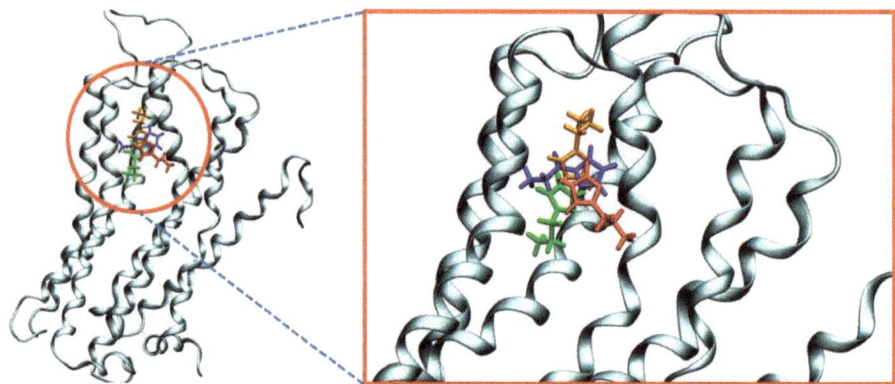

**Figure 2.** Overlap of four most favorable histamine binding poses within the $H_2$ receptor as predicted by molecular docking that differ only in the ligand orientation. The computed binding free energies are $-7.4$ kcal mol$^{-1}$ (blue), $-7.1$ kcal mol$^{-1}$ (green), $-6.8$ kcal mol$^{-1}$ (orange) and $-6.7$ kcal mol$^{-1}$ (red).

Apart from being positioned in the same binding pocket, a closer analysis of the predicted binding poses shows that all histamine molecules are located in the area between third and fifth transmembrane α-helices, in line with many earlier literature reports on the binding of $H_2$ receptor ligands [28–30]. This provides some credence to the obtained results, which is further promoted by the calculated binding affinities. Namely, Figure 2 shows that the most favorable pose is associated with the binding energy of $-7.4$ kcal mol$^{-1}$, which very well agrees with the experimental value of $-5.9$ kcal mol$^{-1}$ obtained from the measured p$K_i$ value of 4.3 [31]. Lastly, let us briefly mention that we have repeated the identical docking procedure for the less stable N1–H ($N^\pi$) histamine tautomer, and the results showed an analogous placement within the $H_2$ receptor and the identical binding energy of $-7.4$ kcal mol$^{-1}$. Still, due to a described lower stability and the matching lower population of this tautomer relative to its N3–H analogue, N1–H tautomer was not considered further.

## 2.2. Molecular Dynamics Simulation of Histamine in Aqueous Solution

As already described, in aqueous solution, histamine exists almost exclusively (98%) as a monocation protonated at the free ethylamino group (Figure 1) and this protonation form was considered in a 20 Å-thick truncated octahedron simulation box, which involved 3.572 water molecules.

It turned out that histamine is a rather flexible molecule, but the clustering analysis of the obtained structures revealed a predominance of the two types of geometries (Figure 3), termed as *gauche*, in which there exists an intramolecular N–H·····N hydrogen bonding between the protonated amine (N2) as a donor and the imino nitrogen (N1) within the imidazole ring as an acceptor, and *trans*, which is elongated and where such a hydrogen bonding is absent. Interestingly, the results reveal around 73% dominance of the *trans* conformation, which is in an almost perfect agreement with around 80% predicted by other techniques [23,32–35]. It is worth mentioning that two useful geometric parameters, which characterize these two distinct orientations, and which will be used later in the analysis of the conformational preference of histamine within the receptor, are (i) distance between the relevant N1–N2 sites, and (ii) dihedral angle describing the rotation of the ethylamino group around the imidazole ring. In the representative *trans* geometry these are 4.54 Å and 158.8°, respectively, while in *gauche* these are reduced to 3.01 Å and 62.1°, in the same order. Their distribution during MD simulation (Figure S1) also indicates the preference of the *trans* conformation and further demonstrates the suitability of the described two structures as representative.

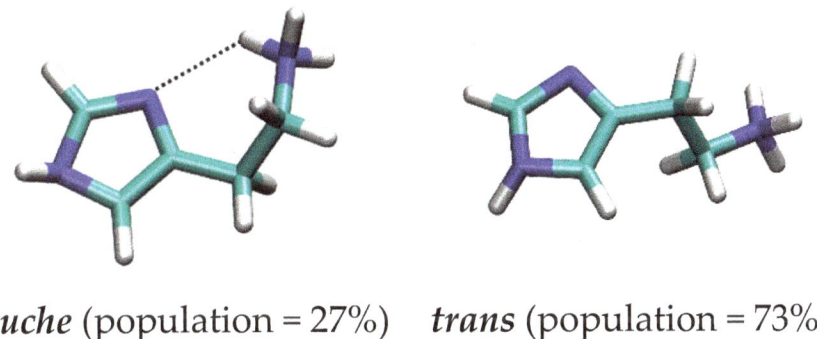

*gauche* (population = 27%)   *trans* (population = 73%)

**Figure 3.** Representative conformations with their population of the histamine monocation in aqueous solution as obtained by the molecular dynamics simulation.

The interactions governing the hydration of histamine also reveal interesting trends. All three nitrogen sites (N1–N3) represent crucial locations to interact with water, with the corresponding RDF displays demonstrating an equal solvent ability to approach them (Figure 4a). Specifically, for all three positions, the predominant interactions are established at the N (histamine)···O (water) distances of around 3 Å, which corresponds to rather strong hydrogen bonds in all cases. The interactions with both N-positions within the imidazole ring show identical patterns, thus indicating that N1 and N3 sites are participating as hydrogen bond acceptor and donor, respectively, with one water molecule in the first hydration shell. The latter is nicely evident in the average number of hydrogen bonding contacts, being 1.2 and 0.7 for N1 and N3, respectively. On the other hand, the interaction with the cationic N2 site is much more frequent, while around three times higher peak for N2 is only partially justified by the fact that the protonated amino group has three equivalent N2–H bonds that can potentially interact with three water molecules at the same time. As a matter of fact, Figure 4b advises that the actual number of hydrogen bonding contacts for this group is predominantly between 2 and 3, with an average value during simulation of 2.2, which is likely due to steric reasons and the exchangeability of individual solvent molecules. In addition, the shape of the RDF curve and a slightly lower distance for the peak maximum for N2 (2.8 Å relative to 2.9 Å for both N1 and N3), also suggests that its interactions with the solvent molecules are stronger than with the other two nitrogen sites. This notion is found in excellent agreement with our earlier report [24], where we utilized the Car-Parrinello molecular dynamics simulation scheme to delineate the experimental IR spectra of histamine in water, which showed a broad feature between 3350 and 2300 cm$^{-1}$ including a mixed contribution from the ring N3–H and the aminoethyl N2–H stretching vibrations, to indicate that the ring amino group absorbs at higher frequencies than the remaining three amino N2–H protons, thus implying the latter forms stronger hydrogen bonding with the surrounding waters.

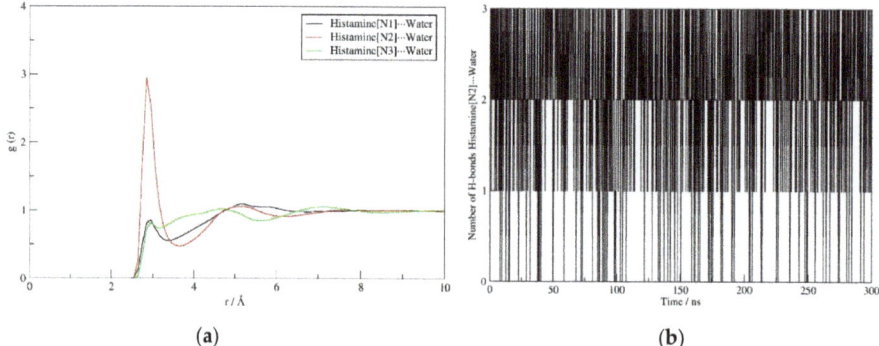

**Figure 4.** RDF displays describing the interaction of the solvent water molecules with N1 (black), N2 (red) and N3 (green) sites on histamine (**a**), and the evolution of the number of N2–H···O(water) hydrogen bonds (**b**) during the molecular dynamics simulation in aqueous solution.

## 2.3. Molecular Dynamics Simulation within the $H_2$ Receptor

Following the presented docking analysis, the obtained four most favorable docking positions, which differ in the orientation of the histamine ligand within the same binding pocket, were solvated in a 10 Å-thick truncated octahedron simulation box involving 18.850 water molecules, and submitted to the MD simulation for the production run of 300 ns. The validity of this approach is justified through the corresponding RMSD graphs, which reveal converged simulation (Figure S2). The obtained trajectories were analyzed by the MM-GBSA protocol in order to obtain the matching binding free energies, $\Delta G_{BIND}$. The mentioned four independent simulations gave $\Delta G_{BIND}$ values between −7.5 and −14.2 kcal mol$^{-1}$, with the trajectory corresponding to the most exergonic binding being employed in further analysis. At this point it is worth to stress that the obtained $\Delta G_{BIND}$ values using this approach are somewhat overestimated in absolute terms. This is a known limitation of the MM-GBSA approach, as extensively discussed in a recent review by Homeyer and Gohlke [36], which also underlined its huge potential in predicting relative binding energies in the biomolecular complexes [36], which is how this approach is utilized here.

The contribution of crucial binding residues is presented in Table 1, while a representative snapshot from the MD trajectory is shown in Figure 5. The specific residues considered for the analysis are those whose favorable contribution exceeds −0.06 kcal mol$^{-1}$ and those with unfavorable contribution over +0.02 kcal mol$^{-1}$, in total 18 residues each. It turns out that the most dominant interaction that histamine establishes within the $H_2$ binding site is that with Asp98, which accounts for almost 40% of the binding energy, being a significant observation. It is established through charge-charge interactions among the protonated amino group on histamine and the anionic carboxylate side chain of Asp98, being highly persistent through the entire simulation, while typically involving N–H·····O hydrogen bond with only one of the carboxylic oxygen atoms (Figure S3). The reason for the latter is the fact that, besides Asp98, the protonated histamine −NH$_3$$^+$ group has the potential to interact with the −OH group on the nearby Tyr250, which is locked in the position through donating a hydrogen bond to the other carboxylic O-atom on Asp98 (Figure 5). The mentioned histamine···Tyr250 interaction is also persistent during simulation (Figure S4, top), evident in a favorable Tyr250 contribution of −0.83 kcal mol$^{-1}$, while Tyr250···Asp98 hydrogen bonding is absent in the beginning, but, once formed, it remains stable during the second half of the trajectory (Figure S4, bottom).

**Table 1.** MM-GBSA binding free energy ($\Delta G_{BIND}$) from the molecular dynamics trajectory and its decomposition on a *per-residue* basis (in kcal mol$^{-1}$). Residues are selected to list all of those with favorable contributions over –0.06 kcal mol$^{-1}$ and unfavorable contributions over +0.02 kcal mol$^{-1}$.

| Histamine $\Delta G_{BIND} = -14.2$ kcal mol$^{-1}$ | | | |
|---|---|---|---|
| Favorable Contributions | | Unfavorable Contributions | |
| Asp98 | −5.45 | Leu193 | +0.02 |
| Val99 | −1.78 | Gly157 | +0.02 |
| Phe254 | −1.60 | Lys166 | +0.02 |
| Thr103 | −1.29 | Lys83 | +0.02 |
| Phe251 | −0.97 | Val273 | +0.02 |
| Thr190 | −0.84 | Asn108 | +0.02 |
| Tyr250 | −0.83 | Lys88 | +0.02 |
| Gly187 | −0.64 | Tyr192 | +0.03 |
| Asp186 | −0.64 | Arg260 | +0.03 |
| Met100 | −0.20 | Ala253 | +0.05 |
| Glu270 | −0.16 | Ser153 | +0.05 |
| Cys102 | −0.14 | Arg161 | +0.06 |
| Glu267 | −0.11 | Lys173 | +0.11 |
| Val178 | −0.10 | Gln177 | +0.11 |
| Tyr94 | −0.08 | Gly183 | +0.13 |
| Val189 | −0.08 | Arg257 | +0.15 |
| Val185 | −0.07 | Leu97 | +0.19 |
| Val255 | −0.06 | Lys175 | +0.29 |

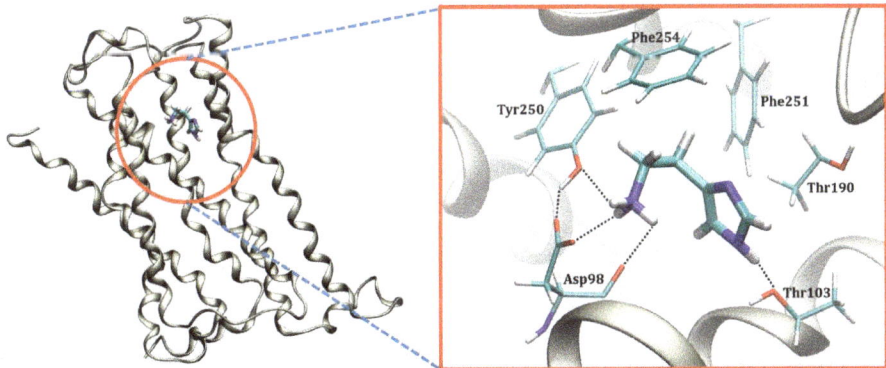

**Figure 5.** Representative position of the histamine monocation within the H$_2$ receptor binding site as obtained from the molecular dynamics simulation.

The imidazole ring is less prone to hydrogen bonding interactions. This holds in particular for its imino N1 nitrogen, for which no particular interactions are observed at all during the entire simulation. In contrast, its amino N3–H group is found in the vicinity of two threonine residues, Thr103 and Thr190. Although N3–H·····O hydrogen bonding interactions with the former are more significant in this respect (Figure S5), thus the higher individual contribution of Thr103 over Thr190 (Table 1), both of these are much less frequent and clearly weaker than those with the –NH$_3^+$ group. Still, the individual contribution of Thr190 is quite notable, −0.84 kcal mol$^{-1}$, not being a consequence of the mentioned hydrogen bonds with the ligand, but rather notable C–H·····π interactions with its imidazole ring (Figure 5). It is very important to underline that this binding pattern is different from the model by Birdsall and co-workers [37] proposed on the basis of the site-directed mutagenesis carried out by Gantz and co-workers [38] that suggested Thr190 to bind histamine on its N1 imino nitrogen through the O–H·····N1 hydrogen bonding. Yet, a closer inspection into the environment around Thr190 shows its –OH group forms a persistent and stable hydrogen bonding with a more distant backbone carbonyl group of Asp186 during 92% of the simulation time (Figure S6) with the average O·····O distance of 2.76 Å, thus ruling out Birdsall's proposal as highly unlikely. To further strengthen this conclusion, we have looked at all other MD trajectories to find an alternative histamine orientation linked with a higher individual Thr190 contribution, which would likely indicate a potential hydrogen bonding connection with the ligand. However, in the case when this was as much as −1.48 kcal mol$^{-1}$ for Thr190, this involved a completely different histamine orientation (Figure S7), yet having as much as 3.3 kcal mol$^{-1}$ less exergonic overall binding free energy (−10.9 kcal mol$^{-1}$), thus being much less relevant. There, the protonated histamine –NH$_3$$^+$ group approaches Asp186 through hydrogen bonding interactions, which makes the latter residue the most relevant for the binding, with an individual contribution of −3.19 kcal mol$^{-1}$. This is, then, followed by Thr190, which, even in this case, does not form hydrogen bonding contacts with histamine, neither with the neighboring –NH$_3$$^+$ group, let alone with its imidazole ring, but rather again interacts through the already described C–H·····π interactions. Let us also mention that such a changed histamine position diminishes the importance of Asp98 and Tyr250 (Figure S7), making the former even disfavoring the binding with the contribution of +0.02 kcal mol$^{-1}$, thus again confirming the insignificance of such ligand binding poses.

Beside the mentioned receptor residues, the rest of the binding pocket is significantly hydrophobic, consisting mostly of aliphatic and aromatic side chains (Figure 5, Table 1). Nevertheless, this still allows histamine to establish a range of additional favorable contacts, including (i) its imidazole ring undergoing the T-shaped π–π stacking interactions with Phe251, and (ii) its ethyl moiety utilizing the C–H·····π interactions with Phe254 (Figure 5). Both of these contacts are rather strong and particularly important, making Phe254 and Phe251 the third and fifth most dominant residues for the histamine binding, with individual contributions reaching −1.60 and −0.97 kcal mol$^{-1}$, respectively (Table 1). This confirms the hydrophobic nature of the H$_2$ receptor binding site, further prompted by a significant contribution of Val99 of −1.78 kcal mol$^{-1}$. Still, likely the most profound evidence for the hydrophobic character of the H$_2$ binding pocket is the conformation of histamine during the binding. As already described, in polar hydrophilic environments, such as the aqueous solution, histamine predominantly assumes the elongated *trans* conformation, which disfavors the intramolecular N2–H·····N1 hydrogen bonding and exposes the protonated –NH$_3$$^+$ group for the interactions with the solvent. In contrast, the increased environment hydrophobicity starts favoring the *gauche* conformation, where the mentioned hydrogen bonding occurs and is allowed by the flexibility of the ethyl linkage. As an illustrative example, in the gas phase, the *gauche* conformer is by as much as 14.8 kcal mol$^{-1}$ more stable [23], and clearly dominates in this paradigmatic hydrophobic media. Along these lines, the conformational preference of the bound histamine reveals interesting trends (Figure 6). The clustering analysis of histamine conformations while inside the H$_2$ receptor shows that, for two thirds of the simulation time, histamine assumes *gauche* conformations, with a mix of structures with and without the N2–H·····N1 hydrogen bonding, while only one third of structures is found in a typical *trans* conformation, thus confirming the hydrophobicity of the H$_2$ receptor interior. Such a distribution of histamine conformations is further evident in the evolution of the corresponding N1–N2 distances and dihedral angles describing the rotation of the ethyl chain (Figure S8), which primarily assume values that support the predominance of the *gauche* orientations.

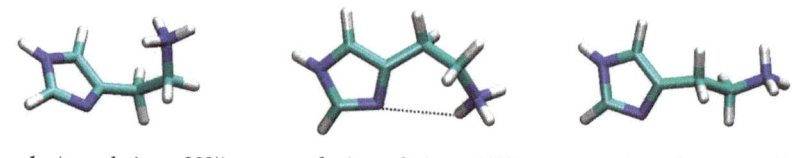

*gauche* (population = 38%)  *gauche* (population = 29%)  *trans* (population = 33%)

**Figure 6.** Representative conformations along with their populations of the histamine monocation at the H$_2$ receptor binding site as revealed by molecular dynamics simulation.

## 2.4. Quantum-Chemical Calculations

In order to computationally evaluate the effect of deuteration on the binding of histamine to the $H_2$ receptor, we have undertaken a series of quantum-chemical calculations at the M06–2X/6–31+G(d) level of theory, employing an implicit quantization of the acidic N–H, O–H and S–H bonds, and utilizing the implicit SMD solvation model with the dielectric constants of $\varepsilon = 78.4$ for aqueous solution and $\varepsilon = 4.0$ for the receptor interior, as already described. Also, we must note that, before reaching the receptor interior, the ligand is present in the aqueous solution, thus our approach is based on evaluating how deuteration affects both environments individually. This results in the separation of the deuteration-induced change in the overall ligand binding affinity, $\Delta\Delta E_{BIND}$, into the contribution arising from the energy of hydration, $\Delta\Delta E_{HYDR}$, and the energy of the interaction with receptor, $\Delta\Delta E_{INTER}$, according to the following equation:

$$\Delta\Delta E_{BIND}(H \rightarrow D) = \Delta\Delta E_{HYDR}(H \rightarrow D) - \Delta\Delta E_{INTER}(H \rightarrow D)$$

In other words, the introduction of deuterium changes not only the geometric parameters of the matching deuterated bonds, but also the energies of the hydrogen bonding interactions in which these bonds participate. It is a fact that deuteration typically reduces the strength of the hydrogen bond, particularly if only one such interaction is considered. Yet, in this case, we are concerned with multiple hydrogen bonds that determine both the hydration and the interaction with the receptor, and since the overall effect ($\Delta\Delta E_{BIND}$) is a difference between these two quantities, it can, at the end, be either positive or negative, depending on the ligand. In line with that, our earlier report on the same receptor showed that deuteration increased the binding of 4-methylhistamine and gave a reduced affinity for 2-methylhistamine, while offered no change for antagonists cimetidine and famotidine [26]. With this in mind, we have extracted the relevant snapshots from the MD simulation in water and $H_2$ receptor, both with and without histamine, and truncated the geometries to clusters involving 36 molecules of water, and receptor residues 98–103, 186–190 and 250–254, respectively. These were submitted to an unconstrained optimization of all geometric parameters, corresponding to the situation with lighter H-nuclei, to be followed by manually shortening by 2.3% and constraining all acidic N–H, O–H and S–H bonds that mirrors deuterated analogues. For histamine, the latter involved all four N–H bonds, three within the protonated N2 amino group and one within the ring N3–H moiety, while for the water molecules this included all O–H bonds. For the receptor fragments, this pertained shortening the O–H bonds in Thr103, Thr190, Tyr250 and Tyr252, and the S–H bond in Cys102. This choice is supported by knowing that threonine, having the least acidic side chain moiety of all three considered residues, spontaneously exchanges all of its –OH protons in $D_2O$ [39], thus justifying the same approach for more acidic tyrosine and cysteine residues.

The hydration energy, $\Delta E_{HYDR}$, is calculated using a reaction scheme depicted in Figure 7 and the obtained values are given in Table 2. This approach relies on transferring histamine from the gas phase into the aqueous solution and forming a hydrated solute-solvent complex. In water, the hydration energy is calculated as −71.63 kcal mol$^{-1}$, indicating that the histamine monocation is well solvated and stabilized in water. This value is reduced in $D_2O$, as a result of modified hydrogen bonding interactions and their strength following deuteration, and assumes −71.20 kcal mol$^{-1}$, a small effect of only 0.43 kcal mol$^{-1}$ in favor of $H_2O$. This implies that the hydration, on its own, works in the direction of promoting the binding of a deuterated system to the receptor.

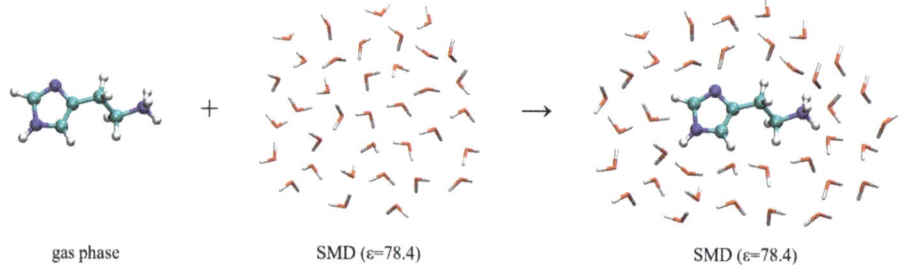

gas phase          SMD ($\varepsilon$=78.4)          SMD ($\varepsilon$=78.4)

**Figure 7.** Computational scheme to calculate the hydration energy of the histamine monocation in the aqueous solution, $\Delta E_{HYDR}$. The selection of dielectric constants is specified in round brackets.

**Table 2.** Calculated deuteration-induced changes in the hydration energy ($\Delta E_{HYDR}$), $H_2$ receptor interaction energy ($\Delta E_{INTER}$), and the overall receptor binding energy ($\Delta E_{BIND,calc}$) as obtained by the (SMD)/M06–2X/6–31+G(d) model (in kcal mol$^{-1}$), the latter compared with the experimentally determined value ($\Delta E_{BIND,exp}$) from ref. [25].

| Ligand | In $H_2O$ | | | In $D_2O$ | | | | |
|---|---|---|---|---|---|---|---|---|
| | $\Delta E_{HYDR}$ | $\Delta E_{INTER}$ | $\Delta E_{BIND}$ | $\Delta E_{HYDR}$ | $\Delta E_{INTER}$ | $\Delta E_{BIND}$ | $\Delta\Delta E_{BIND,calc}$ | $\Delta\Delta E_{BIND,exp}$ |
| Histamine | −71.63 | −82.31 | −10.69 | −71.20 | −82.73 | −11.54 | −0.85 | −0.75 |

On the other hand, the interaction energies with the receptor, $\Delta E_{INTER}$, are estimated through the scheme shown in Figure 8, which considers placing a ligand from the gas phase into the cluster model of the receptor binding site. We have to mention that, despite considering only a truncated receptor model and approximating the rest of its structure with the dielectric constant of $\varepsilon = 4.0$, during the geometry optimization, the structure of histamine and its protonated $-NH_3^+$ group remained as such, although it is positioned in the direct interaction with the $-COO^-$ group from Asp98. In other words, we did not observe a spontaneous histamine···Asp98 proton transfer, which could have occurred due to a limited account of the electrostatic environment that disfavors charge separation. Instead, both the structure and the position of histamine within the binding site remained as described during the MD simulation, which justifies our model and the selection of the most important residues for the cluster-continuum approach. With this in mind, it is important to notice that the interaction energies, $\Delta E_{INTER}$, are consistently higher than $\Delta E_{HYDR}$, which confirms histamine ability to leave the aqueous solution and enter the receptor. In water, this assumes $\Delta E_{INTER} = -82.31$ kcal mol$^{-1}$, which is, interestingly, further increased by 0.42 kcal mol$^{-1}$ to $\Delta E_{INTER} = -82.73$ kcal mol$^{-1}$ upon deuteration. This already indicates that deuterated histamine is better accommodated within the receptor, but the precise magnitude of the resulting effect is interplay between this interaction and histamine placement in the aqueous solution preceding the binding.

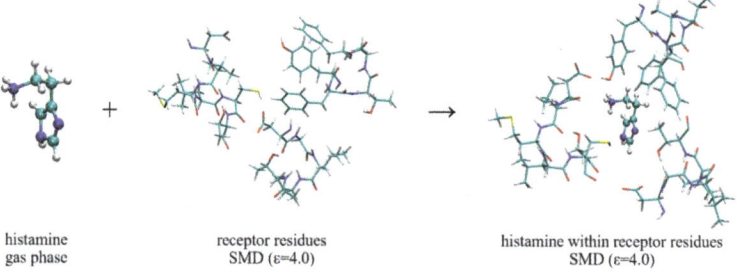

histamine gas phase      receptor residues SMD ($\varepsilon$=4.0)      histamine within receptor residues SMD ($\varepsilon$=4.0)

**Figure 8.** Computational scheme to estimate the interaction energy of the histamine monocation with the $H_2$ receptor, $\Delta E_{INTER}$. The selection of dielectric constants is specified in round brackets.

Combining the mentioned hydration and interaction energies and their differences, one arrives to the overall change in the binding energy for the receptor-ligand recognition, which assumes $\Delta\Delta E_{BIND,calc}$ = −0.85 kcal mol$^{-1}$, being in excellent agreement with the experimentally determined value of −0.75 kcal mol$^{-1}$ [25], thus confirming an increased histamine affinity following deuteration. Both the calculated and computed values indicate around 3–4 times higher histamine affinity following deuteration, which is an interesting observation bearing some pharmacological relevance. The agreement between these sets of data is very impressive, particularly given the simplicity of the computational model used for the implicit deuteration conducted on only a small, but carefully selected part of the receptor molecule, which validates the employed methodology and allows its use in other biological systems as well.

In concluding this section, let us emphasize that receptor activation is a highly complex and dynamic process associated with large conformational changes between receptor states. These are difficult to investigate experimentally, while, at the same time, occurring on time scales that are inaccessible for direct molecular simulations. Still, the results presented here provide convincing support that hydrogen bonding interactions are involved in the receptor activation and firmly advise that deuteration, as the simplest possible structural modification, can have a significant impact for the ligand affinity. This opens the door for the development of perdeuterated drugs, which could have different, yet in some instances more favorable clinical profiles to already marketed substances.

## 3. Computational Details

A homology structure of the H$_2$ receptor was developed earlier [25], which revealed a good agreement with other models reported in the literature [40,41], and was employed here throughout the entire work. The structure of histamine contains an imidazole ring and an aminoethyl side chain, both of which have the ability of accepting a proton, if the medium is acidic enough. According to its p$K_a$ values, 6.0 for the imidazole nitrogen and 9.7 for the aliphatic amino group [42], at physiological pH of 7.4, histamine is predominantly a monocation (96%) protonated at the free amino group. Also, its imidazole ring can exist in two tautomeric forms, 1$H$-imidazole and 3$H$-imidazole (Figure 1), denoted as N$^\pi$–H and N$^\tau$–H, respectively, with plenty of experimental and computational evidence in favor of the latter as the predominant structure in the aqueous solution [23,32–35]. With all this in mind, the structure of the histamine N$^\tau$–monocation was considered in all simulations.

### 3.1. Docking Analysis

The structure of the histamine monocation was optimized with the Gaussian 16 software [43] employing the M06–2X DFT functional with the 6–31+G(d) basis set. To account for the effect of the aqueous solution, during the geometry optimization we included the implicit SMD polarizable continuum model [44] with all parameters for pure water. The molecular docking studies have been done with SwissDock [45], a web server for docking of small molecules on the target proteins based on the EADock DSS engine, taking into account the entire protein surface as potential binding sites for the investigated ligands. Both the preparation of the H$_2$ receptor structure and the visualization of results were performed using the UCSF Chimera program (version 1.14) [46].

### 3.2. Molecular Dynamics Simulation

Several best binding poses of histamine within the H$_2$ receptor, elucidated through the preceding docking analysis, were used for subsequent molecular dynamics simulations. To parametrize histamine, RESP charges were calculated at the HF/6–31G(d) level of theory in Gaussian 16 program [43] to be consistent with the employed GAFF force field, while the protein was modeled using the AMBER ff14SB force field. Such a complex was then solvated in a truncated octahedral box of TIP3P water molecules spanning a 10 Å-thick buffer, neutralized by 12 Cl$^-$ anions, and submitted to the geometry optimization in the AMBER16 program package [47], employing periodic boundary conditions in all directions. An analogous setup, involving a 20 Å-thick buffer of water molecules around isolated histamine monocation, joined by the Cl⁻ counterion, whose position was fixed at the border of the simulation box by a force constant of 30 kcal mol$^{-1}$ and a position restrain between 19–21 Å from histamine, was utilized for the MD simulation pertaining to the aqueous solution. Both approaches were identically repeated to setup analogous simulations concerning the receptor and the aqueous solution without histamine and its counterion. In all instances, optimized systems were gradually heated from 0 to 300 K and equilibrated during 30 ps using NVT conditions, followed by productive and unconstrained MD simulation of 300 ns, employing a time step of 2 fs at a constant pressure (1 atm) and temperature (300 K), the latter held constant using a Langevin thermostat with a collision frequency of 1 ps$^{-1}$. The long-range electrostatic interactions were calculated employing the Particle Mesh Ewald method [48], and were updated in every second step, while the nonbonded interactions were truncated at 11.0 Å.

Histamine binding free energies, $\Delta G_{BIND}$, within the H$_2$ binding pocket were calculated using the established MM-GBSA protocol [49,50] available in AmberTools16 [47], and in line with our earlier reports [51,52]. MM-GBSA is widely used for calculating the binding free energies from snapshots of the MD trajectory with an estimated standard error of 1–3 kcal mol$^{-1}$ [49]. For that purpose, 3000 snapshots collected from the last 30 ns of the corresponding MD trajectories were utilized. The calculated MM-GBSA binding free energies were decomposed into a specific residue contribution on a *per-residue* basis according to the established procedure [53,54]. This protocol evaluates contributions to $\Delta G_{BIND}$ arising from each amino acid residue and identifies the nature of the energy change in terms of the interaction and solvation energies or entropic contributions.

### 3.3. Quantum-Chemical Calculations

Following the MD analysis, which identified residues dominating the histamine binding, we took a representative snapshot and extracted positions of the bound histamine and the surrounding residues 98–103, 186–190 and 250–254. The same residues were pulled out from the H$_2$ receptor MD simulation without histamine. In this way, the cluster representation of the receptor binding site consisted of the following residues Asp98, Val99, Met100, Leu101, Cys102, Thr103, Asp186, Gly187, Leu188, Val189, Thr190, Tyr250, Phe251, Thr252, Ala253 and Phe254, which were considered in their typical protonation forms according to the PROPKA 3.1 analysis [55] carried out on the entire homology structure. From the MD simulation in water, we extracted the position of the nearest 36 water molecules within 4 Å from histamine, which allowed for a spherical solvent layer involving histamine first solvation shell. An analogous cluster with the same number of waters was taken out from the MD simulation of a plain aqueous solution. In this way, we obtained the starting geometries for the quantum-chemical calculations involving the H$_2$ receptor and the aqueous solution, both with and without histamine. These were submitted to a full geometry optimization at the M06-2X/6-31+G(d) level in Gaussian 16 [43]. Total molecular electronic energies were extracted without thermal corrections, so the results reported here correspond to differences in electronic energies. The effect of the rest of the receptor environment was considered through the implicit SMD solvation using a dielectric constant of $\varepsilon = 4.0$, as suggested by Himo and co-workers [56], and a dielectric constant of $\varepsilon = 78.4$ for the aqueous solution, in line with our previous reports [25,26,51]. In addition, such a truncated cluster-continuum model of the entire protein turned out to be very useful in rationalizing various aspects of the catalytic activity [57], selectivity [58] and inhibition [51] of the monoamine oxidase family of enzymes, and is broadly used by different groups to describe various biological phenomena [59–63], which justifies its use here.

Lastly, although the literature presents a number of methods for the quantization of nuclear motion, relevant for studying the H/D isotope substitution, these are limited to only a few degrees of freedom. Yet, these are not applicable here, since we have many critical protons directly involved in the H$_2$ receptor-ligand recognition and water hydration. As such, we employed an approximate empirical treatment of the nuclear quantum effects based on the mentioned experimental work by Bordallo and co-workers [8], which showed that deuteration reduces the electrostatic attraction in the acidic N–D bonds by 2.3% relative to the matching N–H bonds. With this in mind, we imposed the empirical quantization in the following way. Initially, all systems were fully optimized, thus mirroring the case with lighter H nuclei. After that, all acidic N–H, O–H and S–H bonds were shortened by 2.3% and kept frozen during the optimization of other geometric parameters, thus corresponding to heavier D nuclei, in accordance with our earlier reports [25,26].

## 4. Conclusions

This study relied on a range of computational techniques to demonstrate the significance of the hydrogen bonding and other non-covalent interactions for the binding of histamine to its $H_2$ receptor, and evaluated how these are affected by deuteration. Molecular docking analysis determined histamine binding poses on the homology model of the $H_2$ receptor, while molecular dynamics simulation underlined crucial residues governing the binding. This recognized Asp98 as the most dominant residue, accounting for 40% of the total binding energy, further held in place by Tyr250, which donates hydrogen bonding to Asp98 and accepts it from the histamine $-NH_3^+$ group. In contrast to earlier literature reports, we showed that the significant role of Thr190 is not in the −OH hydrogen bonds, but rather in the C–H···π contacts with the imidazole ring, while the former is persistently involved in the hydrogen bonding with a more distant Asp186. The rest of the binding pocket is hydrophobic, allowing for a range of favorable contacts with Phe254, Phe251 and Val99, but also evident in a clear predominance for the *gauche* histamine conformation within the receptor, unlike the aqueous solution where it is *trans*. Molecular dynamics simulation in the aqueous solution revealed that the first histamine solvation shell involves five water molecules at all three nitrogen sites, yet the interaction with its $-NH_3^+$ groups mostly does not occur with three water molecules at the same time, but is linked with an average of 2.2 such contacts during the entire simulation.

Following molecular dynamics simulation, which identified receptor residues crucial for the binding and a representative cluster of 36 water molecules in the aqueous solution, quantum chemical calculations at the M06–2X/6–31+G(d) level utilized the empirical quantization of the acidic X–H bonds (X = N, O, S) to support the increased histamine affinity upon deuteration. The overall binding was separated in two contributions, that from the interaction with the receptor and the one arising from the interaction with the solvent preceding the binding, which were both modeled through a cluster-continuum approach utilizing the implicit SMD solvation with the dielectric constants of $\varepsilon = 4.0$ for the receptor environment, and $\varepsilon = 78.4$ for the aqueous solution. The used computational setup gave the calculated difference in the binding free energy of −0.85 kcal mol$^{-1}$, being in excellent agreement with the measured value of −0.75 kcal mol$^{-1}$, thus confirming the relevance of hydrogen bonds for the receptor activation.

The results of this study highlight the importance of deuteration for the development of new drugs, as the selective replacement of exchangeable hydrogen atoms with deuterium can increase the duration of action due to their slower decomposition [64,65]. In addition, this can result in different, yet in some instances more beneficial clinical profiles to already marketed solutions, and further progress in this area is highly recommended. Finally, we are convinced that advanced molecular simulations of entire receptors with the inclusion of experimental data will finally lead to a methodology that will be able to discriminate between GPCR agonist and antagonists, which is currently limited to QSAR applications [66].

**Supplementary Materials:** The following are available online, Figures S1–S8 showing various analyses from the molecular dynamics simulation.

**Author Contributions:** Conceptualization, J.M. and R.V.; methodology, J.M. and R.V.; formal analysis, L.H. and R.V.; investigation, L.H.; data curation, L.H.; writing—original draft preparation, R.V.; writing—review and editing, L.H., J.M. and R.V.; visualization, L.H.; supervision, R.V. All authors have read and agreed to the published version of the manuscript.

**Funding:** Part of this research was funded by the Slovenian Research Agency, program group P1–0012.

**Acknowledgments:** L.H. wishes to thank the Croatian Science Foundation for a doctoral stipend through the Career Development Project for Young Researchers. J.M. thanks the Slovenian Research Agency for financial support. L.H. and R.V. would like to thank the Zagreb University Computing Centre (SRCE) for granting computational resources on the ISABELLA cluster.

**Conflicts of Interest:** The authors declare no conflict of interest. The funders had no role in the design of the study; in the collection, analyses, or interpretation of data; in the writing of the manuscript, or in the decision to publish the results.

## References

1. Walter, M.; Stark, H. Histamine receptor subtypes: A century of rational drug design. *Front. Biosci.* **2012**, *4*, 461–488. [CrossRef]
2. Cong, X.; Topin, J.; Golebiowski, J. Class A GPCRs: Structure, function, modeling and structure-based ligand design. *Curr. Pharm. Des.* **2017**, *23*, 4390–4409. [CrossRef] [PubMed]
3. Congreve, M.; Langmead, C.J.; Mason, J.S.; Marshall, F.H. Progress in structure based drug design for G protein-coupled receptors. *J. Med. Chem.* **2011**, *54*, 4283–4311. [CrossRef] [PubMed]
4. Keshelava, A.; Solis, G.P.; Hersch, M.; Koval, A.; Kryuchkov, M.; Bergmann, S.; Katanaev, V.L. High capacity in G protein-coupled receptor signaling. *Nat. Commun.* **2018**, *9*, 876. [CrossRef] [PubMed]
5. Mason, J.S.; Bortolato, A.; Congreve, M.; Marshall, F.H. New insights from structural biology into the druggability of G protein-coupled receptors. *Trends Pharmacol. Sci.* **2012**, *33*, 249–260. [CrossRef]
6. Weis, W.I.; Kobilka, B.K. The Molecular Basis of G Protein–Coupled Receptor Activation. *Annu. Rev. Biochem.* **2018**, *87*, 897–919. [CrossRef]
7. Ubbelohde, A.R.; Gallagher, K.J. Acid-base effects in hydrogen bonds in crystals. *Acta Cryst.* **1955**, *8*, 71–83. [CrossRef]
8. De Souza, J.M.; Freire, P.T.C.; Bordallo, H.N.; Argyriou, D.N. Structural isotopic effects in the smallest chiral amino acid: Observation of a structural phase transition in fully deuterated alanine. *J. Phys. Chem. B* **2007**, *111*, 5034–5039. [CrossRef]
9. Shi, C.; Zhang, X.; Yu, C.-H.; Yao, Y.-F.; Zhang, W. Geometric isotope effect of deuteration in a hydrogen-bonded host-guest crystal. *Nat. Commun.* **2018**, *9*, 481. [CrossRef]
10. Rivera-Rivera, L.A.; Wang, Z.; McElmurry, B.A.; Willaert, F.F.; Lucchese, R.R.; Bevan, J.W.; Suenram, R.D.; Lovas, F.J. A ground state morphed intermolecular potential for the hydrogen bonded and van der Waals isomers in OC:HI and a prediction of an anomalous deuterium isotope effect. *J. Chem. Phys.* **2010**, *133*, 184305. [CrossRef]
11. Mishra, A.K.; Murli, C.; Sharma, S.M. High pressure Raman spectroscopic study of deuterated γ-glycine. *J. Phys. Chem. B* **2008**, *112*, 15867–15874. [CrossRef] [PubMed]
12. Goncalves, R.O.; Freire, P.T.C.; Bordallo, H.N.; Lima, J.A., Jr.; Melo, F.E.A.; Mendes Filho, J.; Argyriou, D.N.; Lima, R.J.C. High-pressure Raman spectra of deuterated L-alanine crystal. *J. Raman Spectrosc.* **2009**, *40*, 958–963. [CrossRef]
13. Smirnov, S.N.; Golubev, N.S.; Denisov, G.S.; Benedict, H.; Schah-Mohammedi, P.; Limbach, H.-H. Hydrogen/deuterium isotope effects on the NMR chemical shifts and geometries of intermolecular low-barrier hydrogen-bonded complexes. *J. Am. Chem. Soc.* **1996**, *118*, 4094–4101. [CrossRef]
14. Yarnell, A.T. Heavy-hydrogen drugs turn heads, again. *Chem. Eng. News* **2009**, *87*, 36–39. [CrossRef]
15. Halford, B. Deuterium switcheroo breathes life into old drugs. *Chem. Eng. News* **2016**, *94*, 32–36.
16. Belleau, B.; Burba, J.; Pindell, M.; Reiffenstein, J. Effect of Deuterium Substitution in Sympathomimetic Amines on Adrenergic Responses. *Science* **1961**, *133*, 102–104. [CrossRef]
17. Falconnet, J.B.; Brazier, J.L.; Desage, M. Synthesis of seven deuteromethyl-caffeine analogues; observation of deuterium isotope effects on CMR analysis. *J. Label. Compd. Radiopharm.* **1986**, *23*, 267–276. [CrossRef]
18. Brazier, J.L.; Ribon, B.; Falconnet, Y.; Cherrah, Y.; Benchekroun, Y. Etude et utilisation des effets isotopiques en pharmacologie. *Therapie* **1987**, *42*, 445–450.
19. Cherrah, Y.; Falconnet, J.B.; Desage, M.; Brazier, J.L.; Zini, R.; Tillement, J.P. Study of deuterium isotope effects on protein binding by gas chromatography/mass spectrometry. Caffeine and deuterated isotopomers. *Biomed. Environ. Mass Spectrom.* **1987**, *14*, 653–657. [CrossRef]

20. Cherrah, Y.; Zini, R.; Falconnet, J.B.; Desage, M.; Tillement, J.P.; Brazier, J.L. Study of deutero-isotopomer-induced inhibition of caffeine and phenobarbitone binding to human serum albumin. *Biochem. Pharmacol.* **1988**, *37*, 1311–1315. [CrossRef]
21. Schmidt, C. First deuterated drug approved. *Nat. Biotechnol.* **2017**, *35*, 493–494. [CrossRef] [PubMed]
22. Toth, G.; Bowers, S.G.; Truong, A.P.; Probst, G. The Role and Significance of Unconventional Hydrogen Bonds in Small Molecule Recognition by Biological Receptors of Pharmaceutical Relevance. *Curr. Pharm. Des.* **2007**, *13*, 3476–3493. [CrossRef] [PubMed]
23. Vianello, R.; Mavri, J. Microsolvation of the histamine monocation in aqueous solution: The effect on structure, hydrogen bonding ability and vibrational spectrum. *New J. Chem.* **2012**, *36*, 954–962. [CrossRef]
24. Stare, J.; Mavri, J.; Grdadolnik, J.; Zidar, J.; Maksić, Z.B.; Vianello, R. Hydrogen bond dynamics of histamine monocation in aqueous solution: Car–Parrinello molecular dynamics and vibrational spectroscopy study. *J. Phys. Chem. B* **2011**, *115*, 5999–6010. [CrossRef] [PubMed]
25. Kržan, M.; Vianello, R.; Maršavelski, A.; Repič, M.; Zakšek, M.; Kotnik, K.; Fijan, E.; Mavri, J. The quantum nature of drug-receptor interactions: Deuteration changes binding affinities for histamine receptor ligands. *PLoS ONE* **2016**, *11*, e0154002. [CrossRef]
26. Kržan, M.; Keuschler, J.; Mavri, J.; Vianello, R. Relevance of Hydrogen Bonds for the Histamine H2 Receptor-Ligand Interactions: A Lesson from Deuteration. *Biomolecules* **2020**, *10*, 196. [CrossRef]
27. Cheng, H.C. The power issue: Determination of $K_B$ or $K_i$ from $IC_{50}$—A closer look at the Cheng–Prusoff equation, the Schild plot and related power equations. *J. Pharmacol. Toxicol. Methods* **2001**, *46*, 61–71. [CrossRef]
28. Singh, V.; Gohil, N.; Ramírez-García, R. New insight into the control of peptic ulcer by targeting the histamine $H_2$ receptor. *J. Cell Biochem.* **2018**, *119*, 2003–2011. [CrossRef]
29. Strasser, A.; Wittmann, H.J. Molecular Modelling Approaches for the Analysis of Histamine Receptors and Their Interaction with Ligands. In *Histamine and Histamine Receptors in Health and Disease. Handbook of Experimental Pharmacology*; Hattori, Y., Seifert, R., Eds.; Springer: Cham, Switzerland, 2017; Volume 241, pp. 31–61.
30. Sun, X.; Li, Y.; Li, W.; Xu, Z.; Tang, Y. Computational investigation of interactions between human $H_2$ receptor and its agonists. *J. Mol. Graph. Model.* **2011**, *29*, 693–701. [CrossRef]
31. Schreeb, A.; Łażewska, D.; Dove, S.; Buschauer, A.; Kieć-Kononiwcz, K.; Stark, H. Histamine $H_4$ receptor ligands. In *Histamine $H_4$ Receptor: A Novel Drug Target for Immunoregulation and Inflammation*; Stark, H., Ed.; Versita: London, UK, 2013; pp. 21–61.
32. Collado, J.A.; Tuñón, I.; Silla, E.; Ramírez, F.J. Vibrational Dynamics of Histamine Monocation in Solution: An Experimental (FT-IR, FT-Raman) and Theoretical (SCRF-DFT) Study. *J. Phys. Chem. A* **2000**, *104*, 2120–2131. [CrossRef]
33. Collado, J.A.; Ramírez, F.J. Vibrational spectra and assignments of histamine dication in the solid state and in solution. *J. Raman Spectrosc.* **2000**, *31*, 925–931. [CrossRef]
34. Drozdzewski, P.; Kordon, E. Isotope effects in the far-infrared spectra of histamine complexes with palladium(II). *Vib. Spectrosc.* **2000**, *24*, 243–248. [CrossRef]
35. Xerri, B.; Flament, J.-P.; Petitjean, H.; Berthomieu, C.; Berthomieu, D. Vibrational modeling of copper-histamine complexes: Metal-ligand IR modes investigation. *J. Phys. Chem. B* **2009**, *113*, 15119–15127. [CrossRef] [PubMed]
36. Homeyer, N.; Gohlke, H. Free energy calculations by the molecular mechanics Poisson–Boltzmann surface area method. *Mol. Inform.* **2012**, *31*, 114–122. [CrossRef] [PubMed]
37. Birdsall, N.J. Cloning and structure-function of the $H_2$ histamine receptor. *Trends Pharmacol. Sci.* **1991**, *12*, 9–10. [CrossRef]
38. Gantz, I.; DelValle, J.; Wang, L.D.; Tashiro, T.; Munzert, G.; Guo, Y.J.; Konda, Y.; Yamada, T. Molecular basis for the interaction of histamine with the histamine $H_2$ receptor. *J. Biol. Chem.* **1992**, *267*, 20840–20843.
39. Ramesh Kumara, G.; Gokul Raj, S.; Saxena, A.; Karnal, A.K.; Raghavalu, T.; Mohan, R. Deuteration effects on structural, thermal, linear and nonlinear properties of L-threonine single crystals. *Mater. Chem. Phys.* **2008**, *108*, 359–363. [CrossRef]
40. Zhang, J.; Qi, T.; Wei, J. Homology Modeling and Antagonist Binding Site Study of the Human Histamine H2 Receptor. *Med. Chem.* **2012**, *8*, 1084–1092.

41. Strasser, A. Molecular modeling and QSAR-based design of histamine receptor ligands. *Expert Opin. Drug Discov.* **2009**, *4*, 1061–1075. [CrossRef]
42. Catalan, J.; Abboud, J.L.M.; Elguero, J. Basicity and Acidity of Azoles. *Adv. Heterocycl. Chem.* **1897**, *41*, 187–274.
43. Frisch, M.J.; Trucks, G.W.; Schlegel, H.B.; Scuseria, G.E.; Robb, M.A.; Cheeseman, J.R.; Scalmani, G.; Barone, V.; Petersson, G.A.; Nakatsuji, H.; et al. *Gaussian 16, Revision C.01*; Gaussian, Inc.: Wallingford, CT, USA, 2016.
44. Marenich, A.V.; Cramer, C.J.; Truhlar, D.G. Universal Solvation Model Based on Solute Electron Density and on a Continuum Model of the Solvent Defined by the Bulk Dielectric Constant and Atomic Surface Tensions. *J. Phys. Chem. B* **2009**, *113*, 6378–6396. [CrossRef] [PubMed]
45. Grosdidier, A.; Zoete, V.; Michielin, O. SwissDock, a protein-small molecule docking web service based on EADock DSS. *Nucleic Acids Res.* **2011**, *39*, W270–W277. [CrossRef] [PubMed]
46. Pettersen, E.F.; Goddard, T.D.; Huang, C.C.; Couch, G.S.; Greenblatt, D.M.; Meng, E.C.; Ferrin, T.E. UCSF Chimera—A Visualization System for Exploratory Research and Analysis. *J. Comput. Chem.* **2004**, *13*, 1605–1612. [CrossRef] [PubMed]
47. Case, D.A.; Betz, R.M.; Cerutti, D.S.; Cheatham, T.E., III; Darden, T.A.; Duke, R.E.; Giese, T.J.; Gohlke, H.; Goetz, A.W.; Homeyer, N.; et al. *Amber 2016*; University of California: San Francisco, CA, USA, 2016.
48. Darden, T.; York, D.; Pedersen, L. Particle Mesh Ewald: An $N·\log(N)$ Method for Ewald Sums in Large Systems. *J. Chem. Phys.* **1993**, *98*, 10089–10092. [CrossRef]
49. Genheden, S.; Ryde, U. The MM/PBSA and MM/GBSA Methods to Estimate Ligand-Binding Affinities. *Expert Opin. Drug Discov.* **2015**, *10*, 449–461. [CrossRef]
50. Hou, T.; Wang, J.; Li, Y.; Wang, W. Assessing the Performance of the MM/PBSA and MM/GBSA Methods. 1. The Accuracy of Binding Free Energy Calculations Based on Molecular Dynamics Simulations. *J. Chem. Inf. Model.* **2011**, *51*, 69–82. [CrossRef]
51. Tandarić, T.; Vianello, R. Computational Insight into the Mechanism of the Irreversible Inhibition of Monoamine Oxidase Enzymes by the Antiparkinsonian Propargylamine Inhibitors Rasagiline and Selegiline. *ACS Chem. Neurosci.* **2019**, *10*, 3532–3542. [CrossRef]
52. Perković, I.; Raić-Malić, S.; Fontinha, D.; Prudêncio, M.; Pessanha de Carvalho, L.; Held, J.; Tandarić, T.; Vianello, R.; Zorc, B.; Rajić, Z. Harmicines—Harmine and Cinnamic Acid Hybrids as Novel Antiplasmodial Hits. *Eur. J. Med. Chem.* **2020**, *187*, 111927. [CrossRef]
53. Gohlke, H.; Kiel, C.; Case, D.A. Insights into Protein-Protein Binding by Binding Free Energy Calculation and Free Energy Decomposition for the Ras-Raf and Ras-RalGDS Complexes. *J. Mol. Biol.* **2003**, *330*, 891–913. [CrossRef]
54. Rastelli, G.; Del Rio, A.; Degliesposti, G.; Sgobba, M. Fast and Accurate Predictions of Binding Free Energies Using MM-PBSA and MM-GBSA. *J. Comput. Chem.* **2010**, *31*, 797–810. [CrossRef]
55. Olsson, M.H.M.; Søndergaard, C.R.; Rostkowski, M.; Jensen, J.H. PROPKA3: Consistent treatment of internal and surface residues in empirical $pK_a$ predictions. *J. Chem. Theory Comput.* **2011**, *7*, 525–537. [CrossRef] [PubMed]
56. Liao, R.Z.; Georgieva, P.; Yu, J.G.; Himo, F. Mechanism of mycolic acid cyclopropane synthase: A theoretical study. *Biochemistry* **2011**, *50*, 1505–1513. [CrossRef] [PubMed]
57. Vianello, R.; Repič, M.; Mavri, J. How are biogenic amines metabolized by monoamine oxidases? *Eur. J. Org. Chem.* **2012**, *36*, 7057–7065. [CrossRef]
58. Maršavelski, A.; Vianello, R. What a difference a methyl group makes: The selectivity of monoamine oxidase B towards histamine and N-methylhistamine. *Chem. Eur. J.* **2017**, *23*, 2915–2925. [CrossRef]
59. Himo, F. Recent trends in quantum chemical modeling of enzymatic reactions. *J. Am. Chem. Soc.* **2017**, *139*, 6780–6786. [CrossRef]
60. Blomberg, M.R.A.; Borowski, T.; Himo, F.; Liao, R.-Z.; Siegbahn, P.E.M. Quantum chemical studies of mechanisms for metalloenzymes. *Chem. Rev.* **2014**, *114*, 3601–3658. [CrossRef]
61. Quesne, M.G.; Borowski, T.; de Visser, S.P. Quantum mechanics/molecular mechanics modeling of enzymatic processes: Caveats and breakthroughs. *Chem. Eur. J.* **2016**, *22*, 2562–2581. [CrossRef]
62. Sousa, S.F.; Ribeiro, A.J.M.; Neves, R.P.P.; Brás, N.F.; Cerqueira, N.M.F.S.A.; Fernandes, P.A.; Ramos, M.J. Application of quantum mechanics/molecular mechanics methods in the study of enzymatic reaction mechanisms. *Wires Comput. Mol. Sci.* **2017**, *7*, e1281. [CrossRef]

63. Quesne, M.G.; Silveri, F.; de Leeuw, N.H.; Catlow, C.R.A. Advances in sustainable catalysis: A computational perspective. *Front. Chem.* **2019**, *7*, 182. [CrossRef]
64. Kaur, S.; Gupta, M. Deuteration as a tool for optimization of metabolic stability and toxicity of drugs. *Glob. J. Pharm. Sci.* **2017**, *1*, 555566.
65. Tung, R.D. Deuterium medicinal chemistry comes of age. *Future Med. Chem.* **2016**, *8*, 491–494. [CrossRef] [PubMed]
66. Don, C.G.; Riniker, S. Scents and sense: In silico perspectives on olfactory receptors. *J. Comput. Chem.* **2014**, *35*, 2279–2287. [CrossRef] [PubMed]

**Sample Availability:** Samples of the compounds are not available from the authors.

**Publisher's Note:** MDPI stays neutral with regard to jurisdictional claims in published maps and institutional affiliations.

© 2020 by the authors. Licensee MDPI, Basel, Switzerland. This article is an open access article distributed under the terms and conditions of the Creative Commons Attribution (CC BY) license (http://creativecommons.org/licenses/by/4.0/).

Article

# Pinacolone-Alcohol Gas-Phase Solvation Balances as Experimental Dispersion Benchmarks

Charlotte Zimmermann, Taija L. Fischer and Martin A. Suhm *

Institut für Physikalische Chemie, Georg-August-Universität Göttingen, Tammannstr. 6, 37077 Göttingen, Germany; czimmer2@gwdg.de (C.Z.); tfische1@gwdg.de (T.L.F.)
* Correspondence: msuhm@gwdg.de; Tel.: +49-551-3933112

Academic Editor: Ilya G. Shenderovich
Received: 25 September 2020; Accepted: 28 October 2020; Published: 3 November 2020

**Abstract:** The influence of distant London dispersion forces on the docking preference of alcohols of different size between the two lone electron pairs of the carbonyl group in pinacolone was explored by infrared spectroscopy of the OH stretching fundamental in supersonic jet expansions of 1:1 solvate complexes. Experimentally, no pronounced tendency of the alcohol to switch from the methyl to the bulkier *tert*-butyl side with increasing size was found. In all cases, methyl docking dominates by at least a factor of two, whereas DFT-optimized structures suggest a very close balance for the larger alcohols, once corrected by CCSD(T) relative electronic energies. Together with inconsistencies when switching from a C4 to a C5 alcohol, this points at deficiencies of the investigated B3LYP and in particular TPSS functionals even after dispersion correction, which cannot be blamed on zero point energy effects. The search for density functionals which describe the harmonic frequency shift, the structural change and the energy difference between the docking isomers of larger alcohols to unsymmetric ketones in a satisfactory way is open.

**Keywords:** dispersion; ketone–alcohol complexes; density functional theory; hydrogen bonds; molecular recognition; vibrational spectroscopy; gas phase; benchmark; pinacolone

## 1. Introduction

In nature, directional hydrogen bonds to carbonyl groups [1,2] are frequent, for instance in proteins, DNA or other biopolymers [3,4]. London dispersion interactions are less directional, but at least as omnipresent [5]. An accurate and detailed theoretical description of these interactions and their cooperation or competition is urgently needed. As in any complex interplay, there is a risk of error cancellation. One may easily get the right answer for the wrong reason. The situation calls for systematic isolation attempts with respect to the different contributions. This can be achieved by the study of a series of small hydrogen-bonded complexes at low temperature in the supersonically expanded gas phase by rotational and vibrational spectroscopy [6–8]. Even at low temperature, anharmonic zero point vibrational energy (ZPVE) still complicates the comparison between electronic structure theory and experimental information on the relative energy of different molecular arrangements [9]. A more direct test of the potential energy landscape would be very desirable.

This has led to the concept of ketone solvation balances, which were introduced for acetophenone and its derivatives in combination with alcohols as hydrogen bond donors [10,11] and tested for other ketones [12,13]. The idea is to have two very comparable lone electron pairs available at the acetophenone oxygen, to which alcohols can either dock from the phenyl or from the alkyl side, with little difference in ZPVE. Besides the intrinsic preference of a docking alcohol for the methyl side due to the more favorable local hydrogen bond geometry [11], the alkyl group of the alcohol will interact dispersively (and by Pauli repulsion) with the two ketone substituents and thus contribute to the preference for one of the docking sides. These secondary interactions through space are able to tip the balance towards phenyl side docking [11]. The comparison of different alcohols and acetophenones thus provides information on London dispersion interactions competing with the electronic and zero-point vibrational local hydrogen bond effects, which still largely govern the position of the alcoholic OH vibration. The latter is used to spectrally discriminate the docking isomers and it also contains further information on the competition of forces, because hydrogen bonds can be distorted by distant interactions of the donor molecule. Experimental information on the docking preference comes from the relative abundance of the docking isomers in the quasi-equilibrium established by cooling collisions in a supersonic jet expansion, down to some conformational freezing temperature $T_c$ (roughly 30 to 150 K, depending on low (1 to 5 kJ mol$^{-1}$) and narrow interconversion barriers [10,14,15]) and can thus only be predicted with a large tolerance.

The results of such studies can be used to benchmark the ability of different density functionals to predict the interplay of hydrogen bonding with distant London dispersion and Pauli repulsion, by simply comparing the predictions to experiment. This can be done strictly at the level of observables, without consulting any energy decomposition models [16–18], although the latter are helpful in the interpretation of the findings. A functional which gives the right answer for the right reason in the popular harmonic approximation for vibrations must be able to predict the splitting of the OH stretching vibrations between the docking isomers (because anharmonic effects by construction largely cancel when comparing the isomers) and the relative abundance of the isomers with a reasonable conformational temperature. As a third test, high level single point wavefunction calculations (for which Hessian calculations to reproduce the spectrum would be too costly) at the optimized DFT minima should confirm the energy predictions in a qualitative sense. If at least one of these three diagnostics fails, the DFT functional performance can be proven to be poor down to a sub-kJ/mol accuracy threshold. This was the case for one out of six pairings of aromatic ketones with alcohols in the first systematic study [11], for the otherwise most successful B3LYP-D3 functional. By using the second-most stable and less compact predicted structure in this particular case, the performance could actually be rescued [11]. This former systematic investigation thus suggests a mildly erroneous preference of the B3LYP-D3 functional (at least for a standard def2-TZVP basis set), and to a lesser extent also the TPSS-D3 approach, for compact structures. Other explored functionals such as M06-2X failed the aromatic ketone balance test in several aspects [11] and need not be considered further.

The hypothesis that B3LYP-D3 and TPSS-D3 show an (almost) acceptable performance for ketone dispersion balances obviously calls for further falsification attempts and this is the task of the present study which involves the purely aliphatic pinacolone (see Figure 1), where the phenyl group in acetophenone is replaced by a *tert*-butyl (*t*Bu) group. This removes aromatic–aliphatic dispersive interactions and brings in more bulky donor-acceptor constellations. Cyclopentanol (CpOH) is introduced as a further, more disk-like and flexible alcohol, in addition to methanol (MeOH) and *tert*-butyl alcohol (*t*BuOH), which have been previously explored with acetophenone [11]. Pinacolone monomer does not have a plane of symmetry [19], but in combination with low planarization barriers (Figures S2–S4) and symmetry-breaking alcohol coordination, this should not lead to additional complications in the analysis. Indeed, there are significant variations of the hydrogen bond angle $\alpha$ and the dihedral angle $\tau$ (see Figure 1) with alcohol substitution. These promise to explore the hydrogen bonding potential of carbonyl groups far away from the intrinsic in plane preference.

In this work, we show that, in alcohol–pinacolone balances, the methyl docking side is consistently preferred. According to exploratory calculations, this may extend to many alcohols beyond the experimentally investigated ones. Further, we show that the predictive quality of the two density functionals which were successful for acetophenone (B3LYP and TPSS) decreases with the size of the alcohol, including significant failures for the largest (CpOH). The proposed assignments and observed trends are discussed and an analysis of dispersion interactions on the docking side preference is presented. We provide initial evidence that some of the superficially satisfactory DFT performance for ketone balances must be fortuitous.

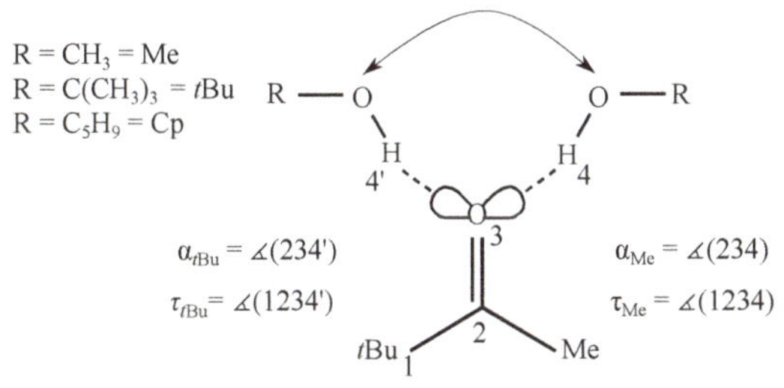

Figure 1. Schematic representation of the two possible docking sides 4 and 4′ in a pinacolone molecule (*t*Bu and Me) with different alcohols (R-OH, with the abbreviations Me for methyl, *t*Bu for *tert*-butyl and Cp for cyclopentyl as R).

## 2. Results and Discussion

We start with the theoretical description of alcohol–pinacolone 1:1 complexes at the level of DFT before comparing to the experimental findings and finally consulting wave-function theory.

### 2.1. Density Functional Predictions

From now on, the abbreviations Pin for the studied ketone pinacolone and MeOH (methanol), *t*BuOH (*tert*-butyl alcohol) and CpOH (cyclopentanol) for the solvating alcohols are consistently used. In Figure 1, two angles $\alpha$ and $\tau$ describing the hydrogen bond geometry are introduced. The hydrogen bond angle between the hydrogen bonded H and the carbonyl group has a local, sp$^2$-explainable preference for ≈120°. The dihedral angle $\tau$ describes the out of plane twist of the docking alcohol OH with respect to the carbonyl plane, with two local preferences near 0° and 180°. Any deviations from these local preferences due to global interactions sensitively affect the hydrogen-bonded OH stretching wavenumber.

As detailed in Table S2 and Figure S1, all six experimentally investigated 1:1 complexes show a narrow distribution for $\alpha$ (115–124°) at the four investigated DFT levels (D3-corrected B3LYP and TPSS with triple and quadruple zeta basis sets). $\tau$ deviates from planarity with increasing size of the alcohol, in steps of roughly 10° from MeOH over *t*BuOH to CpOH. On the *t*Bu docking side of Pin, even MeOH is already displaced by 35–37°, due to the bulkiness of the substituent, whereas the Me docking displacement is less than 10° for MeOH.

The structural trends are reflected in the calculated OH stretching wavenumbers (see Table S4), which are consistently lower for Me docking for all three alcohols, whereas the trend with increasing alcohol size is comparatively weak, relative to the overall hydrogen bond shift. This assists a straightforward interpretation of the experimental spectra.

The energy differences between Me and *t*Bu docking sides fall between 0 and 3 kJ mol$^{-1}$, always preferring the Me side, as shown in Figure 2. The narrow corridor of ±0.2 kJ mol$^{-1}$ in the figure (gray lines) illustrates that it makes almost no difference whether harmonic ZPVE is included or not. The effect of basis size extension is similarly small. This is very favorable for a direct judgement of the DFT functional in terms of the predicted electronic energy difference without worrying about major (anharmonic) zero point energy or basis set effects which can both be quite significant when looking at absolute energies and frequencies [10,20].

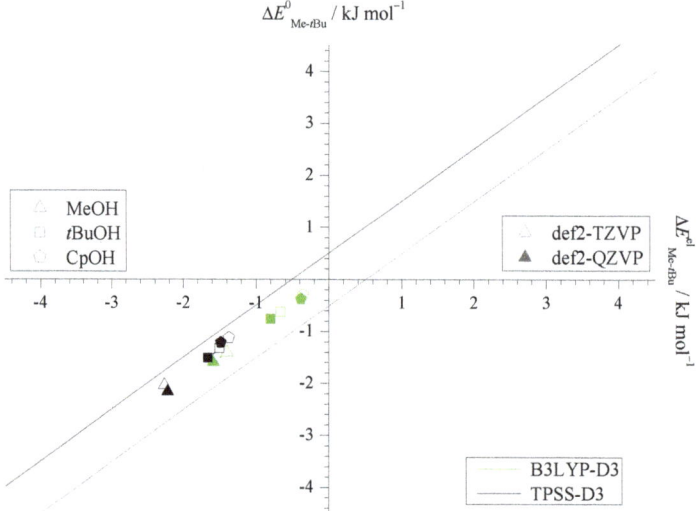

**Figure 2.** Harmonically zero-point corrected energy differences $\Delta E^0_{\text{Me}-t\text{Bu}}$ plotted against the electronic energy differences $\Delta E^{\text{el}}_{\text{Me}-t\text{Bu}}$ referenced to the tBu side, computed at B3LYP-D3 (green) and TPSS-D3 (black) level, each with a def2-TZVP (empty symbols) and def2-QZVP (filled symbols) basis set. The electronic energy differences are seen to be a good approximation to experimentally relevant ZPVE-inclusive differences and the methyl docking side is systematically preferred (see also Table S3).

The predicted spread in docking energy difference of about 2.5 kJ mol$^{-1}$ across the systems promises a large variation of the experimental abundances, but the absence of a sign reversal (corresponding to an absence of data points in the upper right quadrant of Figure 2) despite varying the alcohol size from 1 to 5 carbon atoms is surprising. An explorative search for almost 20 other alcoholic donors (see Table S5) confirms this systematic bias. The steric disadvantage of the tBu side of Pin together with the flexibility of alcohols provides possible explanations. The latter allows the alcohol to dock on the sterically more accessible Me side and at the same time to exploit London dispersion interaction with the tBu side. A good example is benzyl alcohol, where the Me sided structure is almost 2 kJ mol$^{-1}$ more stable, because the benzyl group can still interact favorably with the tBu group of the Pin while the OH group is docking to the Me side of Pin.

Another important feature of carbonyl balances is the feasibility of the isomerization under supersonic jet expansion conditions. A transition state search between the two competing structures for MeOH–Pin yielded an interconversion barrier height of about 3 kJ mol$^{-1}$ when viewed from the tBu docking structure. The interconversion path is distinctly out-of-plane, relaxing the hydrogen bond angle $\alpha$ while switching between small and large $\tau$. This is similar to previous findings for acetophenone [11] and its derivatives and supports a feasible interconversion under supersonic jet conditions, with $T_c$ values significantly below the starting temperature of the expansion. However, the more numerous the contacts between the residue and Pin are, the larger this barrier may become. This is one reason this work focuses on small alcohols to establish the performance of the DFT functionals.

Before switching to experiment, two important observable predictions need to be explored. One is a sufficiently robust infrared cross section ratio for the docking isomers, which is a precondition for reliable experimental abundance determinations from spectral intensities. As shown in Figure S5, the basis set and functional dependences are modest and the trends are smooth, such that this variation and the double-harmonic approximation are not expected to be critical for the theory-experiment comparison.

The most important theoretical assignment aid concerns the predicted positions and differences or splittings of the OH stretching fundamental vibrations. While the harmonic approximation is too crude for absolute predictions, the harmonic Me-tBu differences involve systematic cancellation of anharmonic contributions for similar docking environments. Furthermore, the structural effect of increasing alcohol size is qualitatively similar on both docking sides, as pointed out above, and should translate into relatively uniform wavenumber splittings as a function of the number of alcoholic C atoms. This is illustrated in Figure 3. In all cases, the Me-docking wavenumber is lower, corresponding to a uniformly negative $\Delta\omega_{Me-tBu}$ value and facilitating experimental assignment. The size of the splitting exceeds the spectral resolution and band width [11] by more than an order of magnitude, which is also favorable.

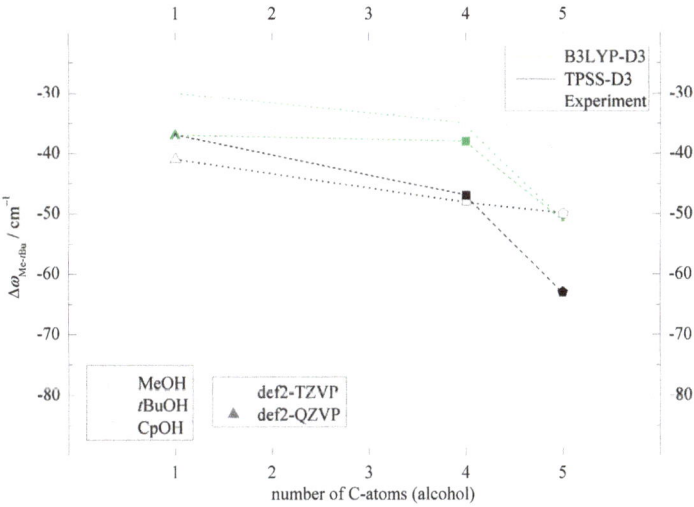

**Figure 3.** Computed OH wavenumber difference between the two docking sides $\Delta\omega_{Me-tBu}$ relative to the number of C-atoms of the corresponding alcohol. This shows that the employed computational methods predict the same spectral trends for MeOH and tBuOH, indicated by dashed lines. For CpOH a somewhat larger discrepancy can be observed, with the smaller basis set TPSS result differing most from the experimental trend (blue) (see Table S4 for details).

With a single exception (TPSS for CpOH for the larger basis set), all predicted harmonic splittings are within $\pm 12\,\text{cm}^{-1}$ of the average value of $-42\,\text{cm}^{-1}$ and there is only a weakly decreasing trend for the splitting with increasing alcohol size. For CpOH, predictions range from $-50$ to $-63\,\text{cm}^{-1}$. These variations are also reflected in the $\tau$ angle (Table S2). They are robust with respect to cross-over re-optimization and indicate a slight TPSS-bistability of the structure depending on the basis set. Anticipating the experimental (anharmonic) result reported in the next section (blue symbols and lines in Figure 3), the larger basis set result appears less likely in absolute numbers but more likely in terms of the trend. Even beyond this outlier, it is clear that the alcohol size trends are not predicted perfectly, thus underscoring the benchmarking potential of this study.

## 2.2. Experimental Results

In Figure 4, the experimental infrared spectra for helium supersonic jet expansions of Pin with MeOH (green), tBuOH (orange) and CpOH (blue) are shown. They feature the rovibrationally broadened alcohol monomer OH stretching bands (MeOH, tBuOH, CpOH), the downshifted hydrogen-bonded homodimer signals ((MeOH)$_2$, (tBuOH)$_2$ and (CpOH)$_2$), as well as the narrow bands of the mixed complexes with docking isomerism $O_{Me}$ and $O_{tBu}$. For MeOH, $O_{Me}$ and $O_{tBu}$

are spectrally downshifted compared to the respective homodimer band, as one might expect from an intrinsically stronger OH···O=C interaction, whereas, in the case of tBuOH and CpOH, they are upshifted. This is already an experimental sign for competition between hydrogen bonding and more global London dispersion interactions. Even when the alcohol is in Me docking position, where there is no sterical crowding, it is displaced out of the ketone plane to maximize interaction with the tBu group (see Table S2). When CpOH is combined with acetone, which lacks the tBu group (see Figure S6), the homodimer and mixed dimer signals actually overlap. This is partly due to less competition from dispersion interaction with the other side of the ketone for the hydrogen bond.

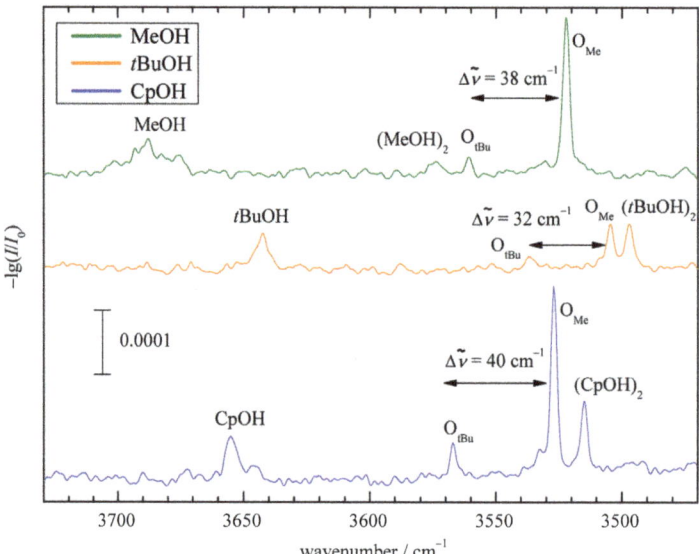

**Figure 4.** FTIR jet OH stretching spectra of Pin with the three alcohols. The 1:1 complexes are marked with O, indexed by the assigned docking preference. Both docking sides are observed. Pin is only a stronger OH shifting partner than the alcohol itself for MeOH.

The more downshifted mixed dimer band $O_{Me}$ in Figure 4 is always significantly stronger and based on the robust DFT predictions for structure (Table S2) and downshift (Figure 3 and Table S4), it must be due to Me docking, as implied by the label. Given that its spectral visibility (Figure S5) is at best twice that of the tBu isomer, it must also be the more stable isomer, in agreement with the DFT computations (Figure 2).

The experimental shift between $O_{Me}$ and $O_{tBu}$ spans a relatively narrow range of 30 to 40 cm$^{-1}$ (Figure 3 and Table S6), which roughly matches the DFT prediction window, except for the TPSS outlier. In many cases, the DFT splitting is somewhat larger than the experimental one, which matches the general overestimation of hydrogen bond shifts by most density functionals. The subtle alcohol substitution trend in the splittings (Figure 3) is not well reproduced, being monotonically decreasing for the DFT predictions and non-monotonic in the experiment, but, considering the superposition of Me and tBu trends, this appears acceptable and does not complicate the spectral assignment.

In Figure 5, the experimentally determined downshifts from the monomer OH fundamental are plotted against the corresponding calculated ones. The fact that all correlation points stay below the diagonal line confirms the systematic overestimation of DFT downshifts, which is more pronounced for TPSS than for B3LYP [11] and only in part due to anharmonicity. The slope of the data points matches the diagonal for methanol (dashed arrows connect isomers), but it becomes flatter for the more bulky alcohols. This indicates that the DFT calculations overestimate the hydrogen bond weakening by bulkiness (dispersion and/or exchange repulsion). Note that non-isomeric acetone docking results [21] (for CpOH, see Figure S6) included in the figure also fit the Pin data for Me docking.

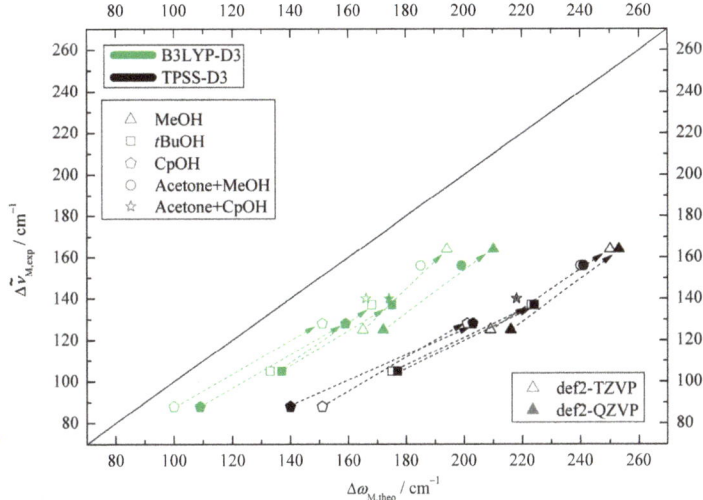

**Figure 5.** Experimental (anharmonic) downshift of the 1:1 complexes $\Delta\tilde{\nu}_{M,exp}$ plotted against the harmonically computed downshifts $\Delta\omega_{M,theo}$ for four computational variants. The harmonic DFT overestimation and the trend between docking sides (dashed arrows from $t$Bu to Me docking) are uniform.

The CpOH–Pin case is also suspicious in terms of the B3LYP energy gap between Me and $t$Bu docking. Based on Figure 4, Me docking should be substantially more stable, even more so if statistically formed conformations freeze rather early in the expansion. However, the predicted energy difference is $\leq 0.5$ kJ mol$^{-1}$ (see Table S3), far too low for such an imbalance. Attempts to rescue the situation in analogy to the acetophenone balance study [11] by searching for metastable minima on the DFT hypersurfaces failed (see Tables S14 and S15 for details). In this context the pseudorotational isomerism of the axial CpOH monomer should be briefly discussed. There are two nearly isoenergetic (B3LYP-D3(BJ,abc)/def2-TZVP energy difference less than 0.1 kJ mol$^{-1}$) and isospectral isomers (about 4 cm$^{-1}$ wavenumber difference, leading to slight shift uncertainty), depending on the position of the axial OH in the envelope conformation. This has caused uncertainty in rotational [22] and has also been addressed in liquid state spectroscopy [23]. However, structure optimizations indicate that this subtle isomerism is relaxed in the complexes with Pin, leading to uniform axial gauche results (or energetically >2 kJ mol$^{-1}$ higher conformations). The problem is thus more fundamental, as the following analysis supports.

For this purpose, the experimental abundance is compared to the predicted B3LYP energy difference for all three investigated systems by calculating a concentration ratio $c_{Me}/c_{tBu}$, which follows from the experimental intensity ratio and the def2-TZVP absorption cross sections (see Table 1). The maximum and minimum values for $I_{Me}/I_{tBu}$ from a Monte Carlo integration program [24] generate a maximum and minimum value for $c_{Me}/c_{tBu}$ which is further transformed to a (semi-)experimental $x_{tBu}$ range. The values confirm that Me docking is strongly preferred for all systems. This result is completely robust with respect to the four theoretical levels, even allowing for possible ZPVE errors of $\pm 0.2$ kJ mol$^{-1}$ and for residual errors in the theoretical cross section ratio (see Tables S3, S7 and S8 for more details).

One should emphasize that the predicted energy imbalance between the two docking isomers is always below 8% (Table S9), so rather small on an absolute scale. Our experiment is thus rather sensitive in detecting errors in this imbalance, making it suitable for benchmarking studies [10].

**Table 1.** Experimental integrated intensity ratios $I_{Me}/I_{tBu}$, B3LYP-D3(BJ,abc)/def2-TZVP cross-section derived docking ratios $c_{Me}/c_{tBu}$ and resulting experimental fractions $x_{tBu}$ for $t$Bu docking. The given ranges represent 95% confidence for $I_{Me}/I_{tBu}$ using an automated statistical evaluation [24] and are carried on to $c_{Me}/c_{tBu}$ and $x_{tBu}$ without including a theoretical cross section ratio uncertainty.

| Donor | $\frac{I_{Me}}{I_{tBu}}$ | $\frac{c_{Me}}{c_{tBu}}$ | $x_{tBu}$ |
|---|---|---|---|
| MeOH  | 5.9–11.8 | 4.5–9.0 | 0.10–0.18 |
| tBuOH | 2.4–4.2  | 1.7–3.0 | 0.25–0.37 |
| CpOH  | 5.5–7.3  | 3.4–4.4 | 0.18–0.23 |

Figure 6a plots the (semi-)experimental fraction of $t$Bu docking (Table 1 and Tables S7 and S8) against the energy difference prediction for the four combinations of functional with basis set. The grey areas indicate qualitative inconsistencies between theoretical prediction and experiment, within the assumptions of uniform anharmonicity and accurate cross section ratio. If Me docking is energetically favorable, $t$Bu should not dominate the expansion and vice versa. Asymmetrical error bars are obtained by taking the mean value for $I_{Me}/I_{tBu}$ as the data point and using the Monte Carlo determined range (see Table 1 and Table S7) as the boundaries.

At first sight, experiment and DFT theory (Figure 6a) are consistent with each other and different DFT levels cannot be discriminated against each other. Even the obvious outliers for CpOH can be accommodated in the allowed (white) area. However, two closer looks at the data reveal deficiencies.

To bring the different theory levels closer together, one can plot the experimental $t$Bu docking abundance gain $\Delta x_{tBu}$ against the theoretical $t$Bu docking energy gain $\Delta\Delta E^0_{tBu}$, when switching from MeOH to a heavier alcohol (Figure S7). One would expect that any energy gain leads to a docking abundance gain, but all DFT methods predict a higher energy gain for CpOH and experiment finds a higher docking abundance gain for $t$BuOH. Clearly, the DFT description is somewhat imbalanced for either CpOH or $t$BuOH or for both.

Another way of analyzing the deficiency is to calculate an effective conformational temperature $T_c$ for each DFT method and pair of isomers from the experimental band integral ratio and the computed IR band strength ratio [10]. Based on the (semi-)experimental concentration ratios $\frac{c_{Me}}{c_{tBu}}$ listed in Table 1 and the computed energy differences $\Delta E^0_{Me-tBu}$ in Figure 6, this can be obtained as

$$T_c \approx -\frac{\Delta E^0_{Me-tBu}}{R \ln \frac{c_{Me}}{c_{tBu}}}$$

with the universal gas constant $R$, if there are no symmetry differences between the docking isomers and the rovibrational partition functions are sufficiently similar due to supersonic jet cooling. $T_c$ should roughly fall in the range of 30 to 150 K [11,15]. This is the case for almost all 12 combinations of system and method, within the respective error bar (Table S10). Only TPSS for $t$BuOH–Pin gives higher $T_c$ values and B3LYP for CpOH–Pin is borderline on the low end. The former could be due to a higher interconversion barrier but the latter is likely due to an overestimated stability of the $t$Bu docking side.

These inconsistencies call for a check with wavefunction theory, which is presented in the next section.

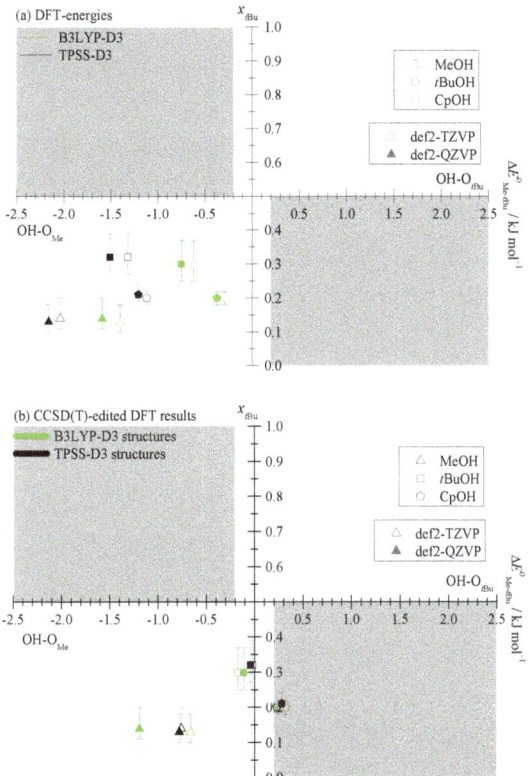

**Figure 6.** Experimental $t$Bu docking fraction $x_{tBu} = c_{tBu}/(c_{tBu} + c_{Me})$ (based on 95% confidence intervals and the mean value for the ratio $I_{Me}/I_{tBu}$ from Table 1 and Tables S7 and S8) plotted against the computed ZPVE corrected energy differences $E^0_{Me-tBu}$. Grey areas indicate inconsistency between experiment and theory, when allowing for an estimated anharmonic ZPVE error of $\pm 0.2$ kJ mol$^{-1}$ and assuming correct cross section ratios form the respective theoretical model. (**a**) DFT energies, where all models predict the correct qualitative docking preference, but the correlation of energy and abundance is non-uniform. (**b**) As in (**a**), but with the electronic energy being replaced by the corresponding DLPNO-CCSD(T) value (see Section 2.3).

### 2.3. DLPNO-CCSD(T) Check

For the large complexes of interest in this work, harmonic frequency analysis and thus zero-point energy calculation is not very practical beyond DFT level. However, single point energies at DLPNO-CCSD(T) level [25] were calculated at the minima obtained for the various DFT methods (with the setting tightPNO, basis sets aug-cc-pVQZ and aug-cc-pVQZ/C, see Table S1). They offer several benefits.

First, they allow judging which of the DFT methods is likely closer to the true minimum by looking at the absolute CCSD(T) energies [9]. In all cases, B3LYP outperforms TPSS but in most B3LYP cases the smaller basis set gives a slightly lower energy. This may be taken as a weak indication that the B3LYP structures are closer to reality, but there could be some compensation between intra- and intermolecular degrees of freedom.

Second, one can replace the DFT electronic energy difference between isomers by the corresponding DLPNO-CCSD(T) difference and keep the structural and ZPVE contributions from the DFT level. This generates a variant of Figure 6a, in which the data points for all larger alcohols now fall close to or into the lower-right grey and thus unphysical region, where major *t*Bu docking is expected but Me docking is predominantly observed (see Figure 6b). Only MeOH stays in the physically meaningful range. The mere fact that DLPNO-CCSD(T) correction leads to such large energy difference changes casts doubt on the quality of the DFT (in particular TPSS) structures. Note that all 12 corrections (see Table S12) promote *t*Bu docking, so the DFT error is highly systematic. For B3LYP, the corrections stay below 1 kJ mol$^{-1}$, for TPSS they always exceed 1 kJ mol$^{-1}$. Because experiment is consistent with a preference for Me docking in all cases, this likely means that the DFT structures for Me docking are relatively far from the best ones, in particular for TPSS. As the CCSD(T) corrections are quite uniform for all three alcohol–pinacolone complexes, it is plausible that the deficiency does not reside so much in the dispersion correction but rather in the functional and its description of differences in hydrogen bonding to the acceptor C=O group.

A third application of DLPNO-CCSD(T) is to provide dispersion contributions to the interaction energy in the LED scheme [16,17] (Table S11). This is a refined way of obtaining such (strictly speaking non-observable) dispersion energies, which is conceptually better than simply evaluating the size of the D3 correction in the complex (Table S13). In the present case, the numbers obtained for both methods are quite similar, but this cannot be generalized except perhaps for large distances, where London dispersion is best defined and LED [16], SAPT [18] or empirical dispersion correction [26] should become comparable as leading corrections to long range electrostatic and inductive interactions. Dispersion always favors *t*Bu docking, by 1.5 to 3.1 kJ mol$^{-1}$ in the LED scheme (1.6 to 2.8 kJ mol$^{-1}$ for D3 corrections). The dependence on the size of the alcohol is quite modest, but CpOH tends to show the largest gains, at least for B3LYP.

Returning to the conformational freezing temperature analysis, now with DLPNO-CCSD(T)-corrected values (Table S10), only MeOH docking yields reasonable $T_c$ values (larger than 30 K). For *t*BuOH docking, B3LYP predicts borderline $T_c$ values and TPSS predictions are far too low. For CpOH, none of the CCSD(T)-corrected DFT results give physical $T_c$ values.

In summary, the DLPNO analysis shows that dispersion-corrected TPSS docking structures are imbalanced, more so than B3LYP structures. It confirms that beyond MeOH, the best isomer energy predictions are inconsistent with experiment or at best borderline (for B3LYP and *t*BuOH docking). Compared to acetophenone [11], Pin is seen to be a more critical test ketone. As it is purely aliphatic, there is likely some error compensation in the apparently more successful mixed aliphatic-aromatic acetophenone case [11].

## 3. Materials and Methods

The spectroscopic data were obtained by probing pulsed supersonic slit jet expansions of Pin+alcohol-seeded helium gas with a synchronized FTIR spectrometer. Specifically, helium (Linde 99.996%) is led through a temperature-controlled gas-flow system, where it passes separate gas saturators filled with the analytes pinacolone (Alfa Aesar > 97%) and alcohol (methanol (Roth $\geq$ 99.9%), *tert*-butyl alcohol (Roth $\geq$ 99.9%) or cyclopentanol (Fluka Chemicals > 99.9%)). The gas mixture is filled into a 67 L reservoir at a pressure of 0.75 bar and pulsed through six magnetic valves into a pre-expansion chamber which is terminated by a 600 mm long and 0.2 mm wide slit nozzle. During about 0.2 s, the gas flows through this slit into a vacuum chamber connected to a buffer volume (23 m$^3$), which is continuously evacuated by a series of pumps with a power of 500 to 2500 m$^3$ h$^{-1}$. The expansion is crossed by a modulated and softly focused IR beam from a Bruker IFS 66v/S FTIR spectrometer with a 150 W tungsten filament, CaF$_2$ optics and a liquid nitrogen cooled InSb detector. The scans are obtained with a resolution of 2 cm$^{-1}$ and are synchronized with the gas pulse. The shown spectra are averaged over 300–425 gas pulses. More details on the experimental setup can be found elsewhere [27]. No evidence was found that more than two structural isomers of the studied 1:1 complexes are formed during the experiment.

To determine the band integral ratios $I_{Me}/I_{tBu}$, an automated statistical evaluation was used, where the main entering parameters include the band positions and band width, which is statistically varied (chosen at $(3.0 \pm 0.5)$ cm$^{-1}$) [24]. The program adds synthetic noise to the spectra, providing statistical error bars for $I_{Me}/I_{tBu}$. The resulting 95% confidence interval was used for further data processing.

DFT calculations were used for assignment purposes and to trigger future benchmarking of their ability to describe the combination of hydrogen bonding and distant London dispersion interactions. Therefore, they were limited to two functionals and two basis sets, but others are invited to find more powerful density functionals for this challenge. The initial structural search (manual and using Crest [28]) was carried out at B3LYP-D3/def2-TZVP level [29–32]. Reoptimization was carried out with a def2-QZVP basis set [32] and with the meta-GGA functional TPSS-D3 [33] using the same def2-TZVP and def2-QZVP basis sets. Three body-inclusive D3 dispersion correction [26] with Becke–Johnson damping [34–37] was always applied. Single point energies were obtained using DLPNO-CCSD(T) [25,38,39] at the DFT-optimized structures. For all these calculations, ORCA version 4.2.1 [40] was used. Further information on computational details can be found in the Supplementary Materials (Table S1). Thermal corrections to the isomer equilibrium were neglected due to the low and mode-dependent temperatures in a jet expansion, with rotational temperatures expected to be on the order 10 K. Vibrational temperatures are on the order of 100 K and conformational temperatures, which depend on the barrier between isomers, are discussed in the main text [10]. The harmonic treatment of the ZPVE is expected to be more than sufficient for this kind of systems and for the achievable accuracy, due to the near-equivalence of the two lone electron pairs [11]. A transition state search for one system was carried out with Woelfling (Turbomole [41,42]) and followed by an optimization with ORCA version 4.2.1 [40].

## 4. Conclusions

Three alcohols of increasing size were combined with pinacolone to determine the hydrogen bonding preference to either the methyl- or the *tert*-butyl-facing lone electron pair of the keto group. As generally predicted for almost two dozen alcohols by dispersion-corrected B3LYP calculations, the methyl side is preferred for methanol, *tert*-butyl alcohol and cyclopentanol. This was qualitatively confirmed by infrared spectroscopy of supersonic jet expansions in combination with approximate IR absorption cross sections. Quantitatively, the DFT predictive power in terms of the spectral splitting decreases with increasing alcohol size. In addition, the observed spectral abundance does not correlate systematically with the predicted energy difference. DLPNO-CCSD(T) energy calculations indicate that B3LYP provides a somewhat better description of the combined hydrogen bond and London dispersion interaction than TPSS. However, in combination with the experiment, they suggest that docking on the methyl side is systematically underrated by both density functionals on the 1 kJ mol$^{-1}$ scale. This only amounts to about 3 % of the total binding energy but is quite significant on the relative energy scale of competitive ketone docking.

Intermolecular energy balances are thus shown to be powerful benchmarking tools to assess the ability of DFT methods to describe hydrogen bonding in competition with London dispersion. The ketone balance variety is particularly useful, as it involves systematically compensating zero-point-energy contributions and therefore allows judging electronic structure predictions in a rather direct way. For acetophenone, only a slight deficiency of the B3LYP functional could be identified [11]. For pinacolone, none of the investigated functionals comes close to describing the spectral splitting and the energetics of the docking isomerism for all three alcohols, but D3-corrected B3LYP performs satisfactorily for methanol docking and borderline for *tert*-butyl alcohol. The qualitative failure of theory to describe the experimentally observed cyclopentanol docking invites studies of related complexes, such as cyclohexanol–pinacolone and cyclopentanol–acetophenone. Larger modifications involve the use of phenol [43] and the switch from the OH chromophore to NH stretching as a probe of the conformational preference.

The goal is to find a density functional which systematically reproduces the harmonic wavenumber splitting between docking isomers within better than about 10 cm$^{-1}$ and which provides a conformational temperature of the correct sign between about 30 and 150 K across a large number of isomeric complexes with low interconversion barrier. Furthermore, DLPNO-CCSD(T) correction should not change the energy difference between the isomers by more than about 0.5 kJ mol$^{-1}$, thus indicating a sufficiently balanced structural description. The best-performing B3LYP-D3/def2-QZVP approach in the present study only fulfills about half of these criteria for the three systems and the corresponding TPSS-D3 calculation even fewer than a quarter. Considering that some of these matches will be fortuitous, this is clearly not a satisfactory state, calling for further experimental and theoretical investigations.

**Supplementary Materials:** The following are available online, Figure S1: Structures, Figures S2–S4: Pinacolone torsion scans, Figure S5: Cross section ratios, Figure S6: CpOH-acetone spectra, Figure S7: Substitution trends, Table S1: Keywords, Table S2: Angles, Table S3: Unimportance of ZPVE, Table S4: OH stretching shifts, Table S5: Other explored donors, Table S6: Experimental band centers, Tables S7 and S8: Band integrals, Table S9: Dissociation energies, Table S10: Conformational temperatures, Table S11: LED analysts, Table S12: CCSD(T) energy differences, Table S13: D3 Analysis, Tables S14 and S15: Higher lying isomers, Tables S16–S21: Cartesian coordinates

**Author Contributions:** Conceptualization, methodology, funding acquisition and supervision, M.A.S.; experimental investigation, C.Z. and T.L.F.; formal analysis, C.Z. and M.A.S.; visualization, data curation and writing–original draft preparation, C.Z.; and writing–review and editing, M.A.S., C.Z. and T.L.F. All authors have read and agreed to the published version of the manuscript.

**Funding:** This research was funded by the Deutsche Forschungsgemeinschaft (DFG, German Research Foundation) grant number 271107160/SPP1807. The APC was co-funded by the Goettingen open access publication funds and by personal membership in the SCS.

**Acknowledgments:** We acknowledge computer time on the GWDG computer cluster as well as the local chemistry cluster— DFG grant number 405832858/INST 186/1294-1 FUGG. Valuable support from the mechanical and electronic workshops of the department is much appreciated, as are discussions with Robert Medel.

**Conflicts of Interest:** The authors declare no conflict of interest. The funders had no role in the design of the study, in the collection, analyses, or interpretation of data, in the writing of the manuscript, or in the decision to publish the results.

## Abbreviations

The following abbreviations are used in this manuscript:

| | |
|---|---|
| MeOH | Methanol |
| tBuOH | tert-Butyl alcohol |
| CpOH | Cyclopentanol |
| Pin | Pinacolone |
| ZPVE | Zero-point vibrational energy |

## References

1. Lommerse, J.P.M.; Price, S.L.; Taylor, R. Hydrogen bonding of carbonyl, ether, and ester oxygen atoms with alkanol hydroxyl groups. *J. Comp. Chem.* **1997**, *18*, 757–774. [CrossRef]
2. Murray-Rust, P.; Glusker, J.P. Directional hydrogen bonding to sp2- and sp3-hybridized oxygen atoms and its relevance to ligand-macromolecule interactions. *J. Am. Chem. Soc.* **1984**, *106*, 1018–1025. [CrossRef]
3. Derewenda, Z.S.; Lee, L.; Derewenda, U. The Occurence of C–H · · · O Hydrogen Bonds in Proteins. *J. Mol. Biol.* **1995**, *252*, 248–262. [CrossRef]
4. Nikolova, E.N.; Stanfield, R.L.; Dyson, H.J.; Wright, P.E. CH···O Hydrogen Bonds Mediate Highly Specific Recognition of Methylated CpG Sites by the Zinc Finger Protein Kaiso. *Biochemistry* **2018**, *57*, 2109–2120. [CrossRef]
5. Wagner, J.P.; Schreiner, P.R. London Dispersion in Molecular Chemistry–Reconsidering Steric Effects. *Angew. Chem. Int. Ed.* **2015**, *54*, 12274–12296. [CrossRef]
6. Schnell, M.; Erlekam, U.; Bunker, P.R.; von Helden, G.; Grabow, J.U.; Meijer, G.; van der Avoird, A. Unraveling the internal dynamics of the benzene dimer: A combined theoretical and microwave spectroscopy study. *Phys. Chem. Chem. Phys.* **2013**, *15*, 10207–10223. [CrossRef]

7. Schwing, K.; Gerhards, M. Investigations on isolated peptides by combined IR/UV spectroscopy in a molecular beam—Structure, aggregation, solvation and molecular recognition. *Int. Rev. Phys. Chem.* **2016**, *35*, 569–677. [CrossRef]
8. Shipman, S.T.; Neill, J.L.; Suenram, R.D.; Muckle, M.T.; Pate, B.H. Structure Determination of Strawberry Aldehyde by Broadband Microwave Spectroscopy: Conformational Stabilization by Dispersive Interactions. *J. Phys. Chem. Lett.* **2011**, *2*, 443–448. [CrossRef]
9. Gottschalk, H.C.; Poblotzki, A.; Suhm, M.A.; Al-Mogren, M.M.; Antony, J.; Auer, A.A.; Baptista, L.; Benoit, D.M.; Bistoni, G.; Bohle, F.; et al. The furan microsolvation blind challenge for quantum chemical methods: First steps. *J. Chem. Phys.* **2018**, *148*, 014301. [CrossRef]
10. Poblotzki, A.; Gottschalk, H.C.; Suhm, M.A. Tipping the Scales: Spectroscopic Tools for Intermolecular Energy Balances. *J. Phys. Chem. Lett.* **2017**, *8*, 5656–5665. [CrossRef]
11. Zimmermann, C.; Gottschalk, H.C.; Suhm, M.A. Three-dimensional docking of alcohols to ketones: An experimental benchmark based on acetophenone solvation energy balances. *Phys. Chem. Chem. Phys.* **2020**, *22*, 2870–2877. [CrossRef]
12. Burevschi, E.; Alonso, E.R.; Sanz, M.E. Binding Site Switch by Dispersion Interactions: Rotational Signatures of Fenchone–Phenol and Fenchone–Benzene Complexes. *Chem. Eur. J.* **2020**, *26*, 11327–11333. [CrossRef]
13. Banerjee, P.; Pandey, P.; Bandyopadhyay, B. Stereo-preference of camphor for H-bonding with phenol, methanol and chloroform: A combined matrix isolation IR spectroscopic and quantum chemical investigation. *Spectrochim. Acta Part A* **2019**, *209*, 186–195. [CrossRef]
14. Gottschalk, H.C.; Altnöder, J.; Heger, M.; Suhm, M.A. Control over the Hydrogen-Bond Docking Site in Anisole by Ring Methylation. *Angew. Chem. Int. Ed.* **2016**, *55*, 1921–1924. [CrossRef] [PubMed]
15. Gottschalk, H.C.; Poblotzki, A.; Fatima, M.; Obenchain, D.A.; Pérez, C.; Antony, J.; Auer, A.A.; Baptista, L.; Benoit, D.M.; Bistoni, G.; et al. The first microsolvation step for furans: New experiments and benchmarking strategies. *J. Chem. Phys.* **2020**, *152*, 164303. [CrossRef] [PubMed]
16. Altun, A.; Neese, F.; Bistoni, G. Local energy decomposition analysis of hydrogen-bonded dimers within a domain-based pair natural orbital coupled cluster study. *Beilstein J. Org. Chem.* **2018**, *14*, 919–929. [CrossRef]
17. Schneider, W.B.; Bistoni, G.; Sparta, M.; Saitow, M.; Riplinger, C.; Auer, A.A.; Neese, F. Decomposition of Intermolecular Interaction Energies within the Local Pair Natural Orbital Coupled Cluster Framework. *J. Chem. Theory Comput.* **2016**, *12*, 4778–4792. [CrossRef]
18. Boese, A.D.; Jansen, G. ZMP-SAPT: DFT-SAPT using ab initio densities. *J. Chem. Phys.* **2019**, *150*, 154101. [CrossRef]
19. Zhao, Y.; Nguyen, H.V.L.; Stahl, W.; Hougen, J.T. Unusual internal rotation coupling in the microwave spectrum of pinacolone. *J. Mol. Spec.* **2015**, *318*, 91–100. [CrossRef]
20. Bende, A. Hydrogen bonding in the urea dimers and adenine–thymine DNA base pair: Anharmonic effects in the intermolecular H-bond and intramolecular H-stretching vibrations. *Theor. Chem. Acc.* **2010**, *125*, 253–268. [CrossRef]
21. Kollipost, F.; Domanskaya, A.V.; Suhm, M.A. Microscopic Roots of Alcohol–Ketone Demixing: Infrared Spectroscopy of Methanol–Acetone Clusters. *J. Phys. Chem. A* **2015**, *119*, 2225–2232. [CrossRef]
22. Carroll, P.B. Laboratory and Astronomical Rotational Spectroscopy. Ph.D. Thesis, California Institute of Technology, Pasadena, CA, USA, 2018. [CrossRef]
23. Abraham, R.J.; Koniotou, R.; Sancassan, F. Conformational analysis. Part 39. A theoretical and lanthanide induced shift (LIS) investigation of the conformations of cyclopentanol and cis- and trans-cyclopentane-1,2-diol. *J. Chem. Soc. Perkin Trans.* **2002**, *2*, 2025–2030. [CrossRef]
24. Karir, G.; Lüttschwager, N.O.B.; Suhm, M.A. Phenylacetylene as a gas phase sliding balance for solvating alcohols. *Phys. Chem. Chem. Phys.* **2019**, *21*, 7831–7840. [CrossRef] [PubMed]
25. Riplinger, C.; Pinski, P.; Becker, U.; Valeev, E.F.; Neese, F. Sparse maps—A systematic infrastructure for reduced-scaling electronic structure methods. II. Linear scaling domain based pair natural orbital coupled cluster theory. *J. Chem. Phys.* **2016**, *144*, 024109. [CrossRef] [PubMed]
26. Grimme, S.; Antony, J.; Ehrlich, S.; Krieg, H. A consistent and accurate ab initio parametrization of density functional dispersion correction (DFT-D) for the 94 elements H-Pu. *J. Chem. Phys.* **2010**, *132*, 154104. [CrossRef]
27. Suhm, M.A.; Kollipost, F. Femtisecond single-mole infrared spectroscopy of molecular clusters. *Phys. Chem. Chem. Phys.* **2013**, *15*, 10702–10721. [CrossRef]

28. Pracht, P.; Bohle, F.; Grimme, S. Automated exploration of the low-energy chemical space with fast quantum chemical methods. *Phys. Chem. Chem. Phys.* **2020**, *22*, 7169–7192. [CrossRef]
29. Becke, A.D. Density-functional exchange-energy approximation with correct asymptotic behavior. *Phys. Rev. A* **1988**, *38*, 3098–3100. [CrossRef]
30. Becke, A.D. Density-functional thermochemistry. III. The role of exact exchange. *J. Chem. Phys.* **1993**, *98*, 5648–5652. [CrossRef]
31. Lee, C.; Yang, W.; Parr, R.G. Development of the Colle-Salvetti correlation-energy formula into a functional of the electron density. *Phys. Rev. B* **1988**, *37*, 785–789. [CrossRef]
32. Weigend, F.; Ahlrichs, R. Balanced basis sets of split valence, triple zeta valence and quadruple zeta valence quality for H to Rn: Design and assessment of accuracy. *Phys. Chem. Chem. Phys.* **2005**, *7*, 3297–3305. [CrossRef]
33. Tao, J.; Perdew, J.P.; Staroverov, V.N.; Scuseria, G.E. Climbing the Density Functional Ladder: Nonempirical Meta–Generalized Gradient Approximation Designed for Molecules and Solids. *Phys. Rev. Lett.* **2003**, *91*, 146401. [CrossRef] [PubMed]
34. Becke, A.D.; Johnson, E.R. A density-functional model of the dispersion interaction. *J. Chem. Phys.* **2005**, *123*, 154101. [CrossRef]
35. Johnson, E.R.; Becke, A.D. A post-Hartree–Fock model of intermolecular interactions. *J. Chem. Phys.* **2005**, *123*, 024101. [CrossRef]
36. Johnson, E.R.; Becke, A.D. A post-Hartree-Fock model of intermolecular interactions: Inclusion of higher-order corrections. *J. Chem. Phys.* **2006**, *124*, 174104. [CrossRef] [PubMed]
37. Grimme, S.; Ehrlich, S.; Goerigk, L. Effect of the damping function in dispersion corrected density functional theory. *J. Comput. Chem.* **2011**, *32*, 1456–1465. [CrossRef]
38. Riplinger, C.; Neese, F. An efficient and near linear scaling pair natural orbital based local coupled cluster method. *J. Chem. Phys.* **2013**, *138*, 034106. [CrossRef]
39. Riplinger, C.; Sandhoefer, B.; Hansen, A.; Neese, F. Natural triple excitations in local coupled cluster calculations with pair natural orbitals. *J. Chem. Phys.* **2013**, *139*, 134101. [CrossRef]
40. Neese, F. Software update: The ORCA program system, version 4.0. *WIREs Comput. Mol. Sci.* **2018**, *8*, e1327. [CrossRef]
41. TURBOMOLE V7.3 2018, a Development of University of Karlsruhe and Forschungszentrum Karlsruhe GmbH, 1989–2007, TURBOMOLE GmbH, Since 2007. Available online: http://www.turbomole.com (accessed on 30 October 2020)
42. Furche, F.; Ahlrichs, R.; Hättig, C.; Klopper, W.; Sierka, M.; Weigend, F. Turbomole. *WIREs Comput. Mol. Sci.* **2014**, *4*, 91–100. [CrossRef]
43. Ebata, T.; Fujii, A.; Mikami, N. Vibrational spectroscopy of small-sized hydrogen-bonded clusters and their ions. *Int. Rev. Phys. Chem.* **1998**, *17*, 331–361. [CrossRef]

**Sample Availability:** All compounds were obtained commercially and samples of the compounds are not available from the authors.

**Publisher's Note:** MDPI stays neutral with regard to jurisdictional claims in published maps and institutional affiliations.

© 2020 by the authors. Licensee MDPI, Basel, Switzerland. This article is an open access article distributed under the terms and conditions of the Creative Commons Attribution (CC BY) license (http://creativecommons.org/licenses/by/4.0/).

Article

# Halogenated Diazabutadiene Dyes: Synthesis, Structures, Supramolecular Features, and Theoretical Studies

Valentine G. Nenajdenko [1,*], Namiq G. Shikhaliyev [2], Abel M. Maharramov [2], Khanim N. Bagirova [2], Gulnar T. Suleymanova [2], Alexander S. Novikov [3], Victor N. Khrustalev [4,5] and Alexander G. Tskhovrebov [4,6,*]

1. M. V. Lomonosov Moscow State University, 1, Leninskie Gory, 119991 Moscow, Russia
2. Department of Organic Chemistry, Baku State University, Z. Xalilov 23, Baku 1148, Azerbaijan; namiqst@gmail.com (N.G.S.); amaharramov@bsu.edu.az (A.M.M.); stella_stand@icloud.com (K.N.B.); gumusqiz91.sg@gmail.com (G.T.S.)
3. Saint Petersburg State University, Universitetskaya Nab. 7/9, 199034 Saint Petersburg, Russia; a.s.novikov@spbu.ru
4. Peoples' Friendship University of Russia, 6 Miklukho-Maklaya, 117198 Moscow, Russia; vnkhrustalev@gmail.com
5. N.D. Zelinsky Institute of Organic Chemistry, Russian Academy of Sciences, 47 Leninsky Av., 119334 Moscow, Russia
6. N.N. Semenov Federal Research Center for Chemical Physics, Russian Academy of Sciences, Kosygina 4, 119991 Moscow, Russia
* Correspondence: nenajdenko@org.chem.msu.ru (V.G.N.); alexander.tskhovrebov@chph.ras.ru (A.G.T.)

Academic Editors: Ilya G. Shenderovich and Antonio Caballero
Received: 6 October 2020; Accepted: 26 October 2020; Published: 29 October 2020

**Abstract:** Novel halogenated aromatic dichlorodiazadienes were prepared via copper-mediated oxidative coupling between the corresponding hydrazones and $CCl_4$. These rare azo-dyes were characterized using $^1H$ and $^{13}C$ NMR techniques and X-ray diffraction analysis for five halogenated dichlorodiazadienes. Multiple non-covalent halogen···halogen interactions were detected in the solid state and studied by DFT calculations and topological analysis of the electron density distribution within the framework of Bader's theory (QTAIM method). Theoretical studies demonstrated that non-covalent halogen···halogen interactions play crucial role in self-assembly of highly polarizable dichlorodiazadienes. Thus, halogen bonding can dictate a packing preference in the solid state for this class of dichloro-substituted heterodienes, which could be a convenient tool for a fine tuning of the properties of this novel class of dyes.

**Keywords:** non-covalent interactions; crystal engineering; halogen bonding; azo dyes; DFT; QTAIM

---

## 1. Introduction

Halogen bonding (XB) is one of the most intensively investigated areas in modern chemistry [1]. The field currently experiences a renaissance due to exploitation of such weak interactions for a number of functional applications, such as catalysis, drug design, nonlinear optics, reactivity control, and construction of functional supramolecular architectures [2–10] Utilization of non-covalent interactions lies at the foundation of the design supramolecular materials and control of their ultimate architectures [11–14]. XB has recently emerged as a powerful tool for the creation of such materials due to its stability, directionality and reversibility [15–17]. In this context, halogen-halogen interactions received particular attention and were intensively explored both experimentally and theoretically [18–21]. Arguably, XB can be more beneficial than the hydrogen bonding in the construction of functional materials and tuning their properties due to its higher directionality [10,22,23].

Recently, we discovered a novel class of azo-dyes, i.e., dichlorodiazadienes, which can be easily prepared via unprecedented copper-catalyzed reaction between CCl$_4$ with N-substituted hydrazones (Scheme 1) [24]. Currently, very little is known about the chemistry and properties of these dichloro-substituted heterodienes [25–31].

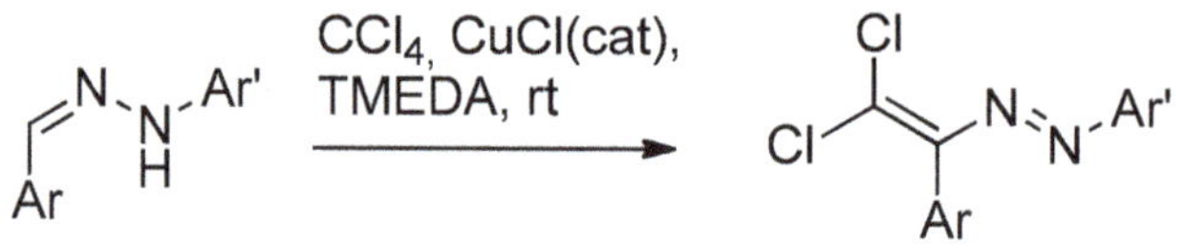

**Scheme 1.** Copper-catalyzed synthesis of dichlorodiazadienes.

Following our interest in construction of supramolecular architectures via non-covalent interactions [32–39] and chemistry of novel diazadienes, we report now the synthesis of halogenated dichlorodiazadienes to demonstrate that dichloro-substituted heterodiene fragment can behave as a strong XB donor/acceptor, what can be used in the design of heterodiene azo-dyes and their self-assembly in the solid state. Incorporation of a halogen atom(s) in the dichloro-dyes' backbone completely changes the way the colorants self-assemble in the crystal. Thus, we show that the XB can dictate a packing preference in the solid state for this class of dichloro-substituted heterodienes. In addition, we performed DFT calculations and topological analysis of the electron density distribution within the formalism of Bader's theory (QTAIM method), which support the presence of intermolecular non-covalent interactions halogen···halogen (Hal···Hal) in the solid state.

## 2. Results and Discussion

The target halogenated azabutadienes **10–18** were synthesized by Cu$^I$-catalyzed reaction between the corresponding hydrazones **1–9** and CCl$_4$ and isolated in up to 82% yield as red crystalline solids (Scheme 2).

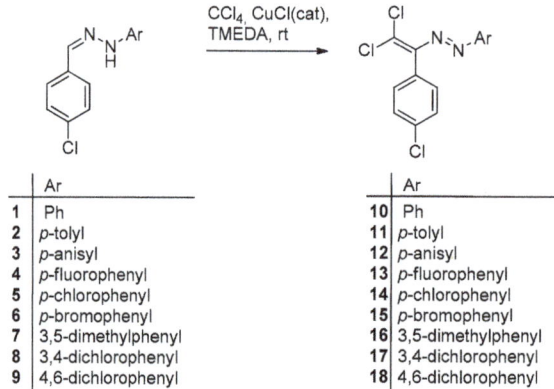

|   | Ar |   | Ar |
|---|---|---|---|
| 1 | Ph | 10 | Ph |
| 2 | p-tolyl | 11 | p-tolyl |
| 3 | p-anisyl | 12 | p-anisyl |
| 4 | p-fluorophenyl | 13 | p-fluorophenyl |
| 5 | p-chlorophenyl | 14 | p-chlorophenyl |
| 6 | p-bromophenyl | 15 | p-bromophenyl |
| 7 | 3,5-dimethylphenyl | 16 | 3,5-dimethylphenyl |
| 8 | 3,4-dichlorophenyl | 17 | 3,4-dichlorophenyl |
| 9 | 4,6-dichlorophenyl | 18 | 4,6-dichlorophenyl |

**Scheme 2.** Copper-catalyzed synthesis of dichlorodiazadienes.

The structure of **10–18** was confirmed by the $^1$H and $^{13}$C NMR spectroscopies and X-ray diffraction analysis for **10**, **13–15**, and **17** (Figures 1–4). $^1$H NMR and $^{13}$C{$^1$H} spectra (CDCl$_3$) are consistent with their solid-state structures. Dyes **10**, **13–15**, and **17** could be easily recrystallized to produce large red crystals, suitable for analysis by single crystal X-ray crystallography. The structural investigations confirmed the formation of azabutadienes. Overall, metrical parameters for **10**, **13–15**, and **17** are similar to those reported for similar azabutadienes [26,29–31]. However, introduction of halogen atoms in the dichloro-dyes' backbone has a dramatic impact on its self-assembly in the crystal. In the crystal packing of **10** (para-chloro substitution at the phenyl, attached the double C=C bond) dye molecules form shifted columns (Figure 1) via π-π interactions. The columns dimerize in the crystal via Cl···Cl attractive interactions between the neighboring dye molecules (type 2 contacts) [23]. The dichloroalkene acts as a donor of the halogen bond here (Figure 1).

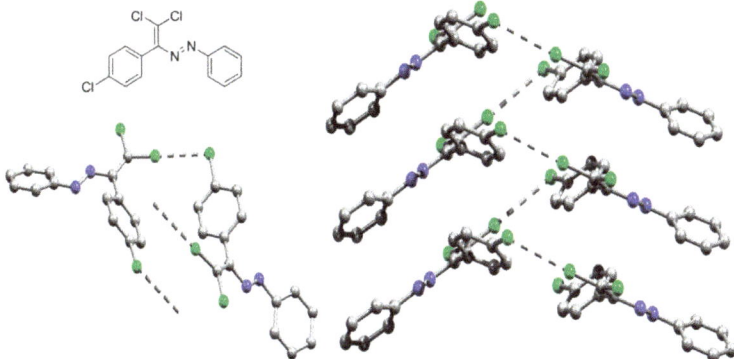

**Figure 1.** Ball-and-stick representation of **10** and its self-assembly via Cl···Cl bonding in the crystal. Blue, green and grey spheres represent nitrogen, chlorine, carbon atoms, respectively. Hydrogen atoms were omitted for clarity.

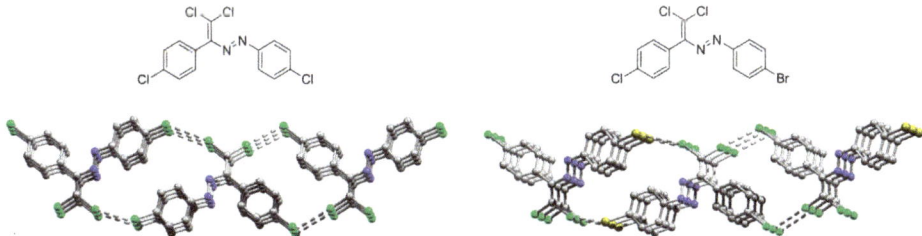

**Figure 2.** Ball-and-stick representations of **14** and **15** and their self-assembly via Cl···Hal bonding in the crystal. Olive-green, blue, green, and grey spheres represent bromine, nitrogen, chlorine, and carbon atoms, respectively. Hydrogen atoms were omitted for clarity.

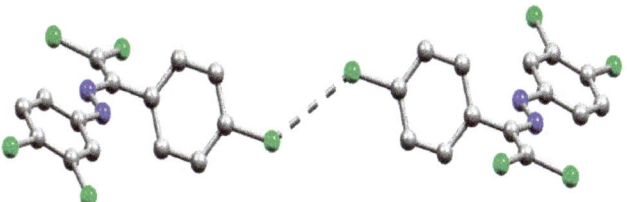

**Figure 3.** Ball-and-stick representation of **17** and its supramolecular dimerization via Cl···Cl type 1 bonding in the crystal. Blue, green, and grey spheres represent nitrogen, chlorine, and carbon atoms, respectively. Hydrogen atoms were omitted for clarity.

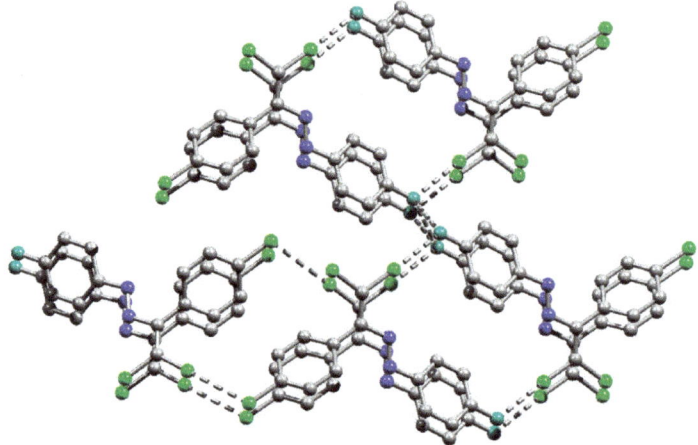

**Figure 4.** Ball-and-stick representation of **13** and its self-assembly via Cl···Cl bonding in the crystal. Blue, green, and grey and cyan spheres represent nitrogen, chlorine, carbon, and fluorine atoms, respectively. Hydrogen atoms were omitted for clarity.

Functionalization of dichloro-dyes with another extra halogen atom (compounds **14** and **15**) does not prevent the formation of columns and supramolecular dimerization via Cl···Cl interactions in the crystal (Figure 2). In addition to this, the columns in the crystal of **14** and **15** interact with another neighboring columns via Cl···Hal (Hal=Cl(**14**), Br(**15**)) type 2 bonding forming 3D supramolecular frameworks (Figure 2).

Introduction of one more halogen atom in the dichloro-dyes' backbone completely changes its self-assembly in the crystal. Remarkably, crystal packing of **17** features only one type of Hal···Hal interaction between the chlorines of the p-cholorophenyl groups (Figure 3), which refer to repulsive type 1 contacts. Halogen atoms, attached to the alkene or dichlorobenzene moieties do not form any halogen bonding. Such a behavior is not very clear at the moment and requires additional studies. One plausible explanation is insufficient nucleophilicity of halogens in **17** for the formation of type 2 contacts.

Finally, when dichloro-dyes are functionalized with the fluorine atom (**13**, para- substitution at the phenyl, attached the double C=C bond, Figure 4), the situation with self-assembly in the crystal is similar to the brominated or chlorinated analogs **14** and **15**. The columns form 3D supramolecular frameworks via Cl···Cl and Cl···F type 2 contacts. An interesting peculiarity of self-assembly of **13** in the crystal is the formation of Cl···F type 1 contacts (Figure 4). Thus, the crystal structure of **13** features a bifurcated XB and a remarkable combination of type 1 and 2 halogen contacts (Figure 4).

Inspection of the crystallographic data suggests the presence of multiple intermolecular non-covalent interactions Hal···Hal in the crystals of **10**, **13–15**, and **17**. Indeed, the observed distances Hal···Hal are shorter than the sum of Bondi's vdW radii for the corresponding atoms [40]. Thus, in addition to structural analysis, a detailed computational studies were desired. In order to understand the nature and quantify energies of various short halogen-halogen contacts the DFT calculations followed by the topological analysis of the electron density distribution within the QTAIM approach [41] were carried out at the ωB97XD/6-311++G ** level of theory for model supramolecular associates containing all types of these noncovalent interactions (see Computational details and Table S1 in the Supplementary Materials). Results of QTAIM analysis summarized in Table 1, the contour line diagrams of the Laplacian of electron density distribution $\nabla^2\rho(r)$, bond paths, and selected zero-flux surfaces as well as visualization of electron localization function (ELF) analysis for selected short halogen-halogen contacts shown in Figure 5 for illustrative purposes.

**Table 1.** Values of the density of all electrons—$\rho(r)$, Laplacian of electron density—$\nabla^2\rho(r)$ and appropriate $\lambda_2$ eigenvalues (with promolecular approximation), energy density—$H_b$, potential energy density—$V(r)$, and Lagrangian kinetic energy—$G(r)$ (a.u.) at the bond critical points (3, −1), corresponding to various short halogen-halogen contacts in **10**, **13–15**, and **17**, and estimated energies for these interactions $E_{int}$ (kcal/mol).

| Halogen–Halogen Contact | $\rho(r)$ | $\nabla^2\rho(r)$ | $\lambda_2$ | $H_b$ | $V(r)$ | $G(r)$ | $E_{int}$ [a] | $E_{int}$ [b] |
|---|---|---|---|---|---|---|---|---|
| **10** Cl···Cl, 3.377 Å (96% from the sum of Bondi's vdW radii) | 0.008 | 0.031 | −0.012 | 0.002 | −0.004 | 0.006 | 1.2 | 1.8 |
| **13** F···F, 2.864 Å (97% from the sum of Bondi's vdW radii) | 0.006 | 0.031 | −0.007 | 0.001 | −0.006 | 0.007 | ≈2 * | ≈2 * |
| **13** F···F, 2.917 Å (99% from the sum of Bondi's vdW radii) | 0.007 | 0.031 | −0.008 | 0.001 | −0.006 | 0.007 | ≈2 * | ≈2 * |
| **13** Cl···F, 2.963 Å (92% from the sum of Bondi's vdW radii) | 0.009 | 0.042 | −0.013 | 0.001 | −0.008 | 0.009 | 2.5 | 2.7 |
| **13** Cl···Cl, 3.405 Å (97% from the sum of Bondi's vdW radii) | 0.007 | 0.029 | −0.011 | 0.002 | −0.004 | 0.006 | 1.2 | 1.8 |
| **14** Cl···Cl, 3.463 Å (99% from the sum of Bondi's vdW radii) | 0.007 | 0.026 | −0.009 | 0.002 | −0.003 | 0.005 | 0.9 | 1.5 |
| **14** Cl···Cl, 3.399 Å (97% from the sum of Bondi's vdW radii) | 0.007 | 0.029 | −0.011 | 0.002 | −0.004 | 0.006 | 1.2 | 1.8 |
| **15** Cl···Br, 3.637 Å (102% from the sum of Bondi's vdW radii) | 0.006 | 0.021 | −0.008 | 0.001 | −0.003 | 0.004 | 0.9 | 1.2 |
| **15** Cl···Cl, 3.394 Å (97% from the sum of Bondi's vdW radii) | 0.007 | 0.030 | −0.011 | 0.002 | −0.004 | 0.006 | 1.2 | 1.8 |
| **17** Cl···Cl, 3.469 Å (99% from the sum of Bondi's vdW radii) | 0.007 | 0.027 | −0.009 | 0.002 | −0.004 | 0.005 | 1.2 | 1.5 |

[a] $E_{int} = 0.49(-V(r))$ (this correlation between the interaction energy and the potential energy density of electrons at the bond critical points (3, −1) was specifically developed for noncovalent interactions involving chlorine atoms) [42]. [b] $E_{int} = 0.47G(r)$ (this correlation between the interaction energy and the kinetic energy density of electrons at the bond critical points (3, −1) was specifically developed for noncovalent interactions involving chlorine atoms) [42]. * There are no generally accepted specific correlations between the interaction energy and the potential or kinetic energy densities of electrons at the bond critical points (3, −1) for F···F noncovalent interactions, but it is clearly expected from values of $V(r)$ and $G(r)$ that strength of these contacts in **13** is approx. 2 kcal/mol.

The QTAIM analysis of **10**, **13–15**, and **17** demonstrates the presence of bond critical points (3, −1) for all weak contacts presented in Table 1. The low magnitude of the electron density (0.006–0.009 a.u.), positive values of the Laplacian of electron density (0.021–0.042 a.u.), and very close to zero positive energy density (0.001–0.002 a.u.) in these bond critical points (3, −1) are typical for halogen-halogen noncovalent interactions [5,39,43]. The balance between the potential and kinetic energy densities of electrons at the bond critical points (3, −1) for studied weak contacts in **10**, **13–15**, and **17** reveals that a covalent contribution is absent in these interactions [44] (Table 1). The Laplacian of electron density is typically decomposed into the sum of contributions along the three principal axes of maximal variation, giving the three eigenvalues of the Hessian matrix ($\lambda_1$, $\lambda_2$ and $\lambda_3$), and the sign of $\lambda_2$ can be utilized to distinguish bonding (attractive, $\lambda_2 < 0$) weak interactions from non-bonding ones (repulsive, $\lambda_2 > 0$) [45,46]. Thus, discussed noncovalent interactions in **10**, **13–15**, and **17** are attractive (Table 1). Overall, it follows from the results of theoretical calculations that all short halogen-halogen contacts in **10**, **13–15**, and **17** are very similar in terms of energies (their estimated strength per one contact vary from 1 to 3 kcal/mol), which correlates well with very close values of minimal and maximal electrostatic surface potentials on halogen atoms in isolated molecules **10**, **13–15**, and **17** (Figure S1 in the Supplementary Materials).

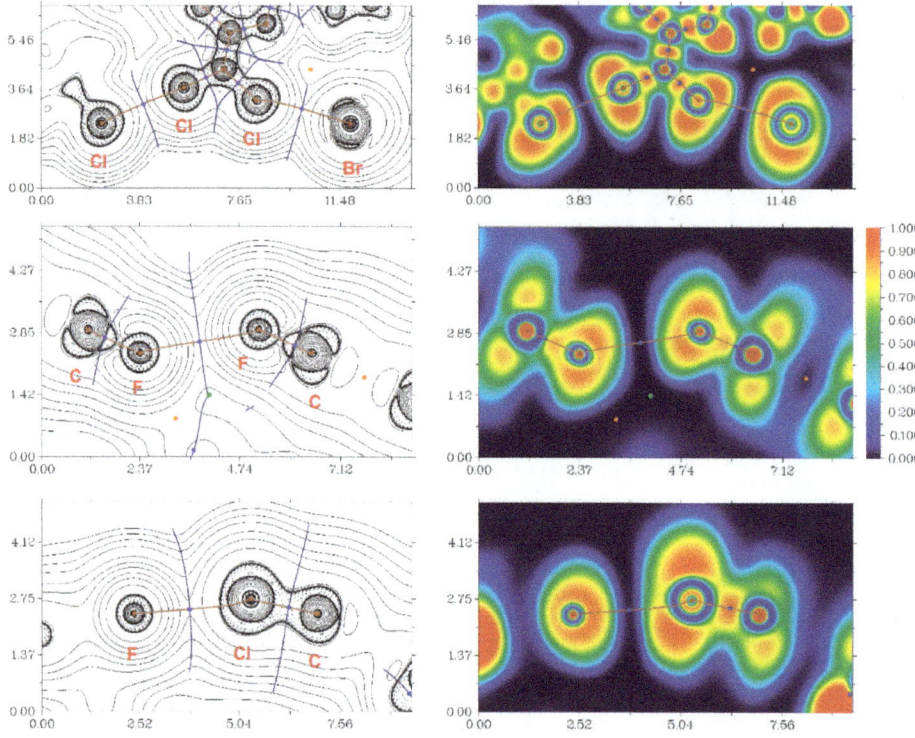

**Figure 5.** Contour line diagrams of the Laplacian of electron density distribution $\nabla^2\rho(r)$, bond paths, and selected zero-flux surfaces (left) and visualization of electron localization function (ELF) analysis (right) for intermolecular Cl···Br and Cl···Cl contacts in **15** (top); F···F (center) and Cl···F (bottom) contacts in **13**. Bond critical points (3, −1) are shown in blue, nuclear critical points (3, −3)—in pale brown, ring critical points (3, +1)—in orange, cage critical points (3, +3)—in light green, length units—Å, bond paths are shown as pale brown lines, and the color scale for the ELF maps is presented in a.u.

To understand what kind of interatomic contacts give the largest contributions in crystal packing, we carried out the Hirshfeld surface analysis for all obtained X-ray structures **10**, **13**–**15**, and **17** (Table 2 and Figure 6). The Hirshfeld surface analysis for the X-ray structures **10**, **13**–**15**, and **17** reveals that in all cases crystal packing determined primarily by interatomic contacts involving chlorine and hydrogen atoms.

**Table 2.** Main partial contributions of different interatomic contacts to the Hirshfeld surfaces of X-ray structures **10**, **13**–**15**, and **17**.

| X-Ray Structure | Contributions of Different Interatomic Contacts to the Hirshfeld Surfaces |
| --- | --- |
| 10 | Cl-H 37.4%, H-H 21.0%, C-H 15.8%, C-C 7.9%, Cl-Cl 5.9%, N-H 5.8%, N-C 3.0%, -Cl-C 2.9%, Cl-N 0.1% |
| 13 | Cl-H 32.6%, H-H 16.2%, C-H 14.6%, C-C 7.4%, Cl-Cl 6.1%, N-H 5.8%, Cl-F 4.1%, N-C 3.1%, Cl-C 3.0%, F-H 2.8%, F-F 2.5%, F-C 1.7% |
| 14 | Cl-H 33.5%, H-H 17.1%, Cl-C 15.7%, C-H 12.6%, C-C 8.5%, N-H 5.2%, Cl-C 4.1%, N-C 3.2%, N-N 0.1% |
| 15 | Cl-H 30.7%, H-H 15.7%, C-H 13.0%, C-C 8.1%, Cl-Cl 5.8%, Br-Cl 5.7%, N-H 5.2%, Br-H 4.6%, Br-Br 3.8%, N-C 3.2%, Cl-C 2.2%, Br-C 1.8%, N-N 0.1% |
| 17 | Cl-H 44.6%, Cl-C 15.6%, H-H 10.5%, Cl-Cl 8.2%, C-H 8.1%, Cl-N 5.1%, C-C 4.4%, N-C 1.9%, N-H 1.5% |

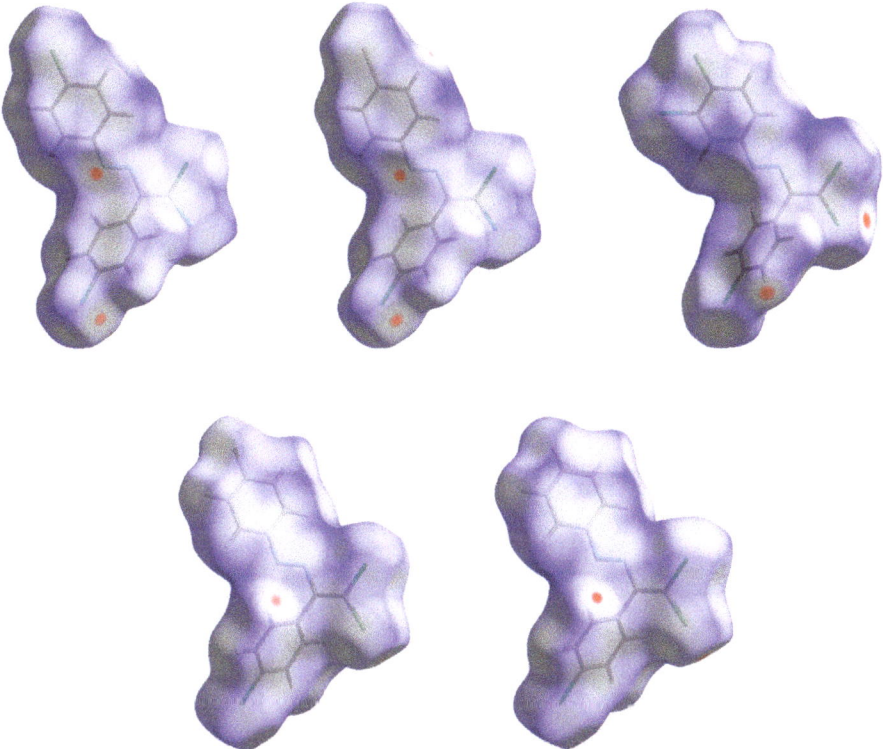

**Figure 6.** Visualization of Hirshfeld surfaces for X-ray structures **14**, **15**, and **17** (top), **13** and **10** (bottom).

## 3. Materials and Methods

General remarks. Unless stated otherwise, all the reagents used in this study were obtained from the commercial sources (Aldrich, TCI-Europe, Strem, ABCR). NMR spectra were recorded on a Bruker Avance 300 ($^1$H: 300 MHz, Karlsruhe, Germany); chemical shifts ($\delta$) are given in ppm relative to TMS, coupling constants (J) in Hz. The solvent signals were used as references (CDCl$_3$: $\delta_C$ = 77.16 ppm; residual CHCl$_3$ in CDCl$_3$: $\delta_H$ = 7.26 ppm; CD$_2$Cl$_2$: $\delta_C$ = 53.84 ppm; residual CHDCl$_2$ in CD$_2$Cl$_2$: $\delta_H$ = 5.32 ppm); $^1$H and $^{13}$C assignments were established using NOESY, HSQC, and HMBC experiments; numbering schemes as shown in the Inserts. IR: Perkin-Elmer Spectrum One spectrometer (Waltham, MA, USA.), wavenumbers ($\tilde{v}$) in cm$^{-1}$. Mass-spectra were obtained on a Bruker micrOTOF spectrometer equipped with electrospray ionization (ESI) source (Bremen, Germany); MeOH, CH$_2$Cl$_2$, or MeOH/CH$_2$Cl$_2$ mixture was used as a solvent. Thermogravimetric analysis (TGA) and differential thermal analysis were determined using a Netzsch TG 209F1 Libra apparatus (Selb, Germany). Solvents were purified by distillation over the indicated drying agents and were transferred under Ar: Et$_2$O (Mg/anthracene), CH$_2$Cl$_2$ (CaH$_2$), hexane (Na/K). Flash chromatography: Merck Geduran® Si 60 (Darmstadt, Germany) (40–63 µm).

The single point calculations based on the experimental X-ray geometries of **10**, **13–15**, and **17** have been carried out at the DFT level of theory using the dispersion-corrected hybrid functional $\omega$B97XD [47] with the help of Gaussian-09 program package ([M. J. Frisch et al., Gaussian-09, Revision C.01, Gaussian, Inc., Wallingford CT, USA, 2010.], full citation for this program is given in the SI). The 6-311++G ** basis sets [48–51] were used for all atoms. The topological analysis of the electron density distribution with the help of the atoms in molecules (QTAIM) method developed by Bader [41] has been performed by using the Multiwfn program (version 3.6, Beijing, China) [52]. The Cartesian atomic coordinates for model supramolecular associates are presented in Table S1, Supporting Information. The Hirshfeld surfaces analysis has been performed by using the CrystalExplorer program (version 17.5, Perth, Australia) [53]. The normalized contact distances ($d_{norm}$) [54] based on Bondi's van der Waals radii [40] were mapped into the Hirshfeld surfaces.

## 3.1. Crystal Structure Determination

X-ray diffraction data for **10, 13–15**, and **17** were collected at the 'RSA' beamline ($\lambda$ = 0.80246 Å) of the Kurchatov Synchrotron Radiation Source. All datasets were collected at 100 K. In total, 720 frames were collected with an oscillation range of 1.0 in the $\varphi$ scanning mode using two different orientations for each crystal. The semi-empirical correction for absorption was applied using the Scala program [55]. The data were indexed and integrated using the utility iMOSFLM from the CCP4 software suite [56,57]. For details, see Table S1. The structures were solved by intrinsic phasing modification of direct methods [58] and refined by a full-matrix least-squares technique on $F^2$ with anisotropic displacement parameters for all non-hydrogen atoms. The hydrogen atoms were placed in calculated positions and refined within the riding model with fixed isotropic displacement parameters [$U_{iso}$(H) = 1.5$U_{eq}$(C) for the methyl groups and 1.2$U_{eq}$(C) for the other groups]. All calculations were carried out using the SHELXTL program [59,60].

Crystallographic data for **10, 13–15**, and **17** have been deposited with the Cambridge Crystallographic Data Center, CCDC 2035010-2035014, respectively. Copies of this information may be obtained free of charge from the Director, CCDC, 12 Union Road, Cambridge CB2 1EZ, UK (fax: +44-1223-336033; e-mail: edeposit@ccdc.cam.ac.uk or www.ccdc.cam.ac.uk).

## 3.2. Synthetic Part

Schiff bases **1–9** were synthesized according to the reported method [20,21]. A mixture of (2-nitrophenyl)hydrazine (10.2 mmol), $CH_3COONa$ (0.82 g) and a corresponding 4-substituted aldehyde (10 mmol) were refluxed with stirring in ethanol (50 mL) for 2 h. The reaction mixture was cooled to room temperature and water (50 mL) was added to give a precipitate of crude product, which was filtered off, washed with diluted ethanol (1:1 with water) and dried in vacuo.

**1.** White solid (69%), mp 118 °C. $^1$H NMR (300 MHz, DMSO-$d_6$) δ 10.46 (s, 1H, NH), 7.85 (s, 1H, CH), 7.66 (d, J = 8.4 Hz, 2H, arom), 7.43 (d, J = 8.4 Hz, 2H, arom), 7.23 (t, J = 7.7 Hz, 2H, arom), 7.09 (d, J = 7.9 Hz, 2H, arom), 6.76 (t, J = 7.2 Hz, 1H, arom). $^{13}$C NMR (75 MHz, DMSO-$d_6$) δ 145.5, 135.4, 135.2, 132.5, 129.5, 129.1, 127.5, 119.41, 112.5.

**2.** White solid (92%), mp 151 °C. $^1$H NMR (300 MHz, DMSO-$d_6$) δ 10.33 (s, 1H, NH), 7.80 (s, 1H, CH), 7.66 (s, 1H, arom), 7.42 (d, J = 8.4 Hz, 2H, arom), 7.00 (q, J = 8.4 Hz, 5H, arom), 2.09 (s, 3H, $CH_3$). $^{13}$C NMR (75 MHz, DMSO-$d_6$) δ 138.6, 130.8, 130.1, 127.7, 125.4, 124.5, 123.3, 122.8, 107.92, 16.1.

**3.** White solid (87%), mp 141 °C. $^1$H NMR (300 MHz, DMSO-$d_6$) δ 10.24 (s, 1H, NH), 7.78 (s, 1H, CH), 7.63 (d, J = 8.5 Hz, 2H, arom), 7.41 (d, J = 8.5 Hz, 2H, arom), 7.01 (d, J = 8.9 Hz, 2H, arom), 6.84 (d, J = 8.9 Hz, 2H, arom), 3.69 (s, 3H, $OCH_3$). $^{13}$C NMR (75 MHz, DMSO-$d_6$) δ 153.2, 139.5, 135.5, 134.2, 132.1, 129.0, 127.3, 115.0, 113.5, 55.7.

4. White solid (77%), mp 135 °C. $^1$H NMR (300 MHz, DMSO-$d_6$)
δ 7.02 (d, 2H, J = 6.0 Hz), 7.22 (t, 2H, J = 9.1 Hz),
7.37 (d, 2H, J = 9.1 Hz), 7.68–7.73(m, 2H), 7.87(s, 1H), 10.49 (s, 1H).
$^{13}$C NMR (75 MHz, DMSO-$d_6$) δ 114.3, 115.9, 116.2, 128.0, 132.1,
132,63, 136,7, 145.0, 109.9.

5. White solid (76%), mp 153 °C. $^1$H NMR (300 MHz, DMSO-$d_6$)
δ 10.58 (s, 1H, NH), 7.85 (s, 1H, CH), 7.67 (d, J = 8.3 Hz, 2H, arom),
7.48–7.38 (m, 2H, arom), 7.25 (d, J = 8.7 Hz, 2H, arom),
7.07 (d, J = 8.7 Hz, 2H, arom). $^{13}$C NMR (75 MHz, DMSO-$d_6$) δ 144.4,
136.3, 135.0, 132.8, 129.3, 129.1, 127.7, 122.6, 113.9.

6. White solid (94%), mp 131 °C. $^1$H NMR (300 MHz, DMSO-$d_6$)
δ 10.59 (s, 1H, NH), 7.85 (s, 1H, CH), 7.67 (d, J = 8.5 Hz, 2H, arom),
7.43 (d, J = 8.5 Hz, 2H, arom), 7.37 (d, J = 8.8 Hz, 2H, arom),
7.03 (d, J = 8.8 Hz, 2H, arom). $^{13}$C NMR (75 MHz, DMSO-$d_6$) δ 144.8,
136.4, 134.9, 132.8, 132.2, 129.1, 127.7, 114.4, 110.2, 39.9.

7. White solid (72%), mp 119 °C. $^1$H NMR (300 MHz, DMSO-$d_6$)
δ 10.30 (s, 1H, NH), 7.81 (s, 1H, CH), 7.65 (d, J = 8.5 Hz, 2H, arom),
7.42 (d, J = 8.5 Hz, 2H, arom), 6.69 (s, 2H, arom), 6.41 (s, 1H, arom),
2.22 (s, 6H, CH$_3$). $^{13}$C NMR (75 MHz, DMSO-$d_6$) δ 145.3, 138.5, 135.3,
135.0, 132.3, 129.1, 127.5, 121.3, 110.3, 21.7.

8. White solid (88%), mp 114 °C. $^1$H NMR (300 MHz, DMSO-$d_6$)
δ 10.16 (s, 1H, NH), 8.28 (s, 1H, CH), 7.69 (d, J = 8.7 Hz, 2H, arom),
7.56 (d, J = 8.7 Hz, 1H, arom), 7.51–7.43 (m, 3H, arom),
7.35–7.17 (m, 1H, arom). $^{13}$C NMR (75 MHz, DMSO-$d_6$) δ 162.3,
140.9, 140.1, 134.6, 133.4, 129.2, 129.0, 128.5, 128.1, 122.8, 117.1, 115.5.

9. White solid (92%), mp 112 °C. $^1$H NMR (300 MHz, DMSO-$d_6$)
δ 10.74 (s, 1H, NH), 7.90 (d, J = 14.3 Hz, 1H), 7.69 (d, J = 8.3 Hz, 2H,
arom), 7.44 (q, J = 8.3, 7.5 Hz, 3H, arom), 7.26 (s, 1H, CH),
7.00 (d, J = 8.4 Hz, 1H, arom). $^{13}$C NMR (75 MHz, DMSO-$d_6$) δ 145.6,
137.6, 134.6, 133.2, 132.1, 131.5, 131.3, 130.2, 129.8, 129.1, 128.0, 120.1,
113.3, 112.8.

### 3.3. Synthesis of Dichlorodiazadiens

A twenty-milliliter screw neck vial was charged with DMSO (10 mL), **1–9** (1 mmol), tetramethylethylenediamine (TMEDA) (295 mg, 2.5 mmol), CuCl (2 mg, 0.02 mmol), and CCl$_4$ (20 mmol, 10 equiv). After 3 h (until TLC analysis showed complete consumption of corresponding Schiff base) reaction mixture was poured into ~0.01 M solution of HCl (100 mL, ~pH = 2), and extracted with dichloromethane (3 × 20 mL). The combined organic phase was washed with water (3 × 50 mL), brine (30 mL), dried over anhydrous Na$_2$SO$_4$ and concentrated in vacuo. The residue was purified by column chromatography on silica gel using appropriate mixtures of hexane and dichloromethane (3/1–1/1).

**10.** Red solid (73%), mp 85 °C. $^1$H NMR (300 MHz, CDCl$_3$) δ 7.71–7.60 (m, 2H, arom), 7.35 (dd, $J$ = 7.6, 3.8 Hz, 4H, arom), 7.28 (s, 1H, arom), 7.03 (d, $J$ = 8.3 Hz, 2H, arom). $^{13}$C NMR (75 MHz, CDCl$_3$) δ 134.6, 131.7, 131.3, 130.7, 129.4, 129.0, 128.4, 127.1, 126.2, 123.1.

**11.** Red solid (79%), mp 90 °C. $^1$H NMR (300 MHz, CDCl$_3$) δ 7.69 (d, $J$ = 8.2 Hz, 2H, arom), 7.42 (d, $J$ = 8.3 Hz, 2H, arom), 7.26 (d, $J$ = 8.2 Hz, 2H, arom), 7.13 (d, $J$ = 8.3 Hz, 2H, arom), 2.42 (s, 3H, CH$_3$). $^{13}$C NMR (75 MHz, CDCl$_3$) δ 162.3, 151.2, 150.9, 142.5, 134.7, 131.4, 131.0, 129.7, 128.4, 123.2, 21.5. Crystals, suitable for X-ray analysis, were obtained by the slow evaporation of saturated hexane/EtOAc (5/1) solution.

**12.** Red solid (72%), mp 96 °C. $^1$H NMR (300 MHz, CDCl$_3$) δ 7.78 (d, $J$ = 9.0 Hz, 2H, arom), 7.42 (d, $J$ = 8.4 Hz, 2H, arom), 7.13 (d, $J$ = 8.4 Hz, 2H, arom), 6.95 (d, $J$ = 9.0 Hz, 2H, arom), 3.88 (s, 3H, OCH$_3$). $^{13}$C NMR (75 MHz, CDCl$_3$) δ 162.7, 162.3, 151.1, 147.2, 134.6, 131.4, 131.2, 128.4, 125.3, 114.2, 55.6.

**13.** Red solid (68%), mp 77 °C. $^1$H NMR (300 MHz, CDCl$_3$) δ 7.81 (dd, $J$ = 8.6, 5.4 Hz, 2H), 7.43 (d, $J$ = 8.3 Hz, 2H), 7.14 (t, $J$ = 8.8 Hz, 4H). $^{13}$C NMR (75 MHz, CDCl$_3$) δ 167.6, 166.4, 151.1, 149.3, 134.8, 131.4, 130.8, 129.6, 128.5, 125.4, 116.2, 115.9. Crystals, suitable for X-ray analysis, were obtained by the slow evaporation of saturated hexane/EtOAc (5/1) solution.

**14.** Red solid (67%), mp 94 °C. $^1$H NMR (300 MHz, CDCl$_3$) δ 7.73 (d, $J$ = 8.6 Hz, 1H), 7.43 (d, $J$ = 8.4 Hz, 2H), 7.12 (d, $J$ = 8.4 Hz, 1H). $^{13}$C NMR (75 MHz, CDCl$_3$) δ 162.3, 151.1, 137.7, 136.5, 134.9, 131.4, 130.6, 129.3, 128.5, 124.4. Crystals, suitable for X-ray analysis, were obtained by the slow evaporation of saturated hexane/EtOAc (5/1) solution.

**15.** Red solid (70%), mp 105 °C. $^1$H NMR (300 MHz, CDCl$_3$) δ 7.69–7.56 (m, 4H, arom), 7.49–7.39 (m, 2H, arom), 7.12 (d, $J$ = 8.4 Hz, 2H, arom). $^{13}$C NMR (75 MHz, CDCl$_3$) δ 151.4, 134.9, 132.3, 131.4, 130.6, 129.8, 128.5, 127.4, 126.3, 124.6. Crystals, suitable for X-ray analysis, were obtained by the slow evaporation of saturated hexane/EtOAc (5/1) solution.

**16.** Red solid (82%), mp 145 °C. $^1$H NMR (300 MHz, CDCl$_3$) δ 7.44 (d, J = 7.7 Hz, 4H, arom), 7.15 (d, J = 7.7 Hz, 3H, arom), 2.40 (s, 6H, CH$_3$). $^{13}$C NMR (75 MHz, CDCl$_3$) δ 157.7, 148.3, 146.7, 134.2, 130.2, 129.0, 126.8, 126.5, 123.9, 116.5, 16.6.

**17.** Red solid (66%), mp 115 °C. $^1$H NMR (300 MHz, CDCl$_3$) δ 7.89 (s, 1H, arom), 7.68–7.61 (m, 1H, arom), 7.54 (d, J = 8.6 Hz, 1H, arom), 7.44 (d, J = 8.3 Hz, 2H, arom), 7.11 (d, J = 8.3 Hz, 2H, arom). $^{13}$C NMR (75 MHz, CDCl$_3$) δ 151.5, 151.4, 135.7, 135.0, 133.5, 131.3, 130.8, 130.4, 129.8, 128.6, 124.5, 122.7. Crystals, suitable for X-ray analysis, were obtained by the slow evaporation of saturated hexane/EtOAc (5/1) solution.

**18.** Red solid (71%), mp 121 °C. $^1$H NMR (300 MHz, CDCl$_3$) δ 7.57 (d, J = 8.7 Hz, 1H, arom), 7.46 (d, J = 2.0 Hz, 1H, arom), 7.37 (d, J = 8.4 Hz, 2H, arom), 7.24 (d, J = 2.3 Hz, 1H, arom), 7.12 (d, J = 8.4 Hz, 2H, arom).

## 4. Conclusions

In summary, **9** novel halogenated dichlorodiazadienes were prepared and fully characterized, while for **5** of them single crystal structures were determined. Solid state structures contained multiple Hal···Hal interactions, which were studied by DFT calculations and topological analysis of the electron density distribution within the framework of Bader's theory (QTAIM method). Calculations showed that the Hal···Hal interactions dictate a packing preference for this newly discovered class of dyes. These results further demonstrate the potential of Hal···Hal bonding in supramolecular engineering and crucial role in the stabilization of the intermolecular networks of dichlorodiazadienes. Further studies into photophysical properties of halogenated dichlorodiazadienes and their applications from our laboratory are underway and will be reported in due course.

**Supplementary Materials:** Figure S1. Visualization of electrostatic surface potentials for **10**, **13–15** and **17** with selected $V_{s,min}/V_{s,max}$ values (in kcal/mol). Table S1: Crystal data and structure refinement for **10**, **13–15** and **17**, Table S2. Values of the density of all electrons—ρ(r), Laplacian of electron density—$\nabla^2$ρ(r) and appropriate $\lambda_2$ eigenvalues (with promolecular approximation), energy density—H$_b$, potential energy density—V(r), and Lagrangian kinetic energy—G(r) (a.u.) at the bond critical points (3, −1), corresponding to Cl···F halogen-halogen contacts in **13**, Table S3. Cartesian atomic coordinates for model supramolecular associates.

**Author Contributions:** Conceptualization, V.G.N. and A.G.T.; writing—review and editing; writing—original draft preparation, V.G.N., A.G.T., A.S.N.; V.G.N. and A.G.T.; software, A.S.N.; investigation, N.G.S.; A.M.M.; K.N.B.; G.T.S.; supervision, V.N.K. All authors have read and agreed to the published version of the manuscript.

**Funding:** This work was performed under the support of the FRCCP RAS State task AAAA-A19-119012990175-9. A.S.N. is grateful to Russian Science Foundation for the support of his theoretical studies (project No. 19-73-00001). We acknowledge the RUDN University Program 5-100. V.G.N. is grateful to RFBR for the support (grant N 18-03-00791).

**Conflicts of Interest:** The authors declare no conflict of interest.

## References

1. Bertani, R.; Sgarbossa, P.; Venzo, A.; Lelj, F.; Amati, M.; Resnati, G.; Pilati, T.; Metrangolo, P.; Terraneo, G. Halogen bonding in metal-organic-supramolecular networks. *Coord. Chem. Rev.* **2010**, *254*, 677–695. [CrossRef]
2. Adonin, S.A.; Gorokh, I.D.; Samsonenko, D.G.; Novikov, A.S.; Korolkov, I.V.; Plyusnin, P.E.; Sokolov, M.N.; Fedin, V.P. Binuclear and polymeric bromobismuthate complexes: Crystal structures and thermal stability. *Polyhedron* **2019**, *159*, 318–322. [CrossRef]

3. Adonin, S.A.; Bondarenko, M.A.; Abramov, P.A.; Novikov, A.S.; Plyusnin, P.E.; Sokolov, M.N.; Fedin, V.P. Bromo- and Polybromoantimonates(V): Structural and Theoretical Studies of Hybrid Halogen-Rich Halometalate Frameworks. *Chem.-A Eur. J.* **2018**, *24*, 10165–10170. [CrossRef]
4. Saha, A.; Rather, S.A.; Sharada, D.; Saha, B.K. C-X···X-C vs C-H···X-C, which one is the more dominant interaction in crystal packing (X = halogen)? *Cryst. Growth Des.* **2018**, *18*, 6084–6090. [CrossRef]
5. Usoltsev, A.N.; Adonin, S.A.; Novikov, A.S.; Samsonenko, D.G.; Sokolov, M.N.; Fedin, V.P. One-dimensional polymeric polybromotellurates(IV): Structural and theoretical insights into halogen···halogen contacts. *CrystEngComm* **2017**, *19*, 5934–5939. [CrossRef]
6. Adonin, S.A.; Gorokh, I.D.; Novikov, A.S.; Abramov, P.A.; Sokolov, M.N.; Fedin, V.P. Halogen Contacts-Induced Unusual Coloring in BiIII Bromide Complex: Anion-to-Cation Charge Transfer via Br···Br Interactions. *Chem.-A Eur. J.* **2017**, *23*, 15612–15616. [CrossRef]
7. Nguyen, H.L.; Horton, P.N.; Hursthouse, M.B.; Legon, A.C.; Bruce, D.W. Halogen Bonding: A New Interaction for Liquid Crystal Formation. *J. Am. Chem. Soc.* **2004**, *126*, 16–17. [CrossRef] [PubMed]
8. Cariati, E.; Cavallo, G.; Forni, A.; Leem, G.; Metrangolo, P.; Meyer, F.; Pilati, T.; Resnati, G.; Righetto, S.; Terraneo, G.; et al. Self-complementary nonlinear optical-phores targeted to halogen bond-driven self-assembly of electro-optic materials. *Cryst. Growth Des.* **2011**, *11*, 5642–5648. [CrossRef]
9. Sun, A.; Lauher, J.W.; Goroff, N.S. Preparation of poly(diiododiacetylene), an ordered conjugated polymer of carbon and iodine. *Science* **2006**, *312*, 1030–1034. [CrossRef]
10. Metrangolo, P.; Resnati, G. Halogen Versus Hydrogen. *Science* **2008**, *321*, 918–919. [CrossRef]
11. Yang, L.; Tan, X.; Wang, Z.; Zhang, X. Supramolecular Polymers: Historical Development, Preparation, Characterization, and Functions. *Chem. Rev.* **2015**, *115*, 7196–7239. [CrossRef]
12. Maharramov, A.M.; Mahmudov, K.T.; Kopylovich, M.N.; Pombeiro, A.J.L. *Non-covalent Interactions in the Synthesis and Design of New Compounds*; John Wiley & Sons Limited: Hoboken, NJ, USA, 2016; ISBN 9781119113874.
13. Berger, G.; Frangville, P.; Meyer, F. Halogen bonding for molecular recognition: New developments in materials and biological sciences. *Chem. Commun.* **2020**, *56*, 4970–4981. [CrossRef] [PubMed]
14. Mahadevi, A.S.; Sastry, G.N. Cooperativity in Noncovalent Interactions. *Chem. Rev.* **2016**, *116*, 2775–2825. [CrossRef] [PubMed]
15. Walsh, R.B.; Padgett, C.W.; Metrangolo, P.; Resnati, G.; Hanks, T.W.; Pennington, W.T. Crystal Engineering through Halogen Bonding: Complexes of Nitrogen Heterocycles with Organic Iodides. *Cryst. Growth Des.* **2001**, *1*, 165–175. [CrossRef]
16. Teyssandier, J.; Mali, K.S.; De Feyter, S. Halogen Bonding in Two-Dimensional Crystal Engineering. *ChemistryOpen* **2020**, *9*, 225–241. [CrossRef]
17. Berger, G.; Soubhye, J.; Meyer, F. Halogen bonding in polymer science: From crystal engineering to functional supramolecular polymers and materials. *Polym. Chem.* **2015**, *6*, 3559–3580. [CrossRef]
18. Saha, B.K.; Rather, S.A.; Saha, A. Interhalogen Interactions in the Light of Geometrical Correction. *Cryst. Growth Des.* **2016**, *16*, 3059–3062. [CrossRef]
19. Bui, T.T.T.; Dahaoui, S.; Lecomte, C.; Desiraju, G.R.; Espinosa, E. The nature of halogen halogen interactions: A model derived from experimental charge-density analysis. *Angew. Chemie-Int. Ed.* **2009**, *48*, 3838–3841. [CrossRef]
20. Yang, H.; Wong, M.W. Application of halogen bonding to organocatalysis: A theoretical perspective. *Molecules* **2020**, *25*, 1045. [CrossRef]
21. Kolář, M.H.; Hobza, P. Computer Modeling of Halogen Bonds and Other σ-Hole Interactions. *Chem. Rev.* **2016**, *116*, 5155–5187. [CrossRef]
22. Priimagi, A.; Cavallo, G.; Forni, A.; Gorynsztejn-Leben, M.; Kaivola, M.; Metrangolo, P.; Milani, R.; Shishido, A.; Pilati, T.; Resnati, G.; et al. Halogen bonding versus hydrogen bonding in driving self-assembly and performance of light-responsive supramolecular polymers. *Adv. Funct. Mater.* **2012**, *22*, 2572–2579. [CrossRef]
23. Cavallo, G.; Metrangolo, P.; Milani, R.; Pilati, T.; Priimagi, A.; Resnati, G.; Terraneo, G. The Halogen Bond. *Chem. Rev.* **2016**, *116*, 2478–2601. [CrossRef]
24. Nenajdenko, V.G.; Shastin, A.V.; Gorbachev, V.M.; Shorunov, S.V.; Muzalevskiy, V.M.; Lukianova, A.I.; Dorovatovskii, P.V.; Khrustalev, V.N. Copper-Catalyzed Transformation of Hydrazones into Halogenated Azabutadienes, Versatile Building Blocks for Organic Synthesis. *ACS Catal.* **2017**, *7*, 205–209. [CrossRef]

25. Shastin, A.V.; Sergeev, P.G.; Lukianova, A.I.; Muzalevskiy, V.M.; Khrustalev, V.N.; Dorovatovskii, P.V.; Nenajdenko, V.G. Dichloro-Substituted 1,2-Diazabuta-1,3-dienes as Highly Reactive Electrophiles in the Reaction with Amines and Diamines: Efficient Synthesis of α-Hydrazo Amidinium Salts. *European J. Org. Chem.* **2018**, *2018*, 4996–5006. [CrossRef]

26. Shastin, A.V.; Tsyrenova, B.D.; Sergeev, P.G.; Roznyatovsky, V.A.; Smolyar, I.V.; Khrustalev, V.N.; Nenajdenko, V.G. Synthesis of a New Family of 1,1-Diazidoethenes: One-Pot Construction of 4-Azido-1,2,3-triazoles via Nitrene Cyclization. *Org. Lett.* **2018**, *20*, 7803–7806. [CrossRef]

27. Tsyrenova, B.; Nenajdenko, V. Synthesis and spectral study of a new family of 2,5-diaryltriazoles having restricted rotation of the 5-aryl substituent. *Molecules* **2020**, *25*, 480. [CrossRef]

28. Sergeev, P.G.; Khrustalev, V.N.; Nenajdenko, V.G. Construction of 6-Aminopyridazine Derivatives by the Reaction of Malononitrile with Dichloro-Substituted Diazadienes. *European J. Org. Chem.* **2020**, *2020*, 4964–4971. [CrossRef]

29. Shikhaliyev, N.Q.; Kuznetsov, M.L.; Maharramov, A.M.; Gurbanov, A.V.; Ahmadova, N.E.; Nenajdenko, V.G.; Mahmudov, K.T.; Pombeiro, A.J.L. Noncovalent interactions in the design of bis-azo dyes. *CrystEngComm* **2019**, *21*, 5032–5038. [CrossRef]

30. Shikhaliyev, N.Q.; Ahmadova, N.E.; Gurbanov, A.V.; Maharramov, A.M.; Mammadova, G.Z.; Nenajdenko, V.G.; Zubkov, F.I.; Mahmudov, K.T.; Pombeiro, A.J.L. Tetrel, halogen and hydrogen bonds in bis(4-((E)-(2,2-dichloro-1-(4-substitutedphenyl)vinyl)diazenyl)phenyl)methane dyes. *Dye. Pigment.* **2018**, *150*, 377–381. [CrossRef]

31. Maharramov, A.M.; Shikhaliyev, N.Q.; Suleymanova, G.T.; Gurbanov, A.V.; Babayeva, G.V.; Mammadova, G.Z.; Zubkov, F.I.; Nenajdenko, V.G.; Mahmudov, K.T.; Pombeiro, A.J.L. Pnicogen, halogen and hydrogen bonds in (E)-1-(2,2-dichloro-1-(2-nitrophenyl)vinyl)-2-(para-substituted phenyl)-diazenes. *Dye. Pigment.* **2018**, *159*, 135–141. [CrossRef]

32. Repina, O.V.; Novikov, A.S.; Khoroshilova, O.V.; Kritchenkov, A.S.; Vasin, A.A.; Tskhovrebov, A.G. Lasagna-like supramolecular polymers derived from the PdII osazone complexes via C(sp2)–H···Hal hydrogen bonding. *Inorganica Chim. Acta* **2020**, *502*, 119378. [CrossRef]

33. Tskhovrebov, A.G.; Novikov, A.S.; Odintsova, O.V.; Mikhaylov, V.N.; Sorokoumov, V.N.; Serebryanskaya, T.V.; Starova, G.L. Supramolecular polymers derived from the PtII and PdII schiff base complexes via C(sp2)–H ··· Hal hydrogen bonding: Combined experimental and theoretical study. *J. Organomet. Chem.* **2019**, *886*, 71–75. [CrossRef]

34. Tskhovrebov, A.G.; Vasileva, A.A.; Goddard, R.; Riedel, T.; Dyson, P.J.; Mikhaylov, V.N.; Serebryanskaya, T.V.; Sorokoumov, V.N.; Haukka, M. Palladium(II)-Stabilized Pyridine-2-Diazotates: Synthesis, Structural Characterization, and Cytotoxicity Studies. *Inorg. Chem.* **2018**, *57*, 930–934. [CrossRef]

35. Mikhaylov, V.N.; Sorokoumov, V.N.; Liakhov, D.M.; Tskhovrebov, A.G.; Balova, I.A. Polystyrene-supported acyclic diaminocarbene palladium complexes in Sonogashira cross-coupling: Stability vs. catalytic activity. *Catalysts* **2018**, *8*, 141. [CrossRef]

36. Tskhovrebov, A.G.; Luzyanin, K.V.; Kuznetsov, M.L.; Sorokoumov, V.N.; Balova, I.A.; Haukka, M.; Kukushkin, V.Y. Substituent R-dependent regioselectivity switch in nucleophilic addition of N-phenylbenzamidine to PdII-and PtII-complexed isonitrile RN≡C giving aminocarbene-like species. *Organometallics* **2011**, *30*, 863–874. [CrossRef]

37. Mikhaylov, V.N.; Sorokoumov, V.N.; Novikov, A.S.; Melnik, M.V.; Tskhovrebov, A.G.; Balova, I.A. Intramolecular hydrogen bonding stabilizes trans-configuration in a mixed carbene/isocyanide PdII complexes. *J. Organomet. Chem.* **2020**, *912*, 121174. [CrossRef]

38. Tskhovrebov, A.G.; Luzyanin, K.V.; Haukka, M.; Kukushkin, V.Y. Synthesis and characterization of cis-(RNC)2PtII species useful as synthons for generation of various (aminocarbene)Pt II complexes. *J. Chem. Crystallogr.* **2012**, *42*, 1170–1175. [CrossRef]

39. Tskhovrebov, A.G.; Novikov, A.S.; Kritchenkov, A.S.; Khrustalev, V.N.; Haukka, M. Attractive halogen···halogen interactions in crystal structure of trans-dibromogold(III) complex. *Zeitschrift Krist. Cryst. Mater.* **2020**. [CrossRef]

40. Bondi, A. Van der Waals volumes and radii of metals in covalent compounds. *J. Phys. Chem.* **1966**, *70*, 3006–3007. [CrossRef]

41. Bader, R.F.W. A Quantum Theory of Molecular Structure and Its Applications. *Chem. Rev.* **1991**, *91*, 893–928. [CrossRef]

42. Bartashevich, E.V.; Tsirelson, V.G. Interplay between non-covalent interactions in complexes and crystals with halogen bonds. *Russ. Chem. Rev.* **2014**, *83*, 1181–1203. [CrossRef]
43. Adonin, S.A.; Gorokh, I.D.; Novikov, A.S.; Samsonenko, D.G.; Yushina, I.V.; Sokolov, M.N.; Fedin, V.P. Halobismuthates with halopyridinium cations: Appearance or non-appearance of unusual colouring. *CrystEngComm* **2018**, *20*, 7766–7772. [CrossRef]
44. Espinosa, E.; Alkorta, I.; Elguero, J.; Molins, E. From weak to strong interactions: A comprehensive analysis of the topological and energetic properties of the electron density distribution involving X-H···F-Y systems. *J. Chem. Phys.* **2002**, *117*, 5529–5542. [CrossRef]
45. Johnson, E.R.; Keinan, S.; Mori-Sánchez, P.; Contreras-García, J.; Cohen, A.J.; Yang, W. Revealing noncovalent interactions. *J. Am. Chem. Soc.* **2010**, *132*, 6498–6506. [CrossRef] [PubMed]
46. Contreras-García, J.; Johnson, E.R.; Keinan, S.; Chaudret, R.; Piquemal, J.-P.; Beratan, D.N.; Yang, W. NCIPLOT: A Program for Plotting Noncovalent Interaction Regions. *J. Chem. Theory Comput.* **2011**, *7*, 625–632. [CrossRef] [PubMed]
47. Chai, J.-D.; Head-Gordon, M. Long-range corrected hybrid density functionals with damped atom–atom dispersion corrections. *Phys. Chem. Chem. Phys.* **2008**, *10*, 6615–6620. [CrossRef]
48. Curtiss, L.A.; McGrath, M.P.; Blaudeau, J.; Davis, N.E.; Binning, R.C.; Radom, L. Extension of Gaussian-2 theory to molecules containing third-row atoms Ga–Kr. *J. Chem. Phys.* **1995**, *103*, 6104–6113. [CrossRef]
49. Francl, M.M.; Pietro, W.J.; Hehre, W.J.; Binkley, J.S.; Gordon, M.S.; DeFrees, D.J.; Pople, J.A. Self-consistent molecular orbital methods. XXIII. A polarization-type basis set for second-row elements. *J. Chem. Phys.* **1982**, *77*, 3654–3665. [CrossRef]
50. Krishnan, R.; Binkley, J.S.; Seeger, R.; Pople, J.A. Self-consistent molecular orbital methods. XX. A basis set for correlated wave functions. *J. Chem. Phys.* **1980**, *72*, 650–654. [CrossRef]
51. McLean, A.D.; Chandler, G.S. Contracted Gaussian basis sets for molecular calculations. I. Second row atoms, Z=11–18. *J. Chem. Phys.* **1980**, *72*, 5639–5648. [CrossRef]
52. Lu, T.; Chen, F. Multiwfn: A multifunctional wavefunction analyzer. *J. Comput. Chem.* **2012**, *33*, 580–592. [CrossRef]
53. Spackman, M.A.; Jayatilaka, D. Hirshfeld surface analysis. *CrystEngComm* **2009**, *11*, 19–32. [CrossRef]
54. McKinnon, J.J.; Jayatilaka, D.; Spackman, M.A. Towards quantitative analysis of intermolecular interactions with Hirshfeld surfaces. *Chem. Commun.* **2007**, 3814–3816. [CrossRef] [PubMed]
55. Evans, P. Scaling and assessment of data quality. *Acta Crystallogr. Sect. D Biol. Crystallogr.* **2006**, *62*, 72–82. [CrossRef]
56. Battye, T.G.G.; Kontogiannis, L.; Johnson, O.; Powell, H.R.; Leslie, A.G.W. iMOSFLM: A new graphical interface for diffraction-image processing with MOSFLM. *Acta Crystallogr. Sect. D Biol. Crystallogr.* **2011**, *67*, 271–281. [CrossRef] [PubMed]
57. Winn, M.D.; Ballard, C.C.; Cowtan, K.D.; Dodson, E.J.; Emsley, P.; Evans, P.R.; Keegan, R.M.; Krissinel, E.B.; Leslie, A.G.W.; McCoy, A.; et al. Overview of the CCP4 suite and current developments. *Acta Crystallogr. Sect. D Biol. Crystallogr.* **2011**, *67*, 235–242. [CrossRef]
58. Sheldrick, G.M. SHELXT—Integrated space-group and crystal-structure determination. *Acta Crystallogr. Sect. A Found. Crystallogr.* **2015**, *71*, 3–8. [CrossRef]
59. Sheldrick, G.M. A short history of SHELX. *Acta Crystallogr. A* **2008**, *64*, 112–122. [CrossRef]
60. Sheldrick, G.M. Crystal structure refinement with SHELXL. *Acta Crystallogr. Sect. C Struct. Chem.* **2015**, *71*, 3–8. [CrossRef]

**Sample Availability:** Samples of the compounds **1–18** are available from the authors.

**Publisher's Note:** MDPI stays neutral with regard to jurisdictional claims in published maps and institutional affiliations.

© 2020 by the authors. Licensee MDPI, Basel, Switzerland. This article is an open access article distributed under the terms and conditions of the Creative Commons Attribution (CC BY) license (http://creativecommons.org/licenses/by/4.0/).

Article

# Inter- vs. Intramolecular Hydrogen Bond Patterns and Proton Dynamics in Nitrophthalic Acid Associates

Kinga Jóźwiak [1], Aneta Jezierska [1], Jarosław J. Panek [1], Eugene A. Goremychkin [2], Peter M. Tolstoy [3], Ilya G. Shenderovich [4] and Aleksander Filarowski [1,*]

[1] Faculty of Chemistry, University of Wrocław 14 F. Joliot-Curie str., 50-383 Wrocław, Poland; kin.joz@o2.pl (K.J.); aneta.jezierska@chem.uni.wroc.pl (A.J.); jaroslaw.panek@chem.uni.wroc.pl (J.J.P.)
[2] Frank Laboratory of Neutron Physics, Joint Institute for Nuclear Research 6 F. Joliot-Curie str., 141980 Dubna, Russia; goremychkin@jinr.ru
[3] Institute of Chemistry, St. Petersburg State University, Universitetskij pr. 26, 198504 St. Petersburg, Russia; peter.tolstoy@spbu.ru
[4] Institute of Organic Chemistry, University of Regensburg, Universitaetstrasse 31, 93053 Regensburg, Germany
* Correspondence: Ilya.Shenderovich@chemie.uni-regensburg.de (I.G.S.); aleksander.filarowski@chem.uni.wroc.pl (A.F.); Tel.: +48-71-375-7229 (A.F.)

Academic Editor: Goar Sánchez
Received: 10 September 2020; Accepted: 12 October 2020; Published: 14 October 2020

**Abstract:** Noncovalent interactions are among the main tools of molecular engineering. Rational molecular design requires knowledge about a result of interplay between given structural moieties within a given phase state. We herein report a study of intra- and intermolecular interactions of 3-nitrophthalic and 4-nitrophthalic acids in the gas, liquid, and solid phases. A combination of the Infrared, Raman, Nuclear Magnetic Resonance, and Incoherent Inelastic Neutron Scattering spectroscopies and the Car–Parrinello Molecular Dynamics and Density Functional Theory calculations was used. This integrated approach made it possible to assess the balance of repulsive and attractive intramolecular interactions between adjacent carboxyl groups as well as to study the dependence of this balance on steric confinement and the effect of this balance on intermolecular interactions of the carboxyl groups.

**Keywords:** proton dynamics; carboxyl group; CPMD; DFT; IINS; IR; Raman; NMR

## 1. Introduction

Hydrogen bonding (H-bonding) and steric effects are important tools of molecular engineering. Under certain conditions, their interplay can stabilize species that otherwise exhibit high chemical reactivity [1–4]. The structural complexity increases when there are either other noncovalent interactions or competing H-bonds. The former is critically important in solids [5–9], at confined geometries [10–12], and in aqueous solutions [13–16]. The latter is characteristic for P=O moiety [17,18], specially designed organic molecules [19,20], but most of all for biomolecules [21,22]. The adjustments of bridging proton positions in H-bonds act as one of the mechanisms governing the chemical properties of macromolecules [23–25] and biosystems [26,27]. Changes of weak specific interactions such as H-bonds can evoke a reorganization on the macroscopic scale. Therefore, many-sided elaborate studies of the conformational phenomena are essential not only for fundamental understanding of H-bond nature but also for a number of practical applications, such as design of materials with the required physicochemical properties [28–30].

The wide variety of effects associated with a competition between intra- and intermolecular H-bonding can be illustrated with salicylic acid. In the simplest case of salicylic acid crystals, the carboxyl groups of the molecules form dimers while their hydroxyl groups form intramolecular H-bonds [31,32]. This structure remains qualitatively valid in an aprotic solution when the dimer is deprotonated [33]. In contrast, when the number of competing interactions increases, the co-crystals of salicylic acid exhibit polymorphism and different solubility [32,34,35]. These changes are critically important for pharmaceutical applications. Besides that, the intramolecular H-bond in salicylic acid derivatives can be controlled through intramolecular steric effects. In the crystalline salicylic acid, the O···O distances of this H-bond are about 2.62 Å [31,32]. In 2-hydroxy-3-nitrobenzoic acid, 6-(cyclohexylmethyl)salicylic acid, and 6-(2-cyclohexylethyl)salicylic acid they are only 2.55, 2.54, and 2.52 Å, respectively [32,36]. Is this a general trend that can be expected for other molecules' structures?

This paper presents the conformational studies of 3- and 4-nitrophthalic acids (3 and 4, Figure 1). These compounds are characterized by the presence of strong intermolecular and intramolecular H-bonds in co-crystals with various organic compounds [37–41]. These bonds might mutually convert one into another in compounds with adjacent carboxyl groups under impact of external factors. The first papers about dimeric formation by carboxyl group were published by Pfeiffer et al. in 1910 [42–44]. The carboxylic acid dimer units (2 × (COOH)) have still attracted attention for researchers involved in H-bonding studies [45–55]. In References [56–58], authors show a strong effect of H-bonds on the conformational state of compounds. The domination of cis conformation of carboxyl group, so-called Z-effect, has been elucidated by Lyssenko et al. [59]. Recently, two polymorphic forms of cinchromeronic acid (the derivative of phthalic acid) have been discovered and studied [60]. It has been shown that the polymorphic forms are caused by the proton transfer and reorientation of the carboxyl groups. Computational studies of H-bonds and stable conformers are important for the development of the conformational polymorphism of the molecular complexes such as benzoic acid with pyridine [61]. Moreover, H-bonded networks of phthalic acids can be used as ligands for metal-organic aggregates [62,63].

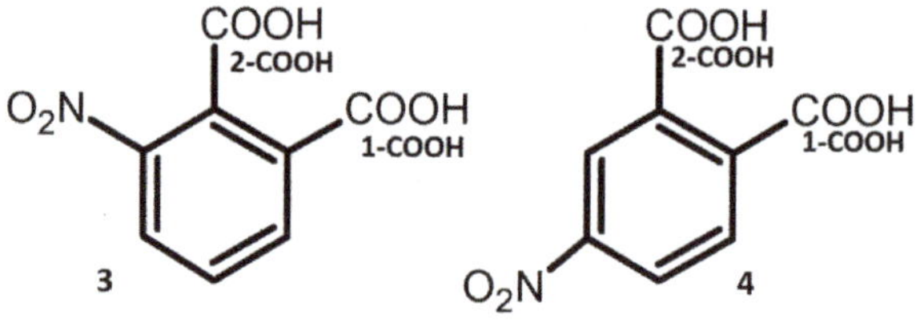

**Figure 1.** Chemical structures of 3-nitrophthalic (3) and 4-nitrophthalic (4) acids.

The main aim of this study was to characterize intramolecular interactions between adjacent carboxyl groups in the presence and absence of intramolecular steric effects and the effect of all these interactions on intermolecular interactions of these carboxyl groups. The nitro substitution was chosen because this moiety is rigid, relatively small, and causes considerable steric strain. Besides static density-functional theory (DFT) computations, this study covers simulations performed using the Car–Parrinello molecular dynamics (CPMD) approach, which supports NMR (Nuclear Magnetic Resonance), IR (infrared), Raman, and IINS (Incoherent Inelastic Neutron Scattering) experimental measurements with the employment of a neutron radiation source.

The outline of the manuscript is as follows. Firstly, the conformational analysis on the basis of static DFT calculations is presented. Next, the proton and functional groups' dynamics were studied by DFT and CPMD calculations. The following part delves into the investigations of conformational equilibrium in the solutions accomplished by NMR spectroscopy as well as IR, Raman, and IINS studies of the compounds in the solid state. Additionally, the spectral analysis on the basis of the experimental and computational results by means of H/D isotopic substitution was performed. The concluding remarks are given in the last section.

## 2. Results and Discussion

### 2.1. DFT Study of H-Bond, Nitro, and Carboxyl Groups' Dynamics of Nitrophthalic Acids

The quantum-mechanical calculations were accomplished at the B3LYP/6-311+G(d,p) level of theory for the detection of the most stable conformer of monomeric **3** and **4**. These calculations show that the most stable conformer does not contain the intramolecular H-bond (Figure 2). Generally, intramolecular H-bonds can significantly decrease the energy of isolated molecules [64]. However, conformers **3(III)**, **3(IV)**, **3(VI)**, **4(IV)**, and **4(V)** with the intramolecular H-bond feature significant steric tensions between the carboxyl groups that increases further if the nitro group is nearby. In consequence, the energies of conformers **3(III)** and **4(IV)** are higher as compared to **3(I)** and **4(I)**.

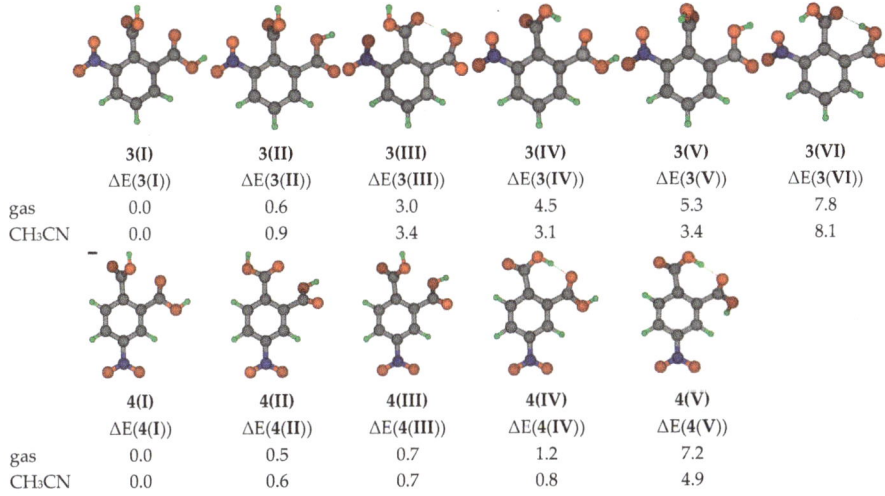

|  | 3(I) ΔE(3(I)) | 3(II) ΔE(3(II)) | 3(III) ΔE(3(III)) | 3(IV) ΔE(3(IV)) | 3(V) ΔE(3(V)) | 3(VI) ΔE(3(VI)) |
|---|---|---|---|---|---|---|
| gas | 0.0 | 0.6 | 3.0 | 4.5 | 5.3 | 7.8 |
| CH$_3$CN | 0.0 | 0.9 | 3.4 | 3.1 | 3.4 | 8.1 |

|  | 4(I) ΔE(4(I)) | 4(II) ΔE(4(II)) | 4(III) ΔE(4(III)) | 4(IV) ΔE(4(IV)) | 4(V) ΔE(4(V)) |
|---|---|---|---|---|---|
| gas | 0.0 | 0.5 | 0.7 | 1.2 | 7.2 |
| CH$_3$CN | 0.0 | 0.6 | 0.7 | 0.8 | 4.9 |

**Figure 2.** Conformers of monomeric **3** (upper row) and **4** (bottom row) and their relative energies (ΔE = $E_{min}$(conformer) − $E_i$(conformer), kcal/mol) obtained at the B3LYP/6-311+G(d,p) level of theory for the gas phase and in acetonitrile (CH$_3$CN). $E_{min}$(conformer) stands for the energy of **3(I)** or **4(I)**. $E_i$(conformer) stands for the energy of the conformer under consideration.

Using the knowledge of the monomer's conformations, the calculations and analysis of the possible structures of hydrogen-bonded dimers were performed (labelled **D3** and **D4** in Figure S1). The most stable conformation of the dimers was obtained when the molecules were arranged orthogonally (**D3(I)**, **D3(II)**, **D4(I)**, and **D4(II)**, Figure S1). However, the planar orientations of the molecules caused only a small increase in energy (**D3(III)**, **D3(IV)**, and **D4(V)**). Indeed, in the crystal of **3**, one carboxyl group of each molecule formed a dimer with the planar orientations of the rings while the other carboxyl group formed a hydrogen-bonded molecular chain between such dimers [37]. Structures in which the second carboxyl group was oriented orthogonally to the intermolecular H-bonded group were energetically beneficial. This result was conditioned by a smaller steric repulsion between carboxyl groups (and the nitro group in case of **3**). Thus, the formation of intramolecular H-bonds in the dimers was unfavorable. Oligomers **D3(IX)** and **D4(VIII)**, in which molecules did not form carboxyl group dimers, exhibited higher energies (Figure S1). However, the energy increase was quite moderate, especially for compound **4**. Moreover, **D3(X)** and **D4(VII)** possessed one intramolecular H-bond each.

For the assessment of dynamic effects associated with the rotations of carboxyl and nitro groups, the corresponding potential energy profiles were calculated at the B3LYP/6-311+G(d,p) level of theory for the monomers of both acids. The DFT calculations of the rotation of the nitro groups (gradual increase of the torsional angles C2C3NO5 in **3** and C3C4NO5 in **4** (Figure S2)) revealed the similarity of the rotational energy barriers for both compounds: 4.8 and 5.8 kcal/mol for **3** and **4**, respectively (Figure S3). These barriers resulted from the disruption of the π-electronic coupling between the nitro group and the benzene ring, which caused energetically disadvantageous configurations at CCNO ≅ 90° (Figure S3). For **3**, one can also observe a small barrier at C2C3NO5 ≅ 180°, caused by the repulsion between the nitro and carboxyl groups (Figure S3). It is noteworthy that the rotation of either nitro or carboxyl group evoked the simultaneous rotation of the neighboring functional groups and, therefore, it led to moderately high energy barriers.

In contrast to the nitro group rotation, the calculations showed a significant difference between the energy barrier heights for the rotation of the carboxyl groups in **3** and **4** (Figure 3). For **3**, these barriers were 5.5–6.5 kcal/mol, which was 4–5 kcal/mol higher than for **4**. This difference resulted from a strong steric effect between three functional groups in **3**. The steric squeezing between carboxyl groups in **4** was weaker than in **3** because the nitro group was in the *meta* position. The energy barriers for the nitro and carboxyl groups' rotation were not very high. Thus, for these compounds a significant dynamics of all functional groups can be expected.

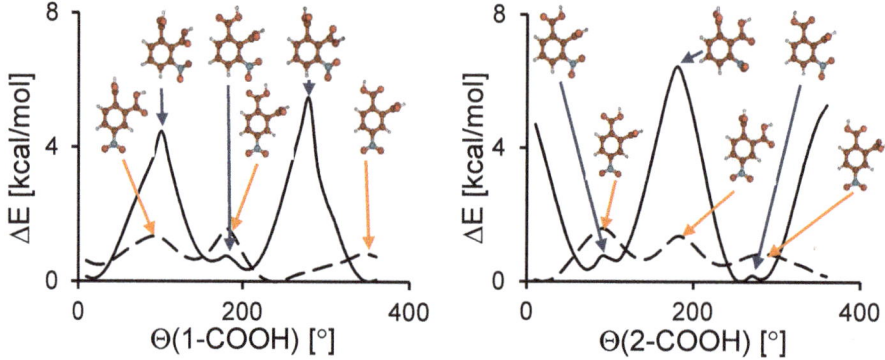

**Figure 3.** Calculated potential energy curves for the carboxyl group rotation in **3** (solid line) and **4** (dashed line).

In order to study the H-bond dynamics, we calculated potential energy profiles for proton transfer in intramolecular H-bonds in **3(III)**, **4(IV)**, **D3(III)**, and **D4(I)** in the gas phase and taking into account the effect of a polar solvent (CH$_3$CN) using the polarizable continuum model (PCM) approach (Figure 4 and Figure S4a,b). The O-H distance in one of the carboxyl groups was gradually elongated while other structural parameters were optimized for each step. The profile of the curves and its numerical values were similar for monomers and dimers. The calculations of the potential energy curves for the intramolecular proton transfer in the monomeric species showed no second minimum in the range of O···H distances 1.4–1.7 Å (Figure 4, curve a). According to the earlier presented analysis [65], which rests upon the experimental and computational data, this result proves the absence of the proton transfer within the intramolecular hydrogen bond, i.e., the absence of a tautomeric equilibrium (Figure 5F). In turn, the intermolecular transfer of one proton within the intermolecular hydrogen bond in dimers of compound **3** induced a transfer of the second proton in the adjacent intermolecular hydrogen bond (Figure 4, curve c). The calculated potential energy curves for dimers **D3(III)** and **D4(I)** turned to be double-well. The energy required for this concerted double proton transfer was about 6.3 kcal/mol for the gas phase (Figure S4b). The use of the PCM approximation for acetonitrile reduced the barrier down to 5.7 kcal/mol (Figure 4). This fact supports the possibility to observe the tautomeric equilibrium with double proton transfer (Figure 5B,C) in an experiment. Taking into account that the PCM approximation strongly underestimates the effect of polar media on hydrogen-bonded systems [66–69], one can expect a fast, concerted proton transfer in phthalic acid dimers in polar solvents. Previously, a double proton transfer in carboxylic acid dimers was experimentally detected in low-temperature NMR spectra (110 K, CDF$_3$/CDF$_2$Cl mixture as solvent) as a triplet splitting of the bridging proton signal for $^{13}$C-labelled acetic acid due to $^2J$(C,H) spin-spin coupling [70]. Though dimers **D3(I)–D3(VII)** are the most stable forms, the formation of oligomers (structure **G**, Figure 5) and complexes of other types (**D** and **E** dimers, Figure 5) is also possible. This fact is also supported by the crystallographic and spectroscopic studies [71–74].

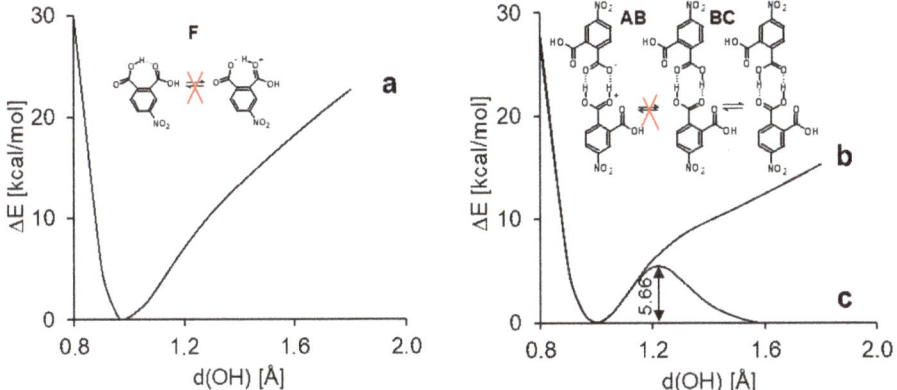

**Figure 4.** The potential energy profile for a gradual displacement of one proton within the H-bond in the **4(II)** monomer (**a**) and the **D4(I)** dimer (**b** and **c**) calculated in the PCM approximation in acetonitrile. The curves **a** and **c** represent a case when all other structural parameters are optimized. The curve **b** represents a case when the position of the adjacent bridged proton is fixed.

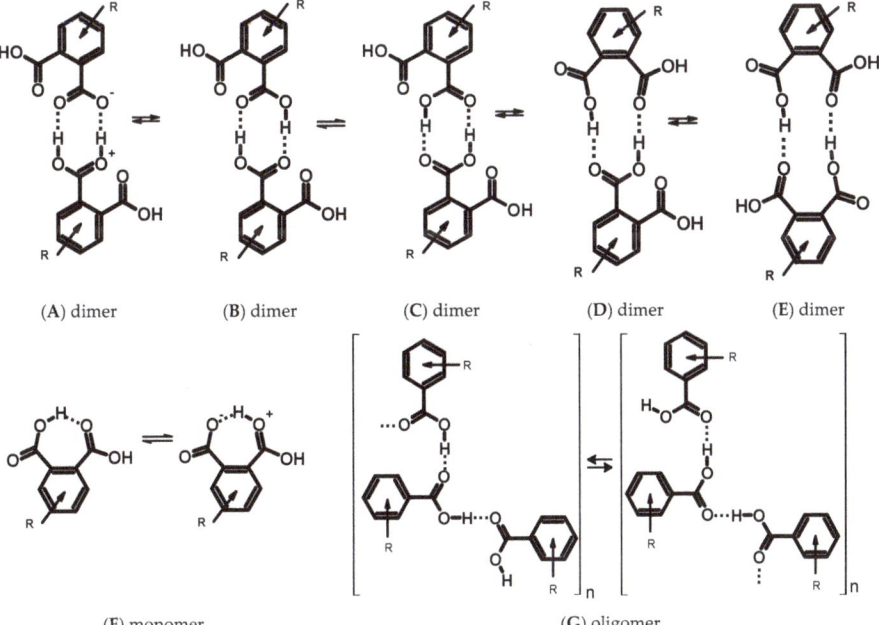

**Figure 5.** Schemes of the prototropic equilibria for the carboxyl aryl derivatives and their intermolecular complexes.

To explore the possibility of a single proton transfer in the dimers (equilibrium **BA**, Figure 5), the calculation was performed at a constant O-H distance of the adjacent hydrogen bond. There was no local minimum on the potential energy curve (Figure 4, curve **b**). Therefore, the formation of a zwitterionic complex (Figure 5, structure **A**) was disadvantageous and there was a poor chance to observe the equilibrium **BA** experimentally (Figure 5). Nevertheless, the profile for the single proton transfer in the dimer was more shallow than that in the monomer.

The calculated H-bond energies ($\Delta E(HB) \approx E_{min}(\text{non-HB}) - E_i(HB)$ [75]) in the studied dimeric complexes were smaller than 7 kcal/mol per H-bond. The estimated values of the energies calculated for the dimers of **3** and **4** correlated well with the energies reported for similar systems. For example, according to the experimental temperature-dependent attenuated total reflection (ATR) IR studies of ibuprofen by Ludwig et al. [76], the enthalpy of the transition between doubly H-bonded cyclic dimers to singly H-bonded linear dimers is equal to −5.07 kcal/mol. The binding energy of a *p*-biphthalate dimer obtained at the B3LYP/6-31+G* approximation is about 12.4 kcal/mol (6.2 kcal/mol per one hydrogen bond) [77]. Such H-bonds are characterized as weak ones. However, the studied dimers exhibited an easy double proton transfer. Such phenomenon is typical for Strong Short H-Bonds (SSHB) [78]. This observation can be rationalized as follows. An elongation of the OH distance results in an increase of the electron density on the adjacent oxygen of the same carboxyl group, thereby it strengthens the basicity of this oxygen. When the OH bond length is ca. 1.25 Å, the basicity of the adjacent oxygen becomes sufficient to evoke a spontaneous transfer of the adjacent proton from the opposite carboxyl group. A further elongation of the OH bond brings about a moderate decrease of the dimer energy, thus creating a double-well potential. Following Gilli's terminology [79], this phenomenon can be called charge flow-assisted hydrogen bond.

## 2.2. Dynamics of Hydrogen Bonding within the Framework of Molecular Dynamics

Molecular dynamics (MD) schemes, which reproduce time evolution of the studied systems, are useful in the investigations of multi-dimensional and complex phenomena [80–82]. In the studied case of phthalic acid derivatives, it was necessary to use the Car–Parrinello MD scheme (CPMD), which is based on the DFT framework and is able to reproduce H-bond properties [83–88]. This section describes how these CPMD simulations illustrate the impact of H-bond strength on the molecular metric parameters.

Table 1 presents statistical data (averages and standard deviations) for the CPMD production runs. After the thermostatted equilibration phase, the data collection without thermostats lasted 24 ps, and only the last 20 ps were taken as the production runs in order to allow the molecules to relax after thermostatting. It was interesting to see that the intramolecular H-bond in **4** was much stronger than in **3**, but the intermolecular bridges of the dimers were of almost the same strength. While the mean and standard deviations of the donor-acceptor distance were lower for a stronger bonding, the opposite was true for the donor-proton bond length. This was a result of increased delocalization of the proton in the stronger bridge, whereas the dynamics of the H-bridge were weaker. The donor-acceptor distances listed in Table 1 indicate that the intermolecular H-bonds in the dimer of **3** were stronger and more delocalized than the intramolecular one in the monomeric **3**. An opposite phenomenon was observed for **4**. This discrepancy can be explained by the difference in the geometry of the structures. For the dimers of **3** and **4**, the geometry of the H-bridges was planar (COH⋯O torsional angle ~0°) and linear (OHO angle ~179°), while, for the monomers, the geometry was neither planar nor linear (COH⋯O and OHO angles were ~64/50° and ~150/160° for **3**/**4**, respectively, Table 1). These deviations from the planarity were caused by a strong electrostatic repulsion between oxygen atoms of the intramolecular H-bonds in the monomers. Moreover, for the monomer of **3**, the phenomenon of non-coplanarity was enhanced by a strong steric repulsion from the nitro group, which led to an additional weakening of the intramolecular H-bond.

**Table 1.** Metric parameters (in Å) for the donor-acceptor (OO) and donor-proton (OH) contacts in the monomers and dimers of **3** and **4**. The CPMD results are given as: Average ± standard deviation.

| Compound | Method | Bridge 1 | | Bridge 2 | | OHO[°] | COH⋯O[°] |
|---|---|---|---|---|---|---|---|
| | | d(OH) | d(OO) | d(OH) | d(OO) | | |
| 3, monomer | CPMD | 0.993 ± 0.022 | 2.728 ± 0.151 | - | - | - | - |
| 3, dimer | - | 1.027 ± 0.032 | 2.634 ± 0.095 | 1.028 ± 0.034 | 2.633 ± 0.091 | - | - |
| 4, monomer | - | 1.005 ± 0.022 | 2.587 ± 0.088 | - | - | - | - |
| 4, dimer | - | 1.028 ± 0.036 | 2.650 ± 0.122 | 1.028 ± 0.037 | 2.653 ± 0.115 | - | - |
| 3, monomer | DFT | 0.978 | 2.670 | - | - | 150.2 | 64.4 |
| 3, dimer | - | 0.999 | 2.679 | 1.001 | 2.660 | 178.9 | 0.2 |
| 4, monomer | - | 0.985 | 2.583 | - | - | 160.0 | 50.1 |
| 4, dimer | - | 1.000 | 2.669 | 0.999 | 2.679 | 178.6 | 0.7 |
| 3, dimer | X-ray [38] | 0.84 | 2.698 | 0.84 | 2.698 | 155.5 | - |
| 3, oligomer | - | 0.84 | 2.681 | - | - | - | - |

Additional insight was provided by the time evolution of the bridge distances, depicted in Figure 6 for the monomers and Figure S5 for the dimers. It was striking that even if the monomer of **4** had the shortest donor-acceptor distance among the studied systems, there were no indications of the proton entering the acceptor side. On the other hand, the intermolecular cyclic dimers of **3** and **4** were typical for carboxylic acids. For **4**, there were numerous instances of the bridge proton being located almost in the middle of the bridge, while for **3** there were just two such cases and one of them was a concerted transfer (occurring at the same time in both bridges). Such synchronicity was less obvious for **4**. This delocalization of the protons in the cyclic dimer of **4** showed that the H-bonding in **4** was stronger than in **3** for both the monomer and the dimer.

It is worth to note that the H-bond in the monomer of **3** was characterized by the greatest dynamics, due to its non-planar structure and, as a consequence, a significant deformation component.

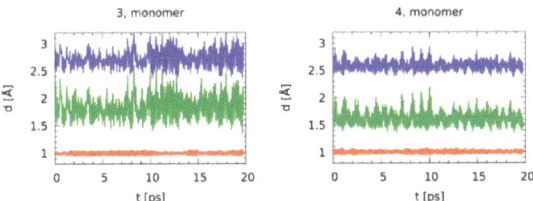

**Figure 6.** Time evolution of the H-bridge metric parameters. The CPMD gas phase simulations of the monomeric **3** and **4**. Red: Donor-proton distance, green: Proton-acceptor distance, blue: Donor-acceptor distance.

### 2.3. NMR Studies of Nitrophthalic Acids

NMR study of H-bonding in solution is challenging due to the short lifetime of H-bonded complexes of low molecular weight. Generally, only a single NMR line is observed for all mobile protons, which represent an average over different, fast interconverting hydrogen-bonded complexes. This problem can be solved using a low-freezing solvent [89]. In this solvent complex, hydrogen-bonded systems can be characterized in great detail [90–92]. However, such experiments are not without their problems. A simplified qualitative analysis is possible when the mole fractions and the individual chemical shifts of different H-bonded complexes are known.

Neither **3** nor **4** was soluble in weakly polar, aprotic solvents. However, their solubility can be increased in the presence of a dissolved base. Possible scenarios of phthalic acid interaction with bases in solution are shown in Figure 7. If the composition of such acid:base complex is 1:1, one carboxyl group of the acid interacts with the base while the other carboxyl group can either form an intramolecular H-bond with the former one (scenario **a**) or remain free (scenario **b**). If the base is in excess, 1:2 acid:base complex can be formed with two near-equal intermolecular H-bonds (scenario c). We did not consider complexes of acid dimers because such complexes are not likely in the presence of the base because the solubility of the acid alone is very low. What are the individual $^1$H chemical shifts of carboxyl protons in these complexes when the base is very strong? In a polar solvent the $^1$H chemical shifts of the carboxyl proton in a 1:1 complex of 2-nitrobenzoic acid with 2,4,6-trimethylpyridine is equal to 16.8 ppm [93]. The length of this H-bond can be elongated due to the steric effects [94,95]. The use of a stronger base can cause both a contraction and a lengthening of the H-bond. The result depends on the position of the bonding proton with respect to the H-bond center. However, the reduction of solvent polarity causes the opposite effect [66–68]. We concluded that the $^1$H chemical shift of the proton in the intermolecular H-bonds in scenarios a, b, and c should have been between 19 and 17 ppm. The $^1$H chemical shift of the proton in the intramolecular H-bond in scenario a was the hardest to estimate. The geometry of this H-bond was forced to adapt to the rigid molecular structure. Most likely, the $^1$H chemical shift of this proton should have been smaller than 15 ppm [33]. The $^1$H chemical shift of the proton of the free carboxyl group in scenario b depended on interaction with CDCl$_3$. At high concentration of 2,6-bis(trifluoromethyl)benzoic acid in dry CDCl$_3$, its mobile proton resonates at 10 ppm. At high concentration of 2,6-bis(trifluoromethyl)benzoic acid in toluene its mobile proton resonates at 8.8 ppm at 300 K and at 7.4 ppm at 354 K. In both solvents the chemical shift depends on the monomer–dimer equilibrium of the acid. We believed that 6 ppm was a safe upper limit for the $^1$H chemical shift of the proton of the free carboxyl group in scenario b. Summarizing the above, the mean $^1$H chemical shifts of the carboxyl protons in scenario a, c, and b were expected to be about 16 ppm, 18 ppm, and below 12 ppm, respectively.

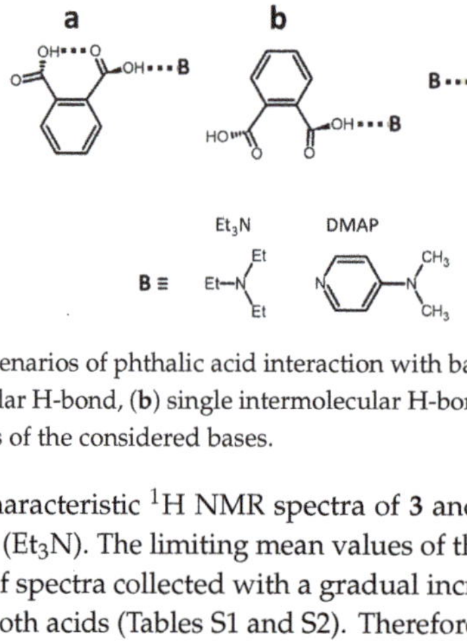

**Figure 7.** Possible scenarios of phthalic acid interaction with bases in nonpolar solution: (a) One intra- and one intermolecular H-bond, (b) single intermolecular H-bond, and (c) two intermolecular H-bonds. Molecular structures of the considered bases.

Figure 8 shows characteristic $^1$H NMR spectra of 3 and 4 in CDCl$_3$ in the presence of a large excess of triethylamine (Et$_3$N). The limiting mean values of the $^1$H chemical shift of carboxyl protons measured using a set of spectra collected with a gradual increase in the mole fraction of 3 or 4 were equal to 14.1 ppm for both acids (Tables S1 and S2). Therefore, the most likely structure of a complex of Et$_3$N with phthalic acids in nonpolar solvents corresponded to scenario a. This result is pretty intuitive while the bulkiness of Et$_3$N significantly increased the entropic cost of the structure shown in scenario c.

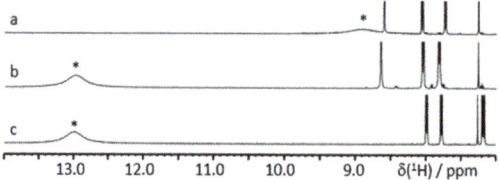

**Figure 8.** Characteristic $^1$H NMR spectra of 3 and 4 in CDCl$_3$ at 300 K in the presence of Et$_3$N. The signals of OH-protons are marked by asterisks. The mole fractions are (a) water:4:Et$_3$N = 1:1.2:35, (b) water:4:Et$_3$N = 1:8.7:35, and (c) water:3:Et$_3$N = 1:8.7:35.

Figure 9a,b shows characteristic $^1$H NMR spectra of **4** in CDCl$_3$ in the presence of a large excess of N,N-dimethylpyridin-4-amine (DMAP). The limiting mean value of the $^1$H chemical shift of carboxyl protons measured using a set of spectra collected with a gradual increase in the mole fraction of **4** was about 18.5 ppm (Table S3). Therefore, the most likely structure of the complex of DMAP with **4** in CDCl$_3$ corresponded to scenario **c**. When the mole fractions of DMAP were only slightly larger than that of **4**, while the total concentration was very low, $^1$H NMR spectra exhibited two separate peaks of different mobile protons (Figure 9c). We attributed the peak at 15.7 ppm to a complex of DMAP with **4** and the peak at 2.4 ppm to water interacting with residual DMAP. The former peak obviously corresponded to the structure in scenario **a**. The mean $^1$H chemical shift in this complex was larger than for Et$_3$N. However, this difference does not mean, obviously, that the intermolecular H-bond in DMAP:**4** was stronger than in Et$_3$N:**4**. Recall that the $^1$H chemical shift of the strongest known H-bond in [FHF]$^-$ is 16.6 ppm [95] while one of the largest $^1$H chemical shifts, of 21.7 ppm, has been measured for a moderately strong H-bond in the proton-bound homodimers of pyridine [33]. More about this issue can be found elsewhere [96]. In contrast, the value of 15.7 ppm in DMAP:**4** can be compared to the value of 18.5 ppm in DMAP:**4**:DMAP. The latter complex has two intermolecular H-bonds while the former has one inter- and one intramolecular H-bond. Therefore, if the effects of mutual influences of the adjacent hydrogen bonds on their geometries in each of these complexes were small, the individual $^1$H chemical shift of the intramolecular H-bond in DMAP:**4** was about 13 ppm.

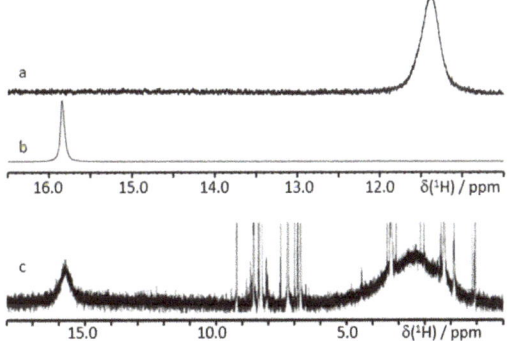

**Figure 9.** Characteristic $^1$H NMR spectra of **4** in CDCl$_3$ at 300 K in the presence of DMAP. The mole fractions are (**a**) water:**4**:DMAP = 1:1.0:260, (**b**) water:**4**:DMAP = 1:3.9:260, and (**c**) water:**4**:DMAP = 1:0.24:0.30.

In contrast to the solution with Et$_3$N, the presence of DMAP did not increase the solubility of **3** in CDCl$_3$. Presumably, **3** did not interact with DMAP by scenario **c** due to a high entropic cost caused by the position of the nitro group. Why did it not interact with DMAP by scenario **a**? We cannot answer this question with certainty.

### 2.4. H-Bonding Vibrational Modes in Carboxyl Dimers

Stretching vibrations of H-bonds have a high diagnostic value for determination of the nature and strength of these bonds [97,98]. Previously, the spectral manifestations of dimerization and isotopic effects on spectroscopic observables were studied for different molecular systems [99–108] including carboxylic acid dimers [109,110]. Upon carboxylic acid dimerization, the structure of the OH stretching band in IR spectra changes most prominently: The narrow band of monomers changes to a broad, intensive, and complex substructured band of dimers shifted to lower wavenumbers.

For a comprehensive spectroscopic investigation of **3** and **4**, we accomplished a study based on IR, Raman, and IINS measurements, as well as DFT, CPMD, and Potential Energy Distribution (PED) calculations. The IR, Raman, and IINS spectra of non-deuterated and deuterated (OH → OD replacement) **3** and **4** are shown in Figures 10 and 11. The experimental spectra were interpreted using the calculated vibrational (DFT) and power spectra (CPMD), and the results of the PED analysis (Tables S4 and S5). More about this issue can be found elsewhere [111–113].

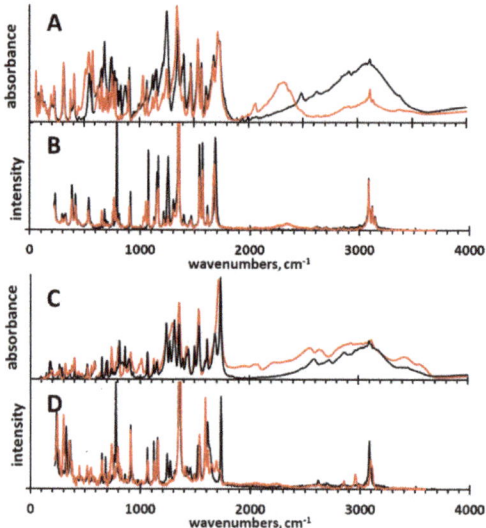

**Figure 10.** Normalized experimental IR and Raman spectra of **3** (**A** and **B**) and **4** (**C** and **D**) (black spectra) and their deuterated (OD) derivatives (red spectra).

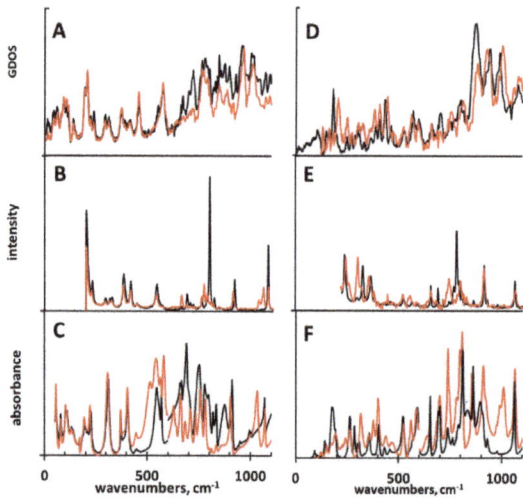

**Figure 11.** Normalized IINS (**A**, **D**), Raman (**B**, **E**), and IR (**C**, **F**) spectra of compounds **3** (**A**—**C**, black spectra) and **4** (**D**–**F**, black spectra) and their deuterated derivatives (red spectra).

According to the crystallographic data [37], the molecules of **3** form H-bonded oligomeric chains of dimers. These H-bonds were almost of the same length (Table 1) and, consequently, were equally strong. Therefore, the stretching vibrations of the OH group ($\nu$(OH)) were within the same spectral range and overlapped in the experimental IR spectra (Figure 10). The shapes of the $\nu$(OH) and $\nu$(OD) bands were very alike to those of carboxylic acid dimer studied experimentally and theoretically by

Flakus et al. [110]. For **3**, the deuteration caused a shift of the ν(OD) band to lower wavenumbers according to the well-established rule with the isotopic spectroscopic ratio ISR = $\delta_{OH}/\delta_{OD}$ = 1.28 [114]. In contrast, for **4** the band ν(OD) expanded strongly; this revealed a complex character of the underlying changes. Deformational (δ(OH)/δ(OD) and γ(OH)/γ(OD)) bands are informative because in **3** and **4** they differed from those observed for intramolecular H-bonds in *ortho*-hydroxy aryl Schiff bases [115–118] and *ortho*-acetophenones [119]. The δ(OH) was a doublet at 1409/1395 cm$^{-1}$ in **3** and a band at 1383 cm$^{-1}$ in **4**. Upon deuteration, these bands disappeared to emerge at 1033/1029 cm$^{-1}$ in **3** and at 1014/995 cm$^{-1}$ in **4**. Thus, the ISR is in the range of 1.36–1.35 for both compounds (Tables S4 and S5). This characteristic behavior of the ISR deviated from that of *ortho*-hydroxy aryl Schiff bases [115] and *ortho*-hydroxy acetophenones [119]. When it comes to the bands assigned to the deformational γ(OH) vibrations, a few bands shifted to the low wavenumbers' region after the deuteration: 876, 835, 819, 796, 752, and 691 cm$^{-1}$ for **3** and 862, 839, and 763 cm$^{-1}$ for **4** in IR/Raman spectra (Tables S4 and S5). The assignments of these bands to the deformational vibrations of the bridging protons was unequivocal because the intensity of the two series of the bands at 895, 875, 845, 826 cm$^{-1}$ and 778, 715, 668 cm$^{-1}$ for **3** and 874 and 704 cm$^{-1}$ for **4** was greatly decreased in the IINS spectra after the deuteration (Figure 11 and Table S4). This phenomenon has been studied in the past [120–125]. As for the emergence of the two series of bands assigned to the deformational vibrations, it can be explained by the presence of the dimers and the oligomers in the solid state (see above). The attribution of two deformational bands to dimers and monomers was suggested by Miyazawa and Pitzer [126] for formic acid in the gas phase and solid nitrogen matrixes. Thus, the two series of the (γ(OH) bands at 876 cm$^{-1}$ (γ(OD) = 627 cm$^{-1}$), 835 cm$^{-1}$ (627 cm$^{-1}$), 819 cm$^{-1}$ (627 cm$^{-1}$), and 796 cm$^{-1}$ (580 cm$^{-1}$) and at 752 cm$^{-1}$ (545 cm$^{-1}$), 690 cm$^{-1}$ (514 cm$^{-1}$), and 874, 704 cm$^{-1}$ are assigned to the deformational vibrations of the carboxyl groups of the dimers and the monomers of **3** and **4**, respectively.

The assignment of γ(OH) can be supported by the previously published d(OO) = f(γ(OH)) correlation [127,128] and the crystallographic data for **3** [37]. The lengths of the H-bridges in the range of 2.65–2.70 Å (d(OO) is calculated by means of the d(OO) = 3.01–4.4 × 10$^{-4}$ γ(OH) correlation, where d(OO) is in Å and γ(OH) is in cm$^{-1}$) matched very well with the experimentally measured ones (d(OO) = 2.698 Å and 2.681 Å [37]). Moreover, the experimentally obtained wavenumber values for the γ(OH) bands can be applied to compare the strength of the H-bonds in **3** and **4**. The obtained results show that the H-bonds in **3** were a bit weaker than in **4** (d(OO) is in the range of 2.65–2.70 Å for **3** and 2.63–2.67 Å for **4**), though the difference in the strength of the hydrogen bonding was not large.

In terms of the H-bond vibrations, the IINS spectroscopy allows one to unequivocally interpret bands (ν$_\sigma$) due to the almost complete disappearance of these bands upon deuteration [115,121]. Based on this phenomenon, two low-intensity bands at 555 and 390 cm$^{-1}$ were assigned to vibrations ν$_\sigma^{asym}$ and ν$_\sigma^{sym}$ of the H-bonds, respectively. Importantly, these bands overlapped with the bands of other vibrations (insensitive to deuteration) both in IINS and IR spectra. However, the changes of the IINS spectra were much clearer than ones of the IR spectra.

The relative strengths of the H-bonds in dimers and monomers of **3** and **4** can be evaluated using atomic velocity power spectra obtained from the CPMD trajectories. The vibrational spectra related to the atomic motion intensity (arbitrary intensities) are presented in Figure 12. The bands of hydroxyl groups are relatively broad (red bars in Figure 12): 3100–3500 cm$^{-1}$ and 2920–3250 cm$^{-1}$ for the monomers of **3** and **4** and 2400–3100 cm$^{-1}$ and 2250–3200 cm$^{-1}$ for the dimers of **3** and **4**. The bands of the dimers are strongly red-shifted as compared to those of the monomers. This shift indicates that the intermolecular H-bonds in the dimers were much stronger than the intramolecular ones in the monomers.

The stretching and bending vibration areas of the dimers of **3** and **4** did overlap (Figure 12, **3d** and **4d**). In contrast, the stretching and bending vibration bands of the monomer of **4** were blue- and red-shifted, respectively, as compared to those of **3** (Figure 12, **3m** and **4m**). Therefore, the strengths of H-bonds in the dimers of **3** and **4** were similar. In contrast, the intramolecular H-bond in the monomer of **3** was weaker than that in the monomer of **4**. The reason for that is that the structure of the monomer of **3** was more bent. This conclusion is consistent with the above interpretation of the experimental data and demonstrates that experimental spectroscopic studies and CPMD simulations greatly enhance each other's results.

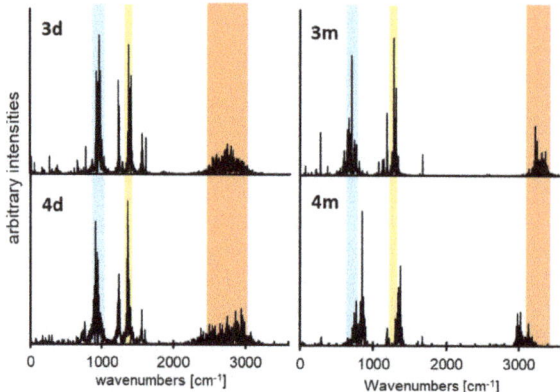

**Figure 12.** Calculated power spectra of atomic velocity–results of the CPMD runs for the monomers of 3 and 4 (**3m** and **4m**) as well as for the dimers of 3 and 4 (**3d** and **4d**). The CPMD power spectra are presented only for the bridged protons vibrational modes. The stretching vibration area is shown in red. The bending vibration areas are shown in blue and yellow.

## 3. Materials and Methods

### 3.1. Compounds and Deuteration

The studied compounds and solvents were purchased from Sigma-Aldrich company and used without further purification. The deuterated sample was prepared by dissolving the product in deuterated methanol ($CH_3OD$). The solution was then heated to 60 °C and refluxed during 30 min. After that, the methanol was removed by evaporation under reduced pressure. This procedure was repeated three times.

### 3.2. Infrared and Raman Measurements

The far and middle infrared (FIR, MIR) absorption measurements were performed using a Bruker Vertex 70v vacuum Fourier Transform spectrometer. The transmission spectra were collected with a resolution of 2 $cm^{-1}$ and with 64 and 32 scans per each spectrum for FIR and MIR, respectively. The FT-FIR spectra (500–50 $cm^{-1}$) were collected for the samples suspended in Apiezon N grease and placed on a polyethylene (PE) disc. The FT-MIR spectra were collected for the samples in a KBr pellet. The Raman spectra of the analyzed samples were obtained using FT-Nicolet Magma 860 spectrophotometer The In:Ga:Ar laser line at 1064 nm was employed for the Raman excitation measurements. The spectra were recorded at the room temperature in the range of 200–3800 $cm^{-1}$ with the spectral resolution of 4 $cm^{-1}$ and with the same number of scans (512/measurement).

### 3.3. Incoherent Inelastic Neutron Scattering (IINS) Measurements

Neutron scattering data were collected at the pulsed IBR-2 reactor at the Joint Institute of Nuclear Research (Dubna) using the time-of-flight inverted geometry spectrometer NERA at 10 K temperature. The spectra were converted from neutron per channel to the scattering function per energy transfer.

At the energy transfer between 5 and 1200 cm$^{-1}$, the relative IINS resolution was estimated to be ca. 3%. The $S(Q, \omega)$ function (scattering law) can be expressed in the form of isotropic harmonic oscillator [129]:

$$S(Q, n\omega) = \frac{(Q^2 \cdot U^2)}{n!} \cdot exp((Q^2 \cdot U^2)) \quad (1)$$

where $Q$ is the momentum transfer and $U^2$ is the mean square displacement defined as

$$U^2 = \frac{\hbar}{2m\omega} = \frac{16.795}{\mu \nu} \quad (2)$$

where $\mu$ is the mass oscillator in amu, $\nu$ is the oscillator energy in cm$^{-1}$, $U^2$ is expressed in Å$^2$, and $n$ is the number of excited states.

### 3.4. NMR Measurements

The $^1$H spectra were recorded at room temperature on a Bruker Avance III 500 MHz spectrometer. CDCl$_3$ was purchased from Sigma-Aldrich and used without further purification. The spectra were measured using the solvent peak as an internal reference, and the chemical shifts were converted to the conventional TMS scale. The number of scans varied between 128 and 256.

### 3.5. Car–Parrinello Molecular Dynamics' Simulations

A dynamical nature of the investigated molecules 3 and 4, with the emphasis on their hydrogen bridges, was studied using Car–Parrinello molecular dynamics (CPMD) [130]. The models of monomers and dimers for the CPMD simulations were constructed on the basis of static DFT gas phase results. The molecular structures were placed in cubic boxes with a = 15 Å for the monomeric forms and a = 22 Å (for compound 3) and a = 25 Å (for compound 4) for dimeric forms. The first-principle molecular dynamics (FPMD) calculations were performed in the gas phase with the empirical van der Waals correction by Grimme (all DFT-D2) [131]. The Perdew–Burke–Ernzerhof (PBE) exchange-correlation DFT functional [132] was applied. The core electrons of the studied monomers and dimers were replaced by norm-conserving pseudopotentials of Troullier–Martins type [133]. The Kohn–Sham orbitals were expanded using the plane-wave basis set with the maximum kinetic energy cutoff of 90 Ry. The Hockney's scheme [134] was used to remove interactions with periodic images and simulate isolated molecule conditions. The orbital coefficients were propagated using the default value of the fictitious orbital mass, 400 a.u., and the nuclear motion timestep was set to 2 a.u. The CPMD simulations were divided into two steps: The equilibration and the production runs. During the equilibration, the ionic temperature was set to 297 K and controlled by Nosé–Hoover thermostat chains with default settings, with each degree of freedom coupled to a separate thermostat ("massive" thermostatting) [135,136]. The Nosé–Hoover thermostat chain was set to 3200 cm$^{-1}$ frequency. The equilibration runs of the CPMD lasted for 50,000 steps for the monomers and dimers. The data collection lasted for 500,000 steps (24 ps) using the NVE microcanonical ensemble (the thermostat chains were detached during the simulations). The obtained trajectories served as a basis for the distance evolution analysis of the bridged proton and the functional groups' dynamic as well as to determine the vibrational features of the investigated compounds from the power spectra of atomic velocity.

The CPMD simulations were carried out using the CPMD 3.17.1 program [137]. The data analysis was performed using locally written utilities and the VMD 1.9.3. program [138]. The graphical presentation of the obtained results was prepared with the Gnuplot graphics package [139], and with the VMD 1.9.3. program [138].

### 3.6. DFT Calculations

This part of the calculations was performed with the Gaussian 09 suite of programs [140] using the density functional theory (DFT) with the three-parameter functional proposed by Becke with the correlation energy according to the Lee–Yang–Parr formula, denoted as B3LYP [141,142]. The triple-zeta split-valence basis set, denoted as 6-311+G(d,p) [143–145] according to the Pople's notation, was applied. The use of diffuse functions is a proper approach for studies of hydrogen bonding [146]. Initially, the geometry optimization was carried out and followed by harmonic frequencies' calculations, confirming that the obtained structures correspond to the minima on the potential energy surface (PES). Next, the one-dimensional reaction path of the bridged proton transfer from donor to the acceptor atom within the intramolecular hydrogen bond was studied. The applied approach was based on stepwise elongation of the O-H distance (with 0.05 or 0.1 Å increments) with full optimization of the remaining structural parameters. The calculations were carried out in the gas phase and with the solvent reaction field using acetonitrile as a solvent. The Polarizable Continuum Model (PCM) method [147] was used to reproduce the solvent influence on the studied molecules. All the performed calculations were conducted for the electronic ground state and without any extra charges on the molecules and dimers. The obtained results were visualized using the MOLDEN software [148].

*3.7. PED Analysis*

The potential energy distribution (PED) of the normal modes was calculated in terms of natural internal coordinates [149] using the Gar2ped program [150].

## 4. Conclusions

The result of the interplay between competing noncovalent interactions in the condensed phase may appear to be quite unexpected. The conformation of carboxyl groups is assumed to be dominantly *cis* due to so-called Z-effect [59]. However, the conformation can be changed in H-bonded associates [56–59]. We herein reported a comprehensive computational and experimental study of this phenomenon using 3-nitrophthalic (**3**) and 4-nitrophthalic acids (**4**) as model systems. It was observed that an intermolecular H-bond interaction between the adjacent carboxyl groups of these molecules became favorable only when one of the groups was involved in a strong intermolecular H-bond. However, even in this case, the spatial distance between the carboxyl groups needed to be increased. If the latter was not possible, for example due to steric hindrances, as in **3**, the intramolecular interaction was energetically unfavorable. As a result, the intramolecular steric hindrances critically affected the solubility, the crystal packing, and the intramolecular proton exchange of phthalic acids.

The structural and energetic parameters of intra- and intermolecular interactions in the monomers, dimers, and aggregates of **3** and **4** were estimated for the gas, liquid, and solid phases.

**Supplementary Materials:** The following are available online, Figure S1: The dimeric forms of compounds **3** and **4** and relative energy values obtained at B3LYP/6-311+G(d,p) level of theory, Figure S2: Structures and atoms numbering of studied compounds **3** and **4**, Figure S3: Calculated potential energy curves for the gradual nitro group rotation of conformers **3(I)** and **4(II)**, Figure S4: Calculated (B3LYP/6-311+G(d,p), PCM approach for acetonitrile (a) and gas phase (b)) potential energy functions by the gradual displacement of one proton for compounds **3** and **4** whereas the remaining parameters were optimized: in the intramolecular hydrogen bond of monomers, in the intermolecular hydrogen bond of dimers and in the intermolecular hydrogen bond of dimers for fixed adjacent bridged proton; Figure S5: Time evolution of the metric parameters of two symmetric hydrogen bridges. The CPMD gas phase simulations of the dimers of **3** and **4**. Donor-proton distance, proton-acceptor distance, donor-acceptor distance; Table S1: $^1$H NMR data for compound **3** in CDCl3 in the presence of N,N-diethylethanamine (Et3N), Table S2: $^1$H NMR data for compound **4** in CDCl3 in the presence of N,N-diethylethanamine (Et3N), Table S3: $^1$H NMR data for compound **4** in CDCl3 in the presence of N,N-dimethylpyridin-4-amine (DMAP), Table S4: Experimental IR, Raman, IINS and calculated DFT (B3LYP/6-311+G(d,p)) spectral data of compound **3** and its mono deuterated (OH→OD) derivative, Table S5: Experimental IR, Raman, IINS and calculated DFT (B3LYP/6-311+G(d,p)) spectral data of compound **4** and its mono deuterated (OH→OD) derivative.

**Author Contributions:** Conceptualization, A.F.; methodology, K.J., A.J., J.J.P., E.A.G., P.M.T., I.G.S. and A.F.; software, K.J., A.J., J.J.P., E.A.G., P.M.T., I.G.S. and A.F.; validation, K.J., A.J., J.J.P., E.A.G., P.M.T., I.G.S. and A.F.; formal analysis, K.J., A.J., J.J.P., E.A.G., P.M.T., I.G.S. and A.F.; investigation, K.J., A.J., J.J.P., E.A.G., P.M.T., I.G.S. and A.F.; resources, K.J., A.J., J.J.P., E.A.G., P.M.T., I.G.S. and A.F.; data curation, K.J., A.J., J.J.P., E.A.G., P.M.T., I.G.S. and A.F.; writing—original draft preparation, A.F.; writing—review and editing, A.F. and I.G.S.; visualization, K.J., A.J., J.J.P., E.A.G., P.M.T., I.G.S. and A.F.; supervision, A.F.; project administration, A.F.; funding acquisition, I.G.S. All authors have read and agreed to the published version of the manuscript.

**Funding:** This research was funded by the Russian Foundation for Basic Research (RFBR, grant no 18-13-00050) and Polish Government Plenipotentiary for the Joint Institute for Nuclear Research n Dubna (75/24/2020; p. 75; date 3 February 2020).

**Acknowledgments:** The authors acknowledge the Wrocław Centre for Networking and Supercomputing Centres (WCSS) for providing computational time and facilities.

**Conflicts of Interest:** The authors declare no conflict of interest.

## References

1. Arp, F.F.; Bhuvanesh, N.; Blümel, J. Di(hydroperoxy)cycloalkane Adducts of Triarylphosphine Oxides: A Comprehensive Study Including Solid-State Structures and Association in Solution. *Inorg. Chem.* **2020**, *59*, 13719–13732. [CrossRef]
2. Arp, F.F.; Bhuvanesh, N.; Blümel, J. Hydrogen peroxide adducts of triarylphosphine oxides. *Dalton Trans.* **2019**, *48*, 14312. [CrossRef] [PubMed]
3. Shenderovich, I.G.; Limbach, H.-H.; Smirnov, S.N.; Tolstoy, P.M.; Denisov, G.S.; Golubev, N.S. H/D isotope effects on the low-temperature NMR parameters and hydrogen bond geometries of $(FH)_2F^-$ and $(FH)_3F^-$ dissolved in $CDF_3/CDF_2Cl$. *Phys. Chem. Chem. Phys.* **2002**, *4*, 5488–5497. [CrossRef]
4. Mauder, D.; Akcakayiran, D.; Lesnichin, S.B.; Findenegg, G.H.; Shenderovich, I.G. Acidity of Sulfonic and Phosphonic Acid-Functionalized SBA-15 under Almost Water-Free Conditions. *J. Phys. Chem. C* **2009**, *113*, 19185–19192. [CrossRef]
5. Melikova, S.M.; Voronin, A.P.; Panek, J.; Frolov, N.E.; Shishkina, A.V.; Rykounov, A.A.; Tretyakov, P.Y.; Vener, M.V. Interplay of pi-stacking and inter-stacking interactions in two-component crystals of neutral closed-shell aromatic compounds: Periodic DFT study. *RSC Adv.* **2020**, *10*, 27899–27910. [CrossRef]
6. Grabowski, S.J. Tetrel Bonds with π-Electrons Acting as Lewis BasesTheoretical Results and Experimental Evidences. *Molecules* **2018**, *23*, 1183. [CrossRef]
7. Grabowski, S.J. Triel Bonds, pi-Hole-pi-Electrons Interactions in Complexes of Boron and Aluminium Trihalides and Trihydrides with Acetylene and Ethylene. *Molecules* **2013**, *20*, 11297–11316. [CrossRef]
8. Shishkina, A.V.; Zhurov, V.V.; Stash, A.I.; Vener, M.V.; Pinkerton, A.A.; Tsirelson, V.G. Noncovalent Interactions in Crystalline Picolinic Acid N-Oxide: Insights from Experimental and Theoretical Charge Density Analysis. *Cryst. Growth Des.* **2013**, *13*, 816–828. [CrossRef]
9. Palusiak, M.; Grabowski, S.J. Do intramolecular halogen bonds exist? Ab initio calculations and crystal structures' evidences. *Struct. Chem.* **2008**, *19*, 5–11. [CrossRef]
10. Gurinov, A.A.; Rozhkova, Y.A.; Zukal, A.; Čejka, J.; Shenderovich, I.G. Mutable Lewis and Brønsted Acidity of Aluminated SBA-15 as Revealed by NMR of Adsorbed Pyridine-$^{15}$N. *Langmuir* **2011**, *27*, 12115–12123. [CrossRef] [PubMed]
11. Shenderovich, I.G. For Whom a Puddle Is the Sea? Adsorption of Organic Guests on Hydrated MCM-41 Silica. *Langmuir* **2020**, *36*, 11383–11392. [CrossRef] [PubMed]
12. Gruenberg, B.; Emmler, T.; Gedat, E.; Shenderovich, I.; Findenegg, G.H.; Limbach, H.-H.; Buntkowsky, G. Hydrogen Bonding of Water Confined in Mesoporous Silica MCM-41 and SBA-15 Studied by 1H Solid-State NMR. *Chem. Eur. J.* **2004**, *10*, 5689–5696. [CrossRef] [PubMed]
13. Kunz, W. Specific ion effects in colloidal and biological systems. *Curr. Opin. Colloid Interface Sci.* **2010**, *15*, 34–39. [CrossRef]
14. Shenderovich, I.G. The Partner Does Matter: The Structure of Heteroaggregates of Acridine Orange in Water. *Molecules* **2019**, *24*, 2816. [CrossRef]
15. Heyne, B. Self-assembly of organic dyes in supramolecular aggregates. *Photochem. Photobiol. Sci.* **2016**, *15*, 1103–1114. [CrossRef]
16. Sharif, S.; Shenderovich, I.G.; González, L.; Denisov, G.S.; Silverman, D.N.; Limbach, H.-H. NMR and Ab initio Studies of Small Complexes Formed between Water and Pyridine Derivatives in Solid and Liquid Phase. *J. Phys. Chem. A* **2007**, *111*, 6084–6093. [CrossRef]
17. Shenderovich, I.G. Effect of Non-Covalent Interactions on the $^{31}$P Chemical Shift Tensor of Phosphine Oxides, Phosphinic, Phosphonic, and Phosphoric Acids and Their Complexes with Lead(II). *J. Phys. Chem. C* **2013**, *117*, 26689–26702. [CrossRef]

18. Tupikina, E.Y.; Bodensteiner, M.; Tolstoy, P.M.; Denisov, G.S.; Shenderovich, I.G. P=O Moiety as an Ambidextrous Hydrogen Bond Acceptor. *J. Phys. Chem. C* **2018**, *122*, 1711–1720. [CrossRef]
19. Nagy, P.I. Competing Intramolecular vs. Intermolecular Hydrogen Bonds in Solution. *Int. J. Mol. Sci.* **2014**, *15*, 19562–19633. [CrossRef]
20. Lesnichin, S.B.; Tolstoy, P.M.; Limbach, H.-H.; Shenderovich, I.G. Counteranion-Dependent Mechanisms of Intramolecular Proton Transfer in Aprotic Solution. *Phys. Chem. Chem. Phys.* **2010**, *12*, 10373–10379. [CrossRef]
21. Benner, S.A. Unusual Hydrogen Bonding Patterns and the Role of the Backbone in Nucleic Acid Information Transfer. *ACS Cent. Sci.* **2016**, *2*, 882–884. [CrossRef] [PubMed]
22. Guthrie, J.P. Short strong hydrogen bonds: Can they explain enzymic catalysis? *Chem. Biol.* **1996**, *3*, 163–170. [CrossRef]
23. Roy, N.; Bruchmann, B.; Lehn, J.-M. DYNAMERS: Dynamic polymers as self-healing materials. *Chem. Soc. Rev.* **2015**, *44*, 3786–3807. [CrossRef] [PubMed]
24. Vyalikh, A.; Emmler, T.; Shenderovich, I.; Zeng, Y.; Findenegg, G.H.; Buntkowsky, G. $^2$H-Solid State NMR and DSC Study of Isobutyric Acid in Mesoporous Silica Materials. *Phys. Chem. Chem. Phys.* **2007**, *9*, 2249–2257. [CrossRef]
25. Li, Z.-T.; Wu, L.-Z. *Hydrogen Bonded Supramolecular Materials*; Springer: Berlin/Heidelberg, Germany, 2015.
26. Chan-Huot, M.; Dos, A.; Zander, R.; Sharif, S.; Tolstoy, P.M.; Compton, S.; Fogle, E.; Toney, M.D.; Shenderovich, I.; Denisov, G.S.; et al. NMR Studies of Protonation and Hydrogen Bond States of Internal Aldimines of Pyridoxal 5′-Phosphate Acid–Base in Alanine Racemase, Aspartate Aminotransferase, and Poly-L-lysine. *J. Am. Chem. Soc.* **2013**, *135*, 18160–18175. [CrossRef]
27. Hynes, J.T.; Klinman, J.P.; Limbach, H.-H.; Schowen, R.L. *Hydrogen-Transfer Reactions*; Wiley-VCH: Weinheim, Germany, 2006.
28. Segura, J.L.; Mancheno, M.J.; Zamora, F. Covalent Organic Frameworks Based on Schiff-base Chemistry: Synthesis, Properties and Potential Applications. *Chem. Soc. Rev.* **2016**, *45*, 5635–5671. [CrossRef]
29. Burrows, A.D. Crystal Engineering Using Multiple Hydrogen Bonds. *Struct. Bonds* **2004**, *108*, 55–96.
30. Pietrzak, M.; Wehling, J.P.; Kong, S.; Tolstoy, P.M.; Shenderovich, I.G.; Lopez, C.; Claramunt, R.M.; Elguero, J.; Denisov, G.S.; Limbach, H.-H. Symmetrization of Cationic Hydrogen Bridges of Protonated Sponges Induced by Solvent and Counteranion Interactions as Revealed by NMR Spectroscopy. *Chem. Eur. J.* **2010**, *16*, 1679–1690. [CrossRef]
31. Woińska, M.; Grabowski, S.; Dominiak, P.M.; Woźniak, K.; Jayatilaka, D. Hydrogen atoms can be located accurately and precisely by x-ray crystallography. *Sci. Adv.* **2016**, *2*, e1600192. [CrossRef]
32. Montis, R.; Hursthouse, M.B. Surprisingly complex supramolecular behaviour in the crystal structures of a family of mono-substituted salicylic acids. *CrystEngComm* **2012**, *14*, 5242–5254. [CrossRef]
33. Golubev, N.S.; Smirnov, S.N.; Schah-Mohammedi, P.; Shenderovich, I.G.; Denisov, G.S.; Gindin, V.A.; Limbach, H.-H. Study of Acid-Base Interaction by Means of Low-Temperature NMR Spectra. Structure of Salicylic Acid. *Russ. J. Gen. Chem.* **1997**, *67*, 1082–1087.
34. Machado, T.C.; Kuminek, G.; Cardoso, S.G.; Rodríguez-Hornedo, N. The role of pH and dose/solubility ratio on cocrystal dissolution, drug supersaturation and precipitation. *Eur. J. Pharm. Sci.* **2020**, *152*, 105422. [CrossRef] [PubMed]
35. Tao, Q.; Chen, J.-M.; Lub, T.-B. Two polymorphs and one hydrate of a molecular salt involving phenazopyridine and salicylic acid. *CrystEngComm* **2013**, *15*, 7852–7855. [CrossRef]
36. Bolla, G.; Sanphui, P.; Nangia, A. Solubility Advantage of Tenoxicam Phenolic Cocrystals Compared to Salts. *Cryst. Growth Des.* **2013**, *13*, 1988–2003. [CrossRef]
37. Glidewell, C.; Low, J.N.; Skakle, J.M.S.; Wardell, J.L. 3-Nitrophthalic acid: C(4) and $R_2^2(8)$ motifs of O-H$\cdots$O hydrogen bonds generate sheets which are linked by C-H$\cdots$O hydrogen bonds. *Acta Cryst.* **2003**, *C59*, o144–o146.
38. Smith, G.; Wermuth, U.D.; Young, D.J.; White, J.M. The 1:1 proton-transfer compounds of 4-(phenyldiazenyl)aniline (aniline yellow) with 3-nitrophthalic, 4-nitrophthalic and 5-nitroisophthalic acids. *Acta Cryst.* **2008**, *C64*, o123–o127. [CrossRef]
39. Smith, G.; Wermuth, U.D. Proton-transfer compounds of isonipecotamide with the aromatic dicarboxylic acids 4-nitrophthalic, 4,5-dichlorophthalic, 5-nitroisophthalic and terephthalic acid. *Acta Cryst.* **2011**, *67*, o259–o264. [CrossRef]

40. Jin, S.; Wang, D.; Du, S.; Linhe, Q.; Fu, M.; Wu, S. Crystal and Molecular Structure of Two Proton Transfer Compounds from Quinolin-8-ol, 4-nitro-phthalic Acid, and 1,5-Naphthalenedisulfonic Acid. *J. Chem. Crystallogr.* **2014**, *44*, 435–441. [CrossRef]
41. Saunders, L.K.; Nowell, H.; Hatcher, L.E.; Shepherd, H.J.; Teat, S.J.; Allan, D.R.; Raithby, P.R.; Wilson, C.C. Exploring short strong hydrogen bonds engineered in organic acid molecular crystals for temperature dependent proton migration behaviour using single crystal synchrotron X-ray diffraction (SCSXRD). *CrystEngComm* **2019**, *21*, 5249–5260. [CrossRef]
42. Pfeiffer, P.; Halperin, O.; Pros, E.; Schwarzkopf, V. Additionsprodukte von Zinnhalogeniden an Carbonylverbindungen, I. Beitrag zur Theorie der Halochromieerscheinungen. *J. Liebegs Ann. Chem.* **1910**, *376*, 285–310. [CrossRef]
43. Pfeiffer, P.; Friedmann, B.; Goldberg, Z.; Pros, E.; Schwarzkopf, V. Beitrag zur Theorie der Halochromieerscheinungen II. Unter experimenteller Mitarbeit. *J. Liebegs Ann. Chem.* **1911**, *383*, 92–155. [CrossRef]
44. Pfeiffer, P. Zur Kenntnis der sauren Salze der Carbonsäuren. *Ber. Deutsch. Chem. Ges.* **1914**, *47*, 1580–1596. [CrossRef]
45. Marechal, Y.; Witkowski, A. Infrared Spectra of H-Bonded Systems. *J. Chem. Phys.* **1968**, *48*, 3637. [CrossRef]
46. Marechal, Y.; Durig, J. *Vibration Spectra and Structure*; Elsevier: Amsterdam/Holland, The Netherlands, 1997.
47. Wójcik, M.J. Infrared Spectra of Hydrogen-Bonded Salicylic Acid and Its Derivatives. Salicylic Acid and Acetylsalicylic Acid. *Chem. Phys. Lett.* **1981**, *83*, 503–507. [CrossRef]
48. Issaoui, N.; Rekik, N.; Oujia, B.; Wójcik, M.J. Theoretical Infrared Line Shapes of H-Bonds within the Strong Anharmonic Coupling Theory. Fermi Resonances Effects. *Int. J. Quant. Chem.* **2010**, *110*, 2583–2602. [CrossRef]
49. Wójcik, M.J.; Szczeponek, K.; Boczar, M. Theoretical Study of Multidimensional Proton Tunnelling in Benzoic Acid Dimer. *Int. J. Mol. Sci.* **2003**, *4*, 422–433. [CrossRef]
50. Kearley, G.J.; Fillaux, F.; Baron, M.-H.; Bennington, S.; Tomkinson, J. A New Look at Proton Transfer Dynamics along the Hydrogen Bonds in Amides and Peptides. *Science* **1994**, *264*, 1285–1289. [CrossRef] [PubMed]
51. Stepanian, S.G.; Reva, I.D.; Radchenko, E.D.; Sheina, G.G. Infrared spectra of benzoic acid monomers and dimers in argon matrix. *Vib. Spectrosc.* **1996**, *11*, 123–133. [CrossRef]
52. Fillaux, F.; Limage, M.H.; Romain, F. Quantum proton transfer and interconversion in the benzoic acid crystal: Vibrational spectra, mechanism and theory. *Chem. Phys.* **2002**, *276*, 181–210. [CrossRef]
53. Shipman, S.T.; Douglass, P.C.; Yoo, H.S.; Hinkle, C.E.; Mierzejewski, E.L.; Pate, B.H. Vibrational dynamics of carboxylic acid dimers in gas and dilute solution. *Phys. Chem. Chem. Phys.* **2007**, *9*, 4572–4586. [CrossRef]
54. Worley, J.D. A Family of Hydrogen Bonds in the Model System Salicylic Acid-Toluene-Water. *J. Chem. Educ.* **1993**, *70*, 417–420. [CrossRef]
55. Zhu, L.; Al-Kaysi, R.O.; Dillon, R.J.; Tham, F.S.; Bardeen, C.J. Crystal Structures and Photophysical Properties of 9-Anthracene Carboxylic Acid Derivatives for Photomechanical Applications. *Cryst. Growth Des.* **2011**, *11*, 4975–4983. [CrossRef]
56. Neumann, M.A.; Craciun, S.; Corval, A.; Johnson, M.R.; Horsewil, A.J.; Benderskii, V.A.; Trommsdorff, H.P. Proton Dynamics and the Tautomerization Potential in Benzoic Acid Crystals. *Ber. Busenges. Phys. Chem* **1998**, *102*, 325–334. [CrossRef]
57. Marushkevich, K.; Khriachtchev, L.; Rasanen, M.; Melavuori, M.; Lundell, J. Dimers of the Higher-Energy Conformer of Formic Acid: Experimental Observation. *J. Phys. Chem. A* **2012**, *116*, 2101–2108. [CrossRef] [PubMed]
58. Marushkevich, K.; Khriachtchev, L.; Lundell, J.; Rasanen, M. cis-trans Formic Acid Dimer: Experimental Observation and Improved Stability against Proton Tunneling. *J. Am. Chem. Soc.* **2006**, *128*, 12060–12061. [CrossRef] [PubMed]
59. Medvedev, M.G.; Bushmarinova, I.S.; Lyssenko, K.A. Z-effect reversal in carboxylic acid associates. *Chem. Commun.* **2016**, *52*, 6593–6596. [CrossRef]
60. Radosavljevic Evans, I.; Howard, J.A.K.; Evans, J.S.O.; Postlethwaite, S.R.; Johnson, M.R. Polymorphism and hydrogen bonding in cinchomeronic acid: A variable temperature experimental and computational study. *CrystEngComm* **2008**, *10*, 1404–1409. [CrossRef]

61. Thomas, L.H.; Craig, G.A.; Gutmann, M.J.; Parkin, A.; Shanklande, K.; Wilson, C.C. Conformational polymorphism of the molecular complex of 3-fluorobenzoic acid with 4-acetylpyridine. *CrystEngComm* **2011**, *13*, 3349–3354. [CrossRef]
62. Kubitschke, J.; Lange, H.; Strutz, H. Carboxylic Acids, Aliphatic. In *Ullmann's Encyclopedia of Industrial Chemistry*; Wiley-VCH: Weinheim, Germany, 2014.
63. Akcakayiran, D.; Mauder, D.; Hess, C.; Sievers, T.K.; Kurth, D.G.; Shenderovich, I.; Limbach, H.-H.; Findenegg, G.H. Carboxylic Acid-Doped SBA-15 Silica as a Host for Metallo-supramolecular Coordination Polymers. *J. Phys. Chem. B* **2008**, *112*, 14637–14647. [CrossRef]
64. Jezierska, A.; Panek, J.J. First-Principle Molecular Dynamics Study of Selected Schiff and Mannich Bases: Application of Two-Dimensional Potential of Mean Force to Systems with Strong Intramolecular Hydrogen Bonds. *J. Chem. Theory Comput.* **2008**, *4*, 375–384. [CrossRef]
65. Sobczyk, L.; Chudoba, D.M.; Tolstoy, T.M.; Filarowski, A. Some Brief Notes on Theoretical and Experimental Investigations of Intramolecular Hydrogen Bonding. *Molecules* **2016**, *21*, 1657. [CrossRef] [PubMed]
66. Shenderovich, I.G.; Denisov, G.S. Adduct under Field—A Qualitative Approach to Account for Solvent Effect on Hydrogen Bonding. *Molecules* **2020**, *25*, 436. [CrossRef] [PubMed]
67. Shenderovich, I.G.; Denisov, G.S. Solvent effects on acid-base complexes. What is more important: A macroscopic reaction field or solute-solvent interactions? *J. Chem. Phys.* **2019**, *150*, 204505. [CrossRef] [PubMed]
68. Shenderovich, I.G. Simplified Calculation Approaches Designed to Reproduce the Geometry of Hydrogen Bonds in Molecular Complexes in Aprotic Solvents. *J. Chem. Phys.* **2018**, *148*, 124313. [CrossRef]
69. Gurinov, A.A.; Denisov, G.S.; Borissova, A.O.; Goloveshkin, A.S.; Greindl, J.; Limbach, H.-H.; Shenderovich, I.G. NMR Study of Solvation Effect on the Geometry of Proton-Bound Homodimers of Increasing Size. *J. Phys. Chem. A* **2017**, *121*, 8697–8705. [CrossRef]
70. Tolstoy, P.M.; Schah-Mohammedi, P.; Smirnov, S.N.; Golubev, N.S.; Denisov, G.S.; Limbach, H.-H. Characterization of Fluxional Hydrogen-Bonded Complexes of Acetic Acid and Acetate by NMR: Geometries and Isotope and Solvent Effects. *J. Am. Chem. Soc.* **2004**, *126*, 5621–5634. [CrossRef]
71. Leiserowitz, L. Molecular Packing Modes. Carboxylic Acids. *Acta Cryst. B* **1976**, *32*, 775–802. [CrossRef]
72. Middlemiss, D.S.; Facchini, M.; Morrison, C.A.; Wilson, C.C. Small energy differences in molecular crystals: A first principles study of tautomerism and dynamics in benzoic acid derivatives. *CrystEngComm* **2007**, *9*, 777–785. [CrossRef]
73. Borissova, A.O.; Lyssenko, K.A.; Gurinov, A.A.; Shenderovich, I.G. Energy Analysis of Competing Non-Covalent Interaction in 1:1 and 1:2 Adducts of Collidine with Benzoic Acids by Means of X-Ray Diffraction. *Z. Phys. Chem.* **2013**, *227*, 775–790. [CrossRef]
74. Voronin, A.P.; Surov, A.O.; Churakov, A.V.; Parashchuk, O.D.; Rykounov, A.A.; Vener, M.V. Combined X-ray Crystallographic, IR/Raman Spectroscopic, and Periodic DFT Investigations of New Multicomponent Crystalline Forms of Anthelmintic Drugs: A Case Study of Carbendazim Maleate. *Molecules* **2020**, *25*, 2386. [CrossRef]
75. Schuster, P. Energy surfaces for hydrogen bonded systems. In *The Hydrogen Bond*; Schuster, P., Zundel, G., Sandorfy, C., Eds.; North-Holland: Amsterdam/Holland, The Netherlands, 1976; Volume II.
76. Emel'yanenko, V.N.; Stange, P.; Feder-Kubis, J.; Verevkin, S.P.; Ludwig, R. Dissecting intermolecular interactions in the condensed phase of ibuprofen and related compounds: The specific role and quantification of hydrogen bonding and dispersion forces. *Phys. Chem. Chem. Phys.* **2020**, *22*, 4896–4904. [CrossRef]
77. Weinhold, F.; Klein, R.A. Anti-Electrostatic Hydrogen Bonds. *Angew. Chem. Int. Ed.* **2014**, *53*, 11214–11217. [CrossRef]
78. Olovsson, I.; Ptasiewicz-Bak, H.; Gustafsson, T.; Majerz, I. Asymmetric hydrogen bonds in centrosymmetric environment: Neutron study of very short hydrogen bonds in potassium hydrogen dichloromaleate. *Acta Cryst.* **2001**, *B57*, 311–316. [CrossRef]
79. Gilli, G.; Gilli, P. Towards an uniced hydrogen-bond theory. *J. Mol. Struct.* **2000**, *552*, 1–15. [CrossRef]
80. Marx, D.; Tuckerman, M.E.; Hutter, J.; Parrinello, M. The Nature of the Hydrated Excess Proton in Water. *Nature* **1999**, *397*, 601–604. [CrossRef]
81. Tuckerman, M.E.; Marx, D.; Klein, M.L.; Parrinello, M. On the quantum nature of the shared proton in hydrogen bonds. *Science* **1997**, *275*, 817–820. [CrossRef] [PubMed]

82. Marx, D.; Chandra, A.; Tuckerman, M.E. Aqueous basic solutions: Hydroxide solvation, structural diffusion, and comparison to the hydrated proton. *Chem. Rev.* **2010**, *110*, 2174–2216. [CrossRef]
83. Dopieralski, P.P.; Perrin, C.L.; Latajka, Z. On the Intramolecular Hydrogen Bond in Solution: Car–Parrinello and Path Integral Molecular Dynamics Perspective. *J. Chem. Theory Comput.* **2011**, *7*, 3505–3513. [CrossRef]
84. Jezierska-Mazzarello, A.; Vuilleumier, R.; Panek, J.J.; Ciccotti, G. Molecular Property Investigations of an ortho-Hydroxy Schiff Base Type Compound with the First-Principle Molecular Dynamics Approach. *J. Phys. Chem. B* **2010**, *114*, 242–253. [CrossRef] [PubMed]
85. Panek, J.J.; Jezierska-Mazzarello, A.B.; Lipkowski, P.; Martyniak, A.; Filarowski, A. Comparison of resonance assisted and charge assisted effects in strengthening of hydrogen bonds in dipyrrins. *J. Chem. Inf. Model.* **2014**, *54*, 86–95. [CrossRef]
86. Brela, M.; Stare, J.; Pirc, G.; Sollner-Dolenc, M.; Boczar, M.; Wójcik, M.J.; Mavri, J. Car–Parrinello Simulation of the Vibrational Spectrum of a Medium Strong Hydrogen Bond by Two-Dimensional Quantization of the Nuclear Motion: Application to 2-Hydroxy-5-nitrobenzamide. *J. Phys. Chem. B* **2012**, *116*, 4510–4518. [CrossRef] [PubMed]
87. Garcia-Viloca, M.; Gonzalez-Lafont, A.; Lluch, J.M. Asymmetry of the Hydrogen Bond of Hydrogen Phthalate Anion in Solution. A QM/MM Study. *J. Am. Chem. Soc.* **1999**, *121*, 9198–9207. [CrossRef]
88. Pirc, G.; Stare, J.; Mavri, J. Car–Parrinello simulation of hydrogen bond dynamics in sodium hydrogen bissulfate. *J. Chem. Phys.* **2010**, *132*, 224506. [CrossRef] [PubMed]
89. Shenderovich, I.G.; Burtsev, A.P.; Denisov, G.S.; Golubev, N.S.; Limbach, H.-H. Influence of the Temperature-Dependent Dielectric Constant on the H/D Isotope Effects on the NMR Chemical Shifts and the Hydrogen Bond Geometry of Collidine-HF Complex in $CDF_3/CDClF_2$ Solution. *Magn. Reson. Chem.* **2001**, *39*, S91–S99. [CrossRef]
90. Mulloyarova, V.V.; Giba, I.S.; Kostin, M.A.; Denisov, G.; Shenderovich, I.G.; Tolstoy, P. Cyclic Trimers of Phosphinic Acids in Polar Aprotic Solvent: Symmetry, Chirality and H/D Isotope Effects on NMR Chemical Shifts. *Phys. Chem. Chem. Phys.* **2018**, *20*, 4901–4910. [CrossRef]
91. Gurinov, A.A.; Lesnichin, S.B.; Limbach, H. H.; Shenderovich, I.G. How Short is the Strongest Hydrogen Bond in the Proton-Bound Homodimers of Pyridine Derivatives? *J. Phys. Chem. A* **2014**, *118*, 10804–10812. [CrossRef]
92. Mulloyarova, V.V.; Ustimchuk, D.O.; Filarowski, A.; Tolstoy, P.M. H/D Isotope Effects on 1H NMR Chemical Shifts in Cyclic Heterodimers and Heterotrimers of Phosphinic and Phosphoric Acids. *Molecules* **2020**, *25*, 1907. [CrossRef]
93. Tolstoy, P.M.; Smirnov, S.N.; Shenderovich, I.G.; Golubev, N.S.; Denisov, G.S.; Limbach, H.-H. NMR studies of solid state, solvent and H/D isotope effects on hydrogen bond geometries of 1:1 complexes of collidine with carboxylic acids. *J. Mol. Struct.* **2004**, *700*, 19–27. [CrossRef]
94. Andreeva, D.V.; Ip, B.; Gurinov, A.A.; Tolstoy, P.M.; Denisov, G.S.; Shenderovich, I.G.; Limbach, H.-H. Geometrical Features of Hydrogen Bonded Complexes Involving Sterically Hindered Pyridines. *J. Phys. Chem. A* **2006**, *110*, 10872–10879. [CrossRef]
95. Golubev, N.S.; Melikova, S.M.; Shchepkin, D.N.; Shenderovich, I.G.; Tolstoy, P.M.; Denisov, G.S. Interpretation of H/D Isotope Effects on NMR Chemical Shifts of [FHF]⁻ Ion Based on Calculations of Nuclear Magnetic Shielding Tensor Surface. *Z. Phys. Chem.* **2003**, *217*, 1549–1563. [CrossRef]
96. Shenderovich, I.G. Maximum Value of the Chemical Shift in the ¹H NMR Spectrum of a Hydrogen-Bonded Complex. *Russ. J. Gen. Chem.* **2006**, *76*, 501–506. [CrossRef]
97. Maréchal, Y. *The Hydrogen Bond and the Water Molecule: The Physics and Chemistry of Water, Aqueous and Bio Media*; Elsevier: Amsterdam, The Netherlands; Oxford, UK, 2007.
98. Nelson, D.L.; Cox, M.M. *Principles of Biochemistry*, 4th ed.; W.H. Freeman: New York, NY, USA, 2005.
99. Flakus, H.T.; Hachuła, B.; Hołaj-Krzak, J.T.; Al-Agel, F.A.; Rekik, N. "Long-distance" H/D isotopic self-organization phenomena in scope of the infrared spectra of hydrogen-bonded terephthalic and phthalic acid crystals. *Spectrochim. Acta A* **2017**, *173*, 65–74. [CrossRef] [PubMed]
100. Flakus, H.T.; Chełmecki, M. Infrared spectra of the hydrogen bond in benzoic acid crystals: Temperature and polarization effects. *Spectrochim. Acta A* **2002**, *58*, 179–196. [CrossRef]
101. Ghalla, H.; Rekik, N.; Michta, A.; Oujia, B.; Flakus, H.T. Theoretical modeling of infrared spectra of the hydrogen and deuterium bond in aspirin crystal. *Spectrochim. Acta A* **2010**, *173*, 37–47. [CrossRef]

102. Flakus, H.T.; Hachuła, B. The source of similarity of the IR spectra of acetic acid in the liquid and solid-state phases. *Vib. Spectrosc.* **2011**, *56*, 170–176. [CrossRef]
103. Flakus, H.T.; Rekik, N.; Jarczyk, A. Polarized IR Spectra of the Hydrogen Bond in 2-Thiopheneacetic Acid and 2-Thiopheneacrylic Acid Crystals: H/D Isotopic and Temperature Effects. *J. Phys. Chem. A* **2012**, *116*, 2117–2130. [CrossRef]
104. Flakus, H.T.; Hachuła, B.; Hołaj-Krzak, J.T. Long-distance inter-hydrogen bond coupling effects in the polarized IR spectra of succinic acid crystals. *Spectrochim. Acta A* **2015**, *142*, 126–134. [CrossRef]
105. Flakus, H.T.; Jarczyk-Jędryka, A. Temperature and H/D Isotopic Effects in the IR Spectra of the Hydrogen Bond in Solid-State 2-Furanacetic Acid and 2-Furanacrylic Acid. *J. Atom. Mol. Opt. Phys.* **2012**, *2012*, 125471. [CrossRef]
106. Flakus, H.T. Vibronic model for H/D isotopic self-organization effects in centrosymmetric dimers of hydrogen bonds. *J. Mol. Struct.* **2003**, *646*, 15–23. [CrossRef]
107. Rodziewicz, P.; Doltsinis, N.L. Formic Acid Dimerization: Evidence for Species Diversity from First Principles Simulations. *J. Phys. Chem. A* **2009**, *113*, 6266–6274. [CrossRef]
108. Majerz, I. Proton Transfer Influence on Geometry and Electron Density in Benzoic Acid–Pyridine Complexes. *Helv. Chim. Acta* **2016**, *99*, 286–295. [CrossRef]
109. Bournay, J.; Marechal, Y. Anamalous isotope effect in the H bonds of acetic acid dimers. *J. Chem. Phys.* **1973**, *59*, 5077–5087. [CrossRef]
110. Vener, M.V.; Kuhn, O.; Bowman, J.M. Vibrational spectrum of the formic acid in the OH stretch region. A model 3D study. *Chem. Phys. Lett.* **2001**, *349*, 562–570. [CrossRef]
111. Kwocz, A.; Panek, J.J.; Jezierska, A.; Hetmańczyk, Ł.; Pawlukojć, A.; Kochel, A.; Lipkowski, P.; Filarowski, A. A molecular roundabout: Triple cycle-arranged hydrogen bonds in light of experiment and theory. *New J. Chem.* **2018**, *42*, 19467–19477. [CrossRef]
112. Filarowski, A.; Koll, A. Intergrated intensity of $\nu_s$(OH) absorption bands in bent hydrogen bonds in *ortho*-dialkylaminomethyl phenols. *Vib. Spectrosc.* **1996**, *12*, 15–24. [CrossRef]
113. Filarowski, A.; Koll, A.; Lipkowski, P.; Pawlukojć, A. Inelastic neutron scattering and vibrational spectra of 2-(*N*-methyl-α-iminoethyl)-phenol and 2-(*N*-methyliminoethyl)-phenol: Experimental and theoretical approach. *J. Mol. Struct.* **2008**, *880*, 97–108. [CrossRef]
114. Zeegers-Huyskens, T. Influence of the nature of the hydrogen bond on the isotopic ratio $\nu_{AH}/\nu_{AD}$. *J. Mol. Struct.* **1992**, *217*, 239–252. [CrossRef]
115. Pająk, J.; Maes, G.; De Borggraeve, W.M.; Boens, N.; Filarowski, A. Matrix-isolation FT-IR and theoretical investigation of the vibrational properties of the sterically hindered *ortho*-hydroxy acylaromatic *Schiff* bases. *J. Mol. Struct.* **2007**, *844–845*, 83–93. [CrossRef]
116. Grzegorzek, J.; Filarowski, A.; Mielke, Z. The photoinduced isomerization and its implication in the photo-dynamical processes in two simple Schiff bases isolated in solid argon. *Phys. Chem. Chem. Phys.* **2011**, *13*, 16596–16605. [CrossRef]
117. Majerz, I.; Pawlukojć, A.; Sobczyk, L.; Dziembowska, T.; Grech, E.; Szady-Chełmieniecka, A. The infrared, Raman and inelastic neutron scattering studies on 5-nitro-N-salicylideneethylamine. *J. Mol. Struct.* **2000**, *552*, 243–247. [CrossRef]
118. Grzegorzek, J.; Mielke, Z.; Filarowski, A. C=N–N=C conformational isomers of 2'-hydroxyacetophenone azine: FTIR matrix isolation and DFT study. *J. Mol. Struct.* **2010**, *976*, 371–376. [CrossRef]
119. Pająk, J.; Maes, G.; De Borggraeve, W.M.; Boens, N.; Filarowski, A. Matrix-isolation FT-IR and theoretical investigation of the competitive intramolecular hydrogen bonding in 5-methyl-3-nitro-2-hydroxyacetophenone. *J. Mol. Struct.* **2008**, *880*, 86–96. [CrossRef]
120. Mitchell, P.C.H.; Parker, S.F.; Ramirez-Cuesta, A.J.; Tomkinson, J. *Vibrational Spectroscopy with Neutrons, with Applications in Chemistry, Biology, Materials Science and Catalysis*; World Scientific: Singapore, 2005.
121. Marques, M.P.M.; Batista de Carvalho, L.A.E.; Valero, R.; Machado, N.F.L.; Parker, S.F. An inelastic neutron scattering study of dietary phenolic acids. *Phys. Chem. Chem. Phys.* **2014**, *16*, 7491–7500. [CrossRef] [PubMed]
122. Fontaine-Vive, F.; Johnson, M.R.; Kearley, G.J.; Cowan, J.A.; Howard, J.A.K.; Parker, S.F. Phonon driven proton transfer in crystals with short strong hydrogen bonds. *J. Chem. Phys.* **2006**, *124*, 234503. [CrossRef]
123. Johnson, M.R.; Trommsdorff, H.P. Vibrational spectra of crystalline formic and acetic acid isotopologues by inelastic neutron scattering and numerical simulations. *Chem. Phys.* **2009**, *355*, 118–122. [CrossRef]

124. Beran, G.J.O.; Chronister, E.L.; Daemen, L.L.; Moehlig, A.R.; Mueller, L.J.; Oomens, J.; Rice, A.; Santiago-Dieppa, D.R.; Tham, F.S.; Theel, K.; et al. Vibrations of a chelated proton in a protonated tertiary diamine. *Phys. Chem. Chem. Phys.* **2011**, *13*, 20380–20392. [CrossRef] [PubMed]
125. Kong, S.; Borissova, A.O.; Lesnichin, S.B.; Hartl, M.; Daemen, L.L.; Eckert, J.; Antipin, M.Y.; Shenderovich, I.G. Geometry and Spectral Properties of the Protonated Homodimer of Pyridine in the Liquid and Solid States. A Combined NMR, X-ray Diffraction and Inelastic Neutron Scattering Study. *J. Phys. Chem. A* **2011**, *115*, 8041–8048. [CrossRef]
126. Miyazawa, T.; Pitzer, K.S. Internal Rotation and Infrared Spectra of Formic Acid Monomer and Normal Coordinate Treatment of Out-of-Plane Vibrations of Monomer, Dimer, and Polymer. *J. Chem. Phys.* **1959**, *30*, 1076–1086. [CrossRef]
127. Novak, A. Hydrogen bonding in solids correlation of spectroscopic and crystallographic data. *Struct. Bonding* **1974**, *18*, 177–216.
128. Howard, J.; Tomkinson, J.; Eckert, J.; Goldstone, J.A.; Taylor, A.D. Inelastic neutron scattering studies of some intramolecular hydrogen bonded complexes: A new correlation of γ(OHO) vs. R (OO). *J. Chem. Phys.* **1983**, *78*, 3150–3155. [CrossRef]
129. Mitchell, P.C.H.; Parker, S.F.; Ramirez-Cuesta, A.J.; Tomkinson, J. *Series on Neutron Techniques and Applications, Vibrational Spectroscopy with Neutrons*; World Scientific Publishing Co. Pte. Ltd.: Singapore, 2005.
130. Car, R.; Parrinello, M. Unified Approach for Molecular Dynamics and Density-Functional Theory. *Phys. Rev. Lett.* **1985**, *55*, 2471–2474. [CrossRef] [PubMed]
131. Grimme, S. Semiempirical GGA-type Density Functional Constructed with a Long-Range Dispersion Correction. *J. Comput. Chem.* **2006**, *27*, 1787–1799. [CrossRef] [PubMed]
132. Perdew, J.P.; Burke, K.; Ernzerhof, M. Generalized Gradient Approximation Made Simple. *Phys. Rev. Lett.* **1996**, *77*, 3865–3868. [CrossRef]
133. Troullier, N.; Martins, J.L. Efficient pseudopotentials for plane-wave calculations. *Phys. Rev. B* **1991**, *43*, 1993–2006. [CrossRef]
134. Hockney, R.W. The potential calculation and some applications. *Methods Comput. Phys.* **1970**, *9*, 136–211.
135. Nosé, S. A unified formulation of the constant temperature molecular dynamics methods. *J. Chem. Phys.* **1984**, *81*, 511–519. [CrossRef]
136. Hoover, W.G. Canonical dynamics: Equilibrium phase-space distributions. *Phys. Rev. A* **1985**, *31*, 1695–1697. [CrossRef]
137. CPMD, Version 3.17.1; Copyright IBM Corp. (1990–2004) Copyright MPI für Festkörperforschung Stuttgart (1997–2001). Available online: http://www.cpmd.org/ (accessed on 14 October 2003).
138. Humphrey, W.; Dalke, A.; Schulten, K. VMD—Visual Molecular Dynamics. *J. Mol. Graph.* **1996**, *14*, 33–38. [CrossRef]
139. Gnuplot, Version 4.2; An Interactive Plotting Program; Thomas Williams and Colin Kelley. 2007. Available online: https://sourceforge.net/projects/gnuplot/ (accessed on 14 October 2007).
140. Frisch, M.J.; Trucks, G.W.; Schlegel, H.B.; Scuseria, G.E.; Robb, M.A.; Cheeseman, J.R.; Scalmani, G.; Barone, V.; Mennucci, B.; Petersson, G.A.; et al. *Gaussian 09, Revision, D.01*; Gaussian, Inc.: Wallingford, UK, 2009.
141. Becke, A.D. Density-functional thermochemistry. III. The role of exact exchange. *J. Chem. Phys.* **1993**, *98*, 5648–5652. [CrossRef]
142. Lee, C.; Yang, W.; Parr, R.G. Development of the Colle-Salvetti Correlation-Energy Formula into a Functional of the Electron Density. *Phys. Rev. B* **1988**, *37*, 785–789. [CrossRef]
143. McLean, A.D.; Chandler, G.S. Contracted Gaussian basis sets for molecular calculations. I. Second row atoms, Z=11-18. *J. Chem. Phys.* **1980**, *72*, 5639–5648. [CrossRef]
144. Krishnan, R.; Binkley, J.S.; Seeger, R.; Pople, J.A. Self-consistent molecular orbital methods. XX. A basis set for correlated wave functions. *J. Chem. Phys.* **1980**, *72*, 650–654. [CrossRef]
145. Frisch, M.J.; Pople, J.A.; Binkley, J.S. Self-consistent molecular orbital methods 25. Supplementary functions for Gaussian basis sets. *J. Chem. Phys.* **1984**, *80*, 3265–3269. [CrossRef]
146. Scheiner, S. *Hydrogen Bonding: A Theoretical Perspective*; Oxford University Press: New York, NY, USA, 1997.
147. Tomasi, J.; Mennucci, B.; Cammi, R. Quantum mechanical continuum solvation models. *Chem. Soc. Rev.* **2005**, *105*, 2999–3093. [CrossRef]
148. Schaftenaar, G.; Noordik, J.H. Molden: A pre- and post-processing program for molecular and electronic structures. *J. Comput.-Aided Mol. Des.* **2000**, *14*, 123–134. [CrossRef]

149. Pulay, P.; Fogarasi, G.; Pang, F.; Boggs, J.E. Systematic ab initio gradient calculation of molecular geometries, force constants, and dipole moment derivatives. *J. Am. Chem. Soc.* **1979**, *101*, 2550–2560. [CrossRef]
150. Martin, J.M.L.; Van Alsenoy, C. *Gar2ped*; University of Antwerp: Antwerpen, Belgium, 1995.

**Sample Availability:** Samples of the compounds are not available from the authors.

**Publisher's Note:** MDPI stays neutral with regard to jurisdictional claims in published maps and institutional affiliations.

© 2020 by the authors. Licensee MDPI, Basel, Switzerland. This article is an open access article distributed under the terms and conditions of the Creative Commons Attribution (CC BY) license (http://creativecommons.org/licenses/by/4.0/).

Article

# Mutual Relations between Substituent Effect, Hydrogen Bonding, and Aromaticity in Adenine-Uracil and Adenine-Adenine Base Pairs[†]

Paweł A. Wieczorkiewicz [1], Halina Szatylowicz [1,*] and Tadeusz M. Krygowski [2]

1 Faculty of Chemistry, Warsaw University of Technology, Noakowskiego 3, 00-664 Warsaw, Poland; pawel.wieczorkiewicz.stud@pw.edu.pl
2 Faculty of Chemistry, University of Warsaw, Pasteura 1, 02-093 Warsaw, Poland; tmkryg@chem.uw.edu.pl
* Correspondence: halina@ch.pw.edu.pl; Tel.: +48-22-234-7755
† Dedicated to Professor Bronisław Marciniak of the Faculty of Chemistry of the Adam Mickiewicz University in Poznań on the occasion of his 70th birthday.

Academic Editors: Ilya G. Shenderovich and Steve Scheiner
Received: 4 July 2020; Accepted: 11 August 2020; Published: 13 August 2020

**Abstract:** The electronic structure of substituted molecules is governed, to a significant extent, by the substituent effect (SE). In this paper, SEs in selected nucleic acid base pairs (Watson-Crick, Hoogsteen, adenine-adenine) are analyzed, with special emphasis on their influence on intramolecular interactions, aromaticity, and base pair hydrogen bonding. Quantum chemistry methods—DFT calculations, the natural bond orbital (NBO) approach, the Harmonic Oscillator Model of Aromaticity (HOMA) index, the charge of the substituent active region (cSAR) model, and the quantum theory of atoms in molecules (QTAIM)—are used to compare SEs acting on adenine moiety and H-bonds from various substitution positions. Comparisons of classical SEs in adenine with those observed in para- and meta-substituted benzenes allow for the better interpretation of the obtained results. Hydrogen bond stability and its other characteristics (e.g., covalency) can be significantly changed as a result of the SE, and its consequences are dependent on the substitution position. These changes allow us to investigate specific relations between H-bond parameters, leading to conclusions concerning the nature of hydrogen bonding in adenine dimers—e.g., H-bonds formed by five-membered ring nitrogen acceptor atoms have an inferior, less pronounced covalent nature as compared to those formed by six-membered ring nitrogen. The energies of individual H-bonds (obtained by the NBO method) are analyzed and compared to those predicted by the Espinosa-Molins-Lecomte (EML) model. Moreover, both SE and H-bonds can significantly affect the aromaticity of adenine rings; long-distance SEs on π-electron delocalization are also documented.

**Keywords:** substituent effect; hydrogen bond; aromaticity; adenine

## 1. Introduction

Fundamental biological importance makes DNA and RNA base pairs important and therefore popular systems for quantum chemical calculations. With the development of quantum chemistry methods and computing power, modeling systems of large sizes became easily accessible [1]. Therefore, since the 1990s many papers on this topic have been published, including articles on hydrogen bonding [2,3], π-stacking interactions [4,5], tautomerization [6], benchmarks of various DFT methods [7], and dispersion models [8], leading to a better understanding of nucleic acid structure and the mechanisms of mutations [9].

Various quantum chemical methods were used to investigate intermolecular interactions in Watson-Crick [10], non-canonical Hoogsteen [11], and adenine-uracil RNA base pairs, as well as in mismatched adenine-adenine base pairs [4,5]. The substituent effects (SEs) on hydrogen bonding were most frequently studied in structurally modified Watson-Crick base pairs (adenine-thymine and guanine-cytosine) [12–18]. Recently, the influence of substituents on the electronic structure of the four most stable purine tautomers and their adenine analogues [19], as well as on the stability of adenine quartets [20] has been presented.

Adenine and uracil belong to five bases constituting DNA or RNA macromolecules that are fundamental to life processes [21]. Undoubtedly, the interactions between all of them, as well as the different types of external influences on their electronic structure, are of great importance for understanding their role in these processes. The influences of the electrophilic or nucleophilic agents belong to this type of interaction; they can cause significant changes in the electronic structure of the bases of DNA or RNA macromolecules. However, the interactions caused by cations or anions are temporary and difficult to systematically study. Their contact with the molecules in question, even if very short, is about 100 times longer than the time during which the electronic structure of the molecule is perturbed and open to some "non-typical" reactions with another reagent. Such situations can cause mutations in the attacked molecule and change its function as an inherent part of DNA or RNA macromolecules. To investigate this problem, instead of free-charged reagents, the attachment of electrophilic or nucleophilic groups (substituents) to a molecular moiety can be used to study their permanent effect on the electronic structure of the substituted molecule. Then, some analogies to a more complex situation can be subject to deeper consideration. The latter type of treatment will be presented in this paper, combined with the studies of mutual interactions of participants in adenine-uracil and adenine-adenine base pairs. In other words, the effect of intramolecular interactions (substituent effect) on individual intermolecular interactions (hydrogen bonds) in substituted base pairs is the subject of this paper. The studied adenine-uracil and adenine-adenine base pairs are shown in Figures 1 and 2 (the separate structures of each dimer are shown in Figures S1–S4 in Supplementary Materials). The naming of the adenine dimers was adopted from the paper of Poater et al. [4] to allow the comparison of the results.

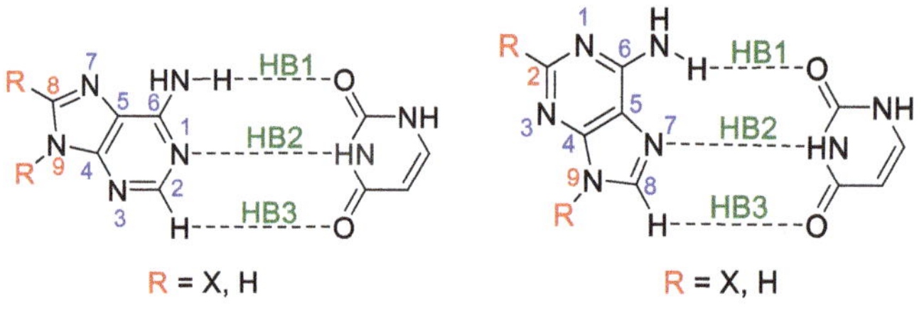

**Figure 1.** Adenine-uracil (**a**) Watson-Crick (abbreviated as WC) and (**b**) Hoogsteen (abbreviated as HG) base pairs with the common numbering of adenine atoms and adopted hydrogen bond numeration. Substitutable positions are marked in red color.

Several methods can be used to estimate the strength of individual hydrogen bonds in base pairs of nucleic acids: the rotational method [22], the compliance constants method [23], the Espinosa-Molins-Lecomte (EML) equation [24] application [25], the atom replacement method [26], the estimation of hydrogen bond energy based on electron density (calculated using the quantum theory of atoms in molecules, QTAIM) [27,28] at the bond critical point (BCP) [29], the application of natural bond orbitals (NBO) [30] method [31], and coordinates interaction approach, [32] as well as the delocalization index [33].

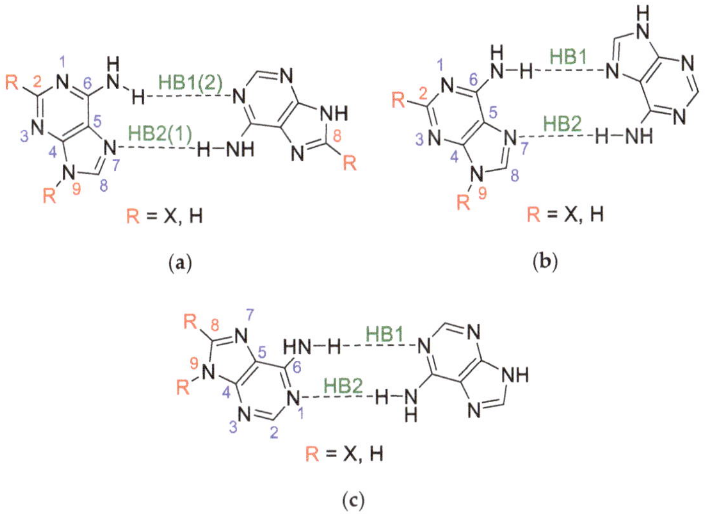

**Figure 2.** Adenine-adenine (**a**) AA2, (**b**) AA3, and (**c**) AA4 base pairs with the common numbering of adenine atoms and adopted hydrogen bond numeration (for AA2 C8-X dimer, numeration in brackets). Substitutable positions are marked in red color.

The most known substituent characteristics are the Hammett constants [34,35]. However, they can only be used to describe the classical substituent effect—how a substituent X affects the properties of a fixed group Y (the so called "reaction site") in a substituted system X-R-Y (R—transmitting moiety). The use of the cSAR (charge of the substituent active region) descriptor [36–38] allows us to study both classical and reverse substituent effects [39]; the latter describes how the electronic properties of substituents X depend on the properties of the moiety R-Y to which they are attached.

In the presented research, the most stable adenine tautomer, 9H, was chosen as a base for the further modification of the molecule. This tautomer has three substitutable hydrogen atoms at the C2, C8, and N9 positions. For each adenine-uracil and adenine-adenine base pair, substituent positions for further analysis were selected to avoid direct intermolecular interactions of a substituent, such as steric interactions of bulky substituents or the formation of a new hydrogen bond. For comparison, the effect of substituents at the C2, C8, or N9 positions in adenine on its physicochemical properties was also considered in detail. Selected substituents that differ in electronic properties (X = NO, $NO_2$, Cl, F, H, Me, OH, $NH_2$) were introduced into the adenine molecule in monomers and dimers.

The strength of the individual hydrogen bonds was characterized using the NBO approach [30,31], the topological parameters within the QTAIM approach [40], the delocalization index [33], and the H-bond lengths.

This work is mainly devoted to the influence of substituents on hydrogen bonding as well as on changes in the electronic structure of adenine, uracil, and their dimers (shown in Figures 1 and 2). The following issues are considered in greater detail:

- The classical substituent effect—i.e., how do substituents in various positions of adenine affect its amino group and individual hydrogen bonds?
- How do substituents affect the π-electron delocalization in adenine rings?
- How does the electronic structure of the substituents, estimated by cSAR(X), depend on the position and kind of a moiety to which they are attached? This means the estimation of the so-called reverse substituent effect.
- How do these characteristics depend on the nature of substituents—i.e., their electron-donating or electron-attracting properties?

- How do these characteristics differ from those estimated for monomers?

## 2. Methods

For all studied systems, geometry optimizations without any symmetry constraints and electronic energy calculations were performed using the Gaussian 16 program [41]. Based on our previous research [42], the DFT-D method was used—namely, B97-D3 dispersion corrected density functional [43,44]—with Dunning's [45] aug-cc-pVDZ basis set. The harmonic vibrational frequencies were calculated at the same level of theory to confirm that all the obtained structures correspond to the minima on the potential energy surface. No imaginary frequencies were found for the obtained series. In the case of asymmetric substituents (NO and OH), a lower energy rotamer was considered. NBO 6.0 software (Theoretical Chemistry Institute, University of Wisconsin, Madison, WI, USA) [46] was used for the NBO calculations.

The substituents were characterized using the cSAR descriptor. It allows a quantitative comparison of the electron-donating/withdrawing effects of different functional groups and correlates with many physicochemical properties [39]. For the X substituent, cSAR is defined as follows (Equation (1)):

$$cSAR(X) = Q_X + Q_I, \quad (1)$$

where $Q_X$ is the sum of the partial charges of the X group atoms, and $Q_I$ is the partial charge of an atom to which a substituent is attached (ipso atom).

Partial charges calculated using the Hirshfeld, [47] Weinhold [30] (NBO), and the Voronoi Deformation Density (VDD) [48] methods were used to select the charge assessment for further investigation. Their values for derivatives of the WC base pair substituted in positions C8-X and N9-X were mutually correlated; the relations between $cSAR(X)$ and $cSAR(X)_{Hir}$ are shown in Figure S5 (Supplementary Materials). Considering both substitution positions (C8 and N9), it can be concluded that, qualitatively, only the VDD and Hirshfeld approaches are nearly equivalent for the estimation of $cSAR(X)$ values. To be able to compare with the results of our previous research [19], only the Hirshfeld charges are used in further discussion and the superscript Hir is omitted.

The interaction energies between monomers A and B were obtained by the supermolecular method (Equations (2)–(4)) [49], using the counterpoise approach [50]:

$$E_{SM} = E_{AB} - (E_A + E_B) + E_{BSSE} = E_{AB}^{int} + E_{AB}^{def}, \quad (2)$$

$$E_{AB}^{int} = E_{AB} - \left(E_A^{dim} + E_B^{dim}\right), \quad (3)$$

$$E_{AB}^{def} = E_A^{dim} - E_A + E_B^{dim} - E_B, \quad (4)$$

where $E_{AB}$ is the electronic energy of a dimer, whereas $E_A$ and $E_B$ are the electronic energies of monomers A and B, and $E_{BSSE}$ is the basis set superposition error (BSSE) energy correction. $E_{AB}^{int}$ and $E_{AB}^{def}$ are the "pure" interaction and deformation energies in the AB dimer, respectively, while $E_A^{dim}$ and $E_B^{dim}$ are the energies of monomers A and B in the dimer geometry, respectively. For the studied systems, the $E_{BSSE}$ values were 0.97–1.12 kcal/mol.

The energy of individual hydrogen bonds was calculated according to the NBO theory [30] as:

$$E_{HB} = E_{n \to \sigma*} - E_{n \to \sigma}, \quad (5)$$

where $E_{n \to \sigma*}$ is the interaction energy between the nonbonding NBO orbital n (lone pair) of an H-bond acceptor atom and an antibonding orbital $\sigma*$ of an H-D bond (where D is a hydrogen bond donor atom), calculated by the NBO 6.0 program from the second-order perturbative analysis of the Fock matrix on an NBO basis. $E_{n \to \sigma}$ is the steric exchange energy between the acceptor's nonbonding Natural Localized Molecular Orbital (NLMO) n and H-D bonding NLMO $\sigma$. The natural steric analysis is accessible in NBO 4.0 and later versions via the STERIC keyword.

The strength of the individual hydrogen bonds was also characterized using the QTAIM topological parameters and delocalization index. The QTAIM calculations were performed with the AIMAII [51] software.

The delocalization index ($\delta$) is a descriptor capable of characterizing both closed-shell and shared-shell interactions [33]. Its value is a measure of the number of electrons delocalized between atoms. $\delta$(A, B) between atoms A and B is calculated within the QTAIM theory, which defines atomic basins in a molecule. Having defined atomic basins, it is possible to calculate $\delta$(A, B) as:

$$\delta(A, B) = 4 \sum_{i,j}^{N/2} S_{ij}(A) S_{ij}(B), \tag{6}$$

where $S_{ij}(A)$ and $S_{ij}(B)$ are the overlaps between orbitals i and j in the atomic basin of A and B, respectively.

Harmonic Oscillator Model of Aromaticity (HOMA) [52] was chosen as an aromaticity descriptor. It is a geometry-based descriptor dependent on the bond lengths of the studied system in comparison to a hypothetical, fully aromatic reference system. HOMA is defined as:

$$\text{HOMA} = 1 - \frac{1}{n} \sum_{i}^{n} \alpha_j \left( d_{\text{opt},j} - d_{j,i} \right)^2, \tag{7}$$

where $n$ is the number of bonds taken into account when carrying out the summation, j means the type of bond (e.g., CC or CN), $\alpha_j$ is an empirical normalization constant, $d_{\text{opt},j}$ is the optimal length of a given bond assumed to be realized for full aromatic systems, and $d_{j,i}$ is an actual bond length in the studied system. The values of HOMA were calculated using the Multiwfn [53] program, with HOMA constants ($\alpha_j$ and optimal bond lengths) taken from Krygowski's paper [54].

## 3. Results and Discussion

The discussion of the results is divided into five parts. The first four concern various aspects of the substituent effect. The last part presents the interrelationships between various characteristics describing the strength of individual hydrogen bonds. The obtained values of the substituent effect descriptors (cSAR, HOMA) and hydrogen bond strength parameters ($E_{HB}$, $d_{HB}$, $\rho_{BCP}$, $\delta$(H,A), $\nabla^2 \rho_{BCP}$, $E_{def}$, $E_{SM}$) for the substituted WC and HG base pairs as well as the adenine dimers are presented in Tables S1–S7 (Supplementary Materials).

### 3.1. Classical Substituent Effect-Intramolecular Interactions

Adenine contains an amino group at the C6 position, which in base pairs is involved in the hydrogen bond, either through the interaction of NH$\cdots$O or through N$\cdots$HN, as shown in Figures 1 and 2. In the studied systems, the substituent also influences the electronic properties of the amino group. The cSAR parameter shows how much these interactions affect its electronic structure. Figure 3 illustrates the ranges of the cSAR(NH$_2$) values (their exact values are listed in Table S1 (Supplementary Materials)). In all cases, the range of the cSAR(NH$_2$) values due to the substituent effect is always greater for the adenine monomer than its dimers, in which the -NH$_2$ group is involved in H-bonding. For adenine in the AA and HG/WC pairs substituted at the C2, C8, or N9 positions, the decrease in the average value ranges compared to the monomer value is equal approximately to 76%, 80% and 56%, respectively. This shows that the H-bonding of the NH$_2$ group in dimers causes its weaker propensity for the substituent effect.

The use of the classical interpretation of the substituent effect allows us to consider how the substituent can affect the properties of the reaction site; in our case, it is the adenine NH$_2$ group. For this purpose, the Hammett-type linear equation is used, in which the slope (a, also known as the reaction constant) describes the sensitivity of the reaction site to the influence of X substituents. Thus, the electronic structure of the NH$_2$ group, involved in H-bonding, is described by the dependences of cSAR(NH$_2$) on cSAR(X), as presented in Table 1 and Figure S6 (Supplementary Materials). In all cases, the regressions have good or at least acceptable determination coefficients ($R^2 > 0.81$).

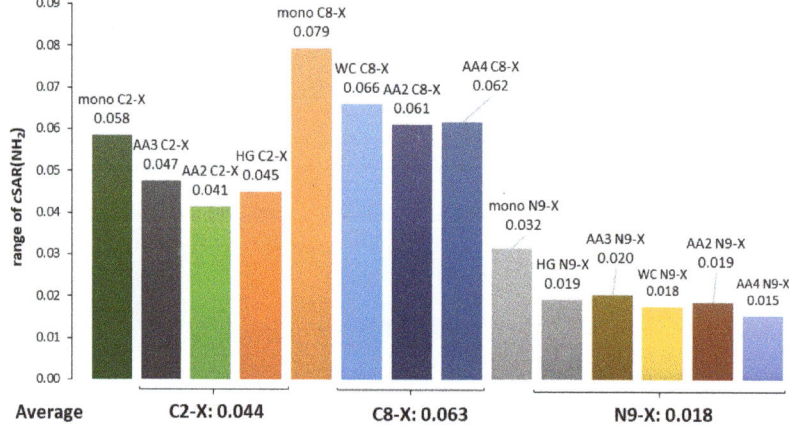

Figure 3. Ranges of cSAR(NH$_2$) variability and their averaged values in substituted adenine monomers and dimers (WC, HG, and AA).

Table 1. Slopes of the obtained linear dependences cSAR(NH$_2$) vs. cSAR(X) and determination coefficients ($R^2$) for the substituted WC and HG base pairs, adenine dimers, and substituted monomers (marked in **bold**, data taken from Ref. [19]) (A-C2-X, A-C8-X, A-N9-X). For asymmetrically substituted AA2 C2-X, C8-X dimer, the cSAR values of underlined monomers are taken.

|  | a | $R^2$ |
|---|---|---|
| **A-C2-X** | −0.222 | 0.913 |
| HG C2-X | −0.183 | 0.882 |
| AA2 C2-X | −0.163 | 0.838 |
| AA3 C2-X | −0.184 | 0.824 |
| **A-C8-X** | −0.297 | 0.989 |
| WC C8-X | −0.219 | 0.991 |
| AA2 C8-X | −0.209 | 0.983 |
| AA4 C8-X | −0.167 | 0.875 |
| **A-N9-X** | −0.166 | 0.902 |
| HG N9-X | −0.084 | 0.817 |
| WC N9-X | −0.087 | 0.835 |
| AA2 N9-X | −0.100 | 0.883 |
| AA3 N9-X | −0.110 | 0.808 |
| AA4 N9-X | −0.079 | 0.813 |
| AA2 C2-X, <u>C8-X</u> | −0.176 | 0.890 |
| AA2 <u>C2-X</u>, C8-X | −0.201 | 0.947 |
| AA3 C2-X, <u>C2-X</u> | −0.182 | 0.818 |
| AA4 C8-X, <u>C8-X</u> | −0.236 | 0.994 |

Looking at the data in Table 1, some observations can be made:

1. The substituents at the C8 position in the five-membered adenine ring in the WC pair have a stronger influence on the electronic structure of NH$_2$ involved in the H-bond than those attached to the C2 in the six-membered ring of the HG pair by the ratio of 0.219/0.183 = 1.20.
2. The substituents attached to N9 in both pairs, WC and HG, affect the electronic structure of NH$_2$ involved in the H-bond by almost the same extent, by the ratio (WC/HG) of 1.03.

3. In all cases of adenine dimers, the effect of substitution at the C2 or C8 position is greater than that at N9 by an average ratio of 1.88.

In the first case, the number of bonds, $n$, between the substituted C atom and another one substituted by the $NH_2$ group involved in H-bonding is equal to three for the WC pair, whereas for the HG pair it is equal to two. According to the documented rule, substituents affect "the reaction site" more strongly from the para position (the number of bonds between functional groups $n = 3$) than from the meta one ($n = 2$) [55,56]; our case follows this rule. For the next point (2), $n = 3$ for both cases, and hence there are almost identical substituent effects from N9 on $NH_2$. Finally (3), the comparisons of the effect of substituents attached to carbon and nitrogen atoms reveal that the influence of substituent at nitrogen is significantly weaker. A possible interpretation is that the lone electron pair at the nitrogen atom can be directly involved in the interaction with the substituent (N-X), thus resulting in its weaker strength of interaction at longer distances.

## 3.2. Classical Substituent Effects-Intermolecular Interactions

The properties of the adenine amino group can also be characterized by the strength of the hydrogen bond in which it is involved. Besides this, substituents may also affect other hydrogen bonds in the system. The following descriptors were used to describe the strength of an individual hydrogen bond: its length ($d_{HB}$); energy ($E_{HB}$); electron density at the H-bond critical point ($\rho_{BCP}$); and delocalization index, $\delta(H,A)$. Their calculated values are collected in Table S7 (Supplementary Materials).

For the WC and HG base pairs, there are three types of H-bonds (see Figure 1 and Figure S1 (Supplementary Materials)):

- (HB1) NH$\cdots$O, in which NH of the amino group is proton-donating towards the oxygen atom of uracil;
- (HB2) N$\cdots$HN, in which the nitrogen atom of the adenine ring is proton accepting from the NH endo group in uracil;
- (HB3) a weak CH$\cdots$O, where the C-H of the adenine ring, C2-H in WC and C8-H in HG, interacts with the carbonyl group of uracil.

In the case of HB3, despite the presence of a bond critical point satisfying the Koch-Popelier [57] criteria for hydrogen bonding, an NBO analysis showed negligible interactions. For this reason, only HB1 and HB2 will be discussed.

Therefore, the following problem should be considered—how the substituents attached to C8 (WC pair), C2 (HG pair), and N9 in both pairs affect the individual H-bond stability. As above, such problems can be presented by a linear correlation of the hydrogen bond strength descriptor versus cSAR(X); its slope describes the sensitivity of H-bonding to the effect of the electron-accepting/donating properties of the substituent expressed by cSAR(X). Figure 4 shows the dependence of the HB1 and HB2 hydrogen bond energy on cSAR(X) for the substituted WC and HG pairs, while the slopes and determination coefficients for all the H-bond descriptors used are summarized in Table 2. A first look at its contents reveals that the slopes of the equations for the HB1 and HB2 bonds differ by a sign. Moreover, in general, regressions have at least good determination coefficients ($R^2 > 0.92$); only in the case of the small variability of hydrogen bond descriptors are they significantly worse. In the further discussion, we consider the dependences of the HB1 and HB2 energies ($E_{HB}$) on the electron-attracting/donating properties of substituents attached to C8, C2 or N9 atoms in WC, HG and AA base pairs. Alternatively, the other relationships presented in Table 2—i.e., $d_{HB}$, $\rho_{BCP}$, and $\delta(H,A)$ on cSAR(X)—lead to similar conclusions.

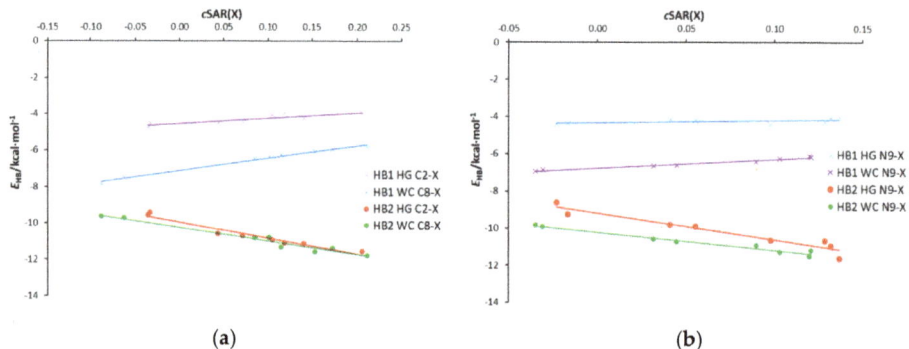

**Figure 4.** Dependence of the HB1 and HB2 energies on cSAR(X) for the WC and HG base pairs substituted at (a) C8 and C2 and (b) N9 positions of the adenine moiety.

**Table 2.** Slopes of the obtained linear relations between the H-bond descriptors (y) and cSAR(X), and the determination coefficients ($R^2$) for the substituted WC and HG base pairs.

| y = a·cSAR(X) + b | | y = $E_{HB}$ | | y = $d_{HB}$ | | y = $\rho_{BCP}$ | | y = $\delta$(H,A) | |
|---|---|---|---|---|---|---|---|---|---|
| Base Pair | HB | a | $R^2$ | a | $R^2$ | a | $R^2$ | a | $R^2$ |
| HG C2-X | 1 | 3.013 | 0.738 | 0.137 | 0.521 | −0.00757 | 0.526 | −0.0215 | 0.566 |
|  | 2 | −8.741 | 0.951 | −0.134 | 0.982 | 0.0172 | 0.984 | 0.0476 | 0.986 |
| WC C8-X | 1 | 6.818 | 0.986 | 0.233 | 0.985 | −0.0175 | 0.980 | −0.0446 | 0.981 |
|  | 2 | −7.487 | 0.966 | −0.143 | 0.968 | 0.0179 | 0.967 | 0.0485 | 0.968 |
| HG N9-X | 1 | 1.344 | 0.314 | 0.100 | 0.329 | −0.00620 | 0.400 | −0.0179 | 0.455 |
|  | 2 | −14.35 | 0.921 | −0.341 | 0.964 | 0.0402 | 0.971 | 0.104 | 0.980 |
| WC N9-X | 1 | 4.927 | 0.971 | 0.182 | 0.929 | −0.0133 | 0.932 | −0.0343 | 0.934 |
|  | 2 | −9.655 | 0.967 | −0.194 | 0.972 | 0.0240 | 0.973 | 0.0645 | 0.970 |

For HB2 in the HG and WC pairs, the slopes are −8.74 and −7.49, respectively. This indicates an increase in the hydrogen bond stability with an increase in the electron-donating power of substituents—i.e., the cSAR(X) values. Thus, in addition to the substituent effect on the amino group, the substituent X also affects the proton-accepting abilities of N atoms of both adenine rings. In these cases, the increase in the cSAR(X) values causes the proton-accepting nitrogen atoms—N1 in the WC pair and N7 in the HG pair—to increase their negative charge. This, in turn, increases the attracting forces towards the hydrogen donating NH group of uracil. A slightly higher slope for HB2 in HG than in the WC base pairs may be explained by a better delocalization way of transmission of the substituent effect. In the case of the HG base pair, the substituent is attached to C2 belonging to a fully aromatic ring, which may be represented by two equivalent canonical structures. In the case of the WC pair, only one (unexcited) canonical structure is possible, i.e., exhibits a lower aromatic character. Hence, there is a worse condition of transmission of the substituent effect.

The HB1 H-bonds represent a different situation. In both cases, an increase in the cSAR(X) values of substituents (i.e., their electron-donating ability) results in a decrease in the HB1 stability expressed by the slopes, which is equal to 3.01 and 6.82 for the HG and WC pairs, respectively. This difference is consistent with the observation of the substituent effect X on the properties of the $NH_2$ group (Figure S6a (Supplementary Materials)), which is involved in the HB1 hydrogen bond. The electron-donating properties of the amino group decrease with the increasing electron-donating ability (more positive cSAR(X) value) of substituent X. Therefore, substitution at the carbon atom (C8 or C2) affects the proton-donating ability of the amino group, decreasing the $E_{HB}$ for the electron-donating X groups. As mentioned above, the difference in the slopes may be explained by a well-known rule—that the substituent effect acting via three bonds (e.g., para-like interactions in benzene derivatives) is more effective than when it operates via two bonds (meta-like interactions). It is usually interpreted that the resonance effect acts more strongly through three than two bonds [55].

The stabilities of HB1 and HB2 in the WC and HG base pairs in the dependence on substitution at N9 are shown in Figure 4b. Despite a similar influence of the X substituent on the properties of the $NH_2$ group (Figure S6a (Supplementary Materials)), its effect on both H-bonds is significantly different. Furthermore, trends in the effect of substituents on hydrogen bonds' stability are similar to those observed for carbon-substituted AU pairs. The only differences are the weaker substituent effect on the HB1 bond and the stronger on the HB2 bond compared to that observed in C-X systems. For the HB1 H-bond, the obtained slopes are 4.93 and 1.34 for the WC and HG pair, respectively (6.82 and 2.71 for the C-X series). In the case of HB2, the resulting slopes are −9.66 and −14.35, respectively, while they are −7.49 and −8.74 for C-X systems, respectively. These results reveal the important role of the lone electron pair at the nitrogen (N9) atom in hydrogen bonding for AU base pairs. The substantial HB2 enhancement by the substituent effect in the HG pair, compared to the WC pair, is due to the fact that the nitrogen atom of the five-membered ring is the proton acceptor of this H-bond. This enhancement is associated with an increase in the charge of this adenine nitrogen atom interacting with the NH group of uracil. This, in turn, is due to the increasing electron-donating power of the substituent—i.e., an increase in the cSAR(X) values.

The results of studies on the effect of substituents attached to C8, C2 or N9 in one adenine on the stability of individual H-bonds in adenine-adenine pairs are presented in Table 3 and Figure S7 (Supplementary Materials). They reveal a similar relationship between the position of the substituent and the stability of H-bonding for adenine dimers, as observed for the WC and HG pairs. The increase in the electron-donating power of substituents causes a decrease in the HB1 stability while strengthening the second H-bond, HB2. This is documented by the positive and negative slope values (Table 3), respectively. However, for AU pairs (both WC and HG), the greater sensitivity of the substituent effect (absolute slope values, Table 2) on the H-bond stability has always been observed for HB2 compared to HB1. This does not apply to mono-substituted adenine dimers. For C8-substituted systems (AA2 and AA4), HB1 is more sensitive than HB2, while the opposite is true for C2-substituted derivatives (AA2 and AA3). This is consistent with the results shown above, where the substituents at the C8 position affect the electronic properties of the adenine amino group more than at the C2 and N9 positions (Table 1). Therefore, they cause greater changes in HB1 stability, which can be expressed by the ratio of slopes $a$ in Table 3. Considering the hydrogen bonding energy, the C8/C2 ratio in AA2 pairs is 1.68, and for C8/N9 in AA4 pairs it is 1.44.

In the case of N9-substituted systems, much larger changes in HB2 strength than in HB1 were noted for the AA2 and AA3 dimers, in which the nitrogen atom of the five-membered ring (N7) is the proton acceptor of this H bond. For AA4 N9-substituted derivatives, HB1 is slightly more sensitive to the effect of substituents than HB2; the slopes are 5.128 and −4.715, respectively (Table 3). For C8 analogs, they are 7.362 and −3.998, respectively. Interestingly, these systems are characterized by the smallest effect of substituents on the electronic structure of the amino group (Table 1) and the strongest intermolecular interactions between adenine-adenine base pairs. In this case, the amino group (proton donor) and the N1 nitrogen atom, which is a proton acceptor and adjacent to the C6 carbon atom, are involved in hydrogen bonds (Figure 2c). A comparison of the structure of the $N1-C6-NH_2$ part in the monomer and dimer shows a significant increase in resonance effects in this fragment. Additionally, in AA4 pairs and WC pairs, H-bonds together with six-membered pyrimidine rings form anthracene-like geometry, which results in the additional stabilization of these systems. This is confirmed by the analysis of the aromaticity of a quasi-aromatic ring consisting of H-bonds [58], which can be described using the QTAIM theory by the total electron energy density at the ring critical point ($H_{RCP}$). The obtained differences between the average $H_{RCP}$ values (in a.u.) in the base pairs are: $\Delta H(AA4 - AA2) = 0.00034$, $\Delta H(AA4 - AA3) = 0.00061$, and $\Delta H(WC - HG) = 0.00038$; the $H_{RCP}$ values for the AA4 and WC dimers are 0.00115 and 0.00135, respectively. It is therefore consistent with the fact that the quasi-aromatic hydrogen bonding systems in AA4 and WC pairs show the strongest H-bonds.

**Table 3.** Slopes of the obtained linear relations between the H-bond descriptors (y) and $cSAR(X)$, and the determination coefficients ($R^2$) for the substituted adenine dimers.

| y = a·cSAR(X) + b | | y = $E_{HB}$ | | y = $d_{HB}$ | | y = $\rho_{BCP}$ | | y = $\delta(H, A)$ | |
|---|---|---|---|---|---|---|---|---|---|
| Base Pair | HB | a | $R^2$ | a | $R^2$ | a | $R^2$ | a | $R^2$ |
| AA2 C2-X | 1 | 3.589 | 0.811 | 0.200 | 0.825 | −0.0156 | 0.825 | −0.0426 | 0.835 |
|  | 2 | −3.794 | 0.987 | −0.193 | 0.993 | 0.0152 | 0.994 | 0.0439 | 0.994 |
| AA2 C8-X | 1 | 6.036 | 0.987 | 0.260 | 0.992 | −0.0218 | 0.985 | −0.0571 | 0.986 |
|  | 2 | −3.036 | 0.963 | −0.161 | 0.958 | 0.0126 | 0.965 | 0.0369 | 0.964 |
| AA3 C2-X | 1 | 2.172 | 0.433 | 0.137 | 0.223 | −0.0096 | 0.262 | −0.0261 | 0.291 |
|  | 2 | −4.843 | 0.762 | −0.276 | 0.579 | 0.0184 | 0.610 | 0.0547 | 0.672 |
| AA4 C8-X | 1 | 7.362 | 0.990 | 0.284 | 0.993 | −0.0252 | 0.989 | −0.0426 | 0.835 |
|  | 2 | −3.998 | 0.976 | −0.189 | 0.976 | 0.0155 | 0.974 | 0.0439 | 0.994 |
| AA2 N9-X | 1 | 1.925 | 0.472 | 0.122 | 0.687 | −0.0096 | 0.712 | −0.0267 | 0.718 |
|  | 2 | −6.930 | 0.978 | −0.370 | 0.986 | 0.0276 | 0.975 | 0.0834 | 0.968 |
| AA3 N9-X | 1 | 1.187 | 0.187 | 0.097 | 0.501 | −0.0068 | 0.575 | −0.0187 | 0.491 |
|  | 2 | −5.778 | 0.792 | −0.413 | 0.943 | 0.0261 | 0.931 | 0.0831 | 0.908 |
| AA4 N9-X | 1 | 5.128 | 0.963 | 0.220 | 0.960 | −0.0190 | 0.964 | −0.0495 | 0.961 |
|  | 2 | −4.715 | 0.949 | −0.225 | 0.967 | 0.0185 | 0.967 | 0.0529 | 0.968 |

Three types of double substituted adenine-adenine pairs were considered in which the same substituent is attached to both adenines. Two of them are symmetric (AA3 C2-X, C2-X, and AA4 C8-X, C8-X) and one is asymmetric (AA2 C2-X, C8-X). The relationships between the energy of individual H-bonds and cSAR(X) are shown in Figure S8. It documents the more complex nature of mutual interactions between the substituents affecting the H-bonding in question. This is manifested by the slopes of linear equations and determination coefficients ($R^2$). However, it is again clearly shown that stronger interactions come out from substitution at the C8 position.

### 3.3. Substituent Effect on the Transmitting Moiety

Adenine molecule consists of five-membered (AD5) and six-membered (AD6) heterocyclic rings. These rings act as transmitters of the substituent effect between X and the "reaction center"—the amino group. In the studied most stable 9H tautomer, both rings have $6\pi$ electrons, so they satisfy the Hückel's 4n + 2 rule [59]. In the AD6 ring, each atom provides a $1\pi$ electron to the delocalized structure, while in the case of AD5, four atoms provide a $1\pi$ electron, and the N9 atom provides $2\pi$ electrons for the overall delocalization. Hence, the substituent effect on the $\pi$-electron delocalization may not be equivalent. The influence of the substituent on the electronic structure of adenine rings can be expressed using aromaticity indices. For this purpose, the HOMA index was used, which can characterize local aromaticity. The calculated HOMA index values of both adenine rings are summarized in Tables S1–S3 (Supplementary Materials), while Table S6 (Supplementary Materials) contains the HOMA values of uracil from the WC and HG pairs. The effect of substituents on HOMA for six- and five-membered adenine rings was approximated by slope values $\Delta HOMA/\Delta cSAR(X)$, where $\Delta HOMA = HOMA(NH_2)$

− HOMA(NO$_2$), and Δ$c$SAR(X) = $c$SAR(NH$_2$) − $c$SAR(NO$_2$). The obtained results for the studied base pairs and monomers are shown in Figure 5.

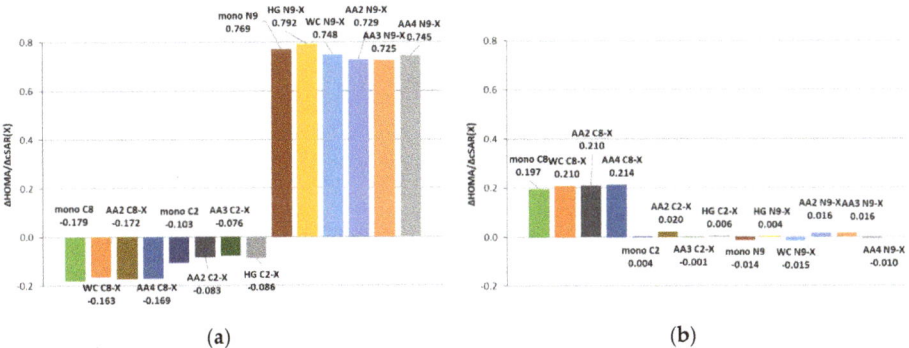

**Figure 5.** Influence of the substituents on Harmonic Oscillator Model of Aromaticity (HOMA) values for (a) five-membered ring and (b) six-membered ring in the studied base pairs, approximated by the slope ΔHOMA/Δ$c$SAR(X) values, where ΔHOMA = HOMA(NH$_2$) − HOMA(NO$_2$) and Δ$c$SAR(X) = $c$SAR(NH$_2$) − $c$SAR(NO$_2$).

The differences in the HOMA values between the substituted and unsubstituted adenine molecules (Tables S8 and S9 (Supplementary Materials)) show that the substitution at C8 affects the electronic structure of the AD6 ring more than the C2 and N9 substitution. Besides, for the C8 substitution, as the electron-donating effect of the substituent described by $c$SAR(X) increases, the HOMA value of the AD6 ring also increases. Moreover, the intermolecular interactions in pairs slightly enhance the effect of substitution from the C8 position on the aromaticity of the six-membered ring compared to the monomer. In the case of C2 and N9 substitution, HOMA does not change monotonically with $c$SAR(X).

The aromaticity of the AD5 ring of adenine has the highest sensitivity to N9 substitution, and the HOMA values increase with the electron-donating ability of the substituent. This can be explained by an interaction between the substituent at the N9 position and the lone electron pair of the N9 atom, which is important for the delocalization in the AD5 ring. An electron-accepting group, such as NO$_2$, may withdraw electrons from the N9 lone electron pair and disturb electron delocalization within the AD5 ring. The substitution at positions C8 and C2 causes smaller changes in the AD5 HOMA value, although in these cases the effect is opposite—the HOMA value decreases with the increasing electron-donating power of the substituent, i.e., $c$SAR(X) value.

Furthermore, in WC and HG pairs, the substituents in adenine also affect the aromaticity of uracil (Table S6 (Supplementary Materials)). Again, the strongest effect is observed from the position C8 than from N9, and the smallest from the position C2. Therefore, it can be concluded that long-distance substituent effects are documented.

### 3.4. Reverse Substituent Effect

The reverse substituent effect quantitatively describes how substituted moiety affects the electronic properties of the substituent X in a system such as R-X or, more generally, X-R-Y. Schematically, it may be presented as in Figure 6.

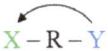

**Figure 6.** Reverse substituent effect in the X-R-Y system.

It should be emphasized that the reverse substituent effect has been known from the beginning, hence there are different substituent constants σ for the para and meta positions [34], similarly to substituent constants σ⁺ and σ⁻ for electron-accepting and -donating reaction sites, respectively [60]. The use of cSAR(X) allows a quantitative description of the electron-donating or accepting properties of the substituent in any system. This gives additional insight into understanding the substituent effect dependent on various substituted moieties.

In our case, we consider how the amino group, variously involved in H-bonding, affects the properties and electronic structure of the substituents expressed by cSAR(X) values. Figure 7 presents the ranges of cSAR(X) values in the dependence on the position and type of moieties to which they are attached. All the data for dimers are compared with the values for adenine monomers.

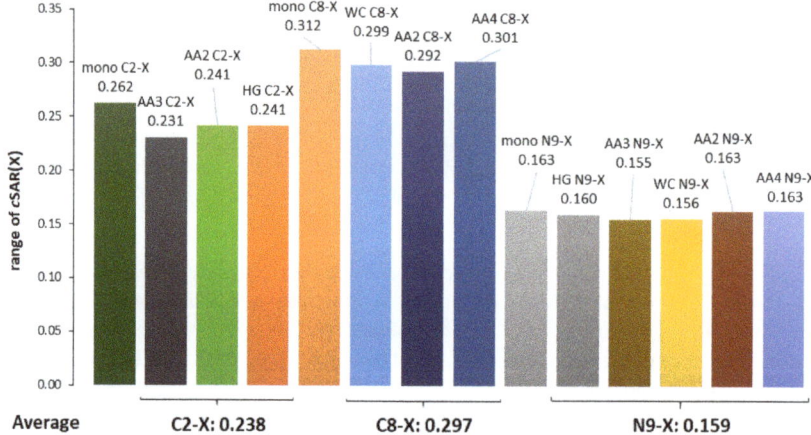

Figure 7. Ranges of cSAR(X) variability and their averaged values for substituents attached to C2, C8, and N9 in adenine monomer and its pairs (WC, HG, and adenine-adenine).

A comparison of the ranges of cSAR(X) values in the adenine monomer and its pairs (WC, HG, and AA) leads to the conclusion that, in the adenine monomer, these ranges are greater than in its pairs for substitution at carbon atoms and almost comparable for substitution at nitrogen. In the case of substitution at C2, the ranges are averaged to 0.238, a value that is significantly lower than for the series substituted at C8 (mean range 0.297), with the ratio of C8/C2 = 1.25. The explanation of this is the same as formerly: the number of bonds between C8, C2 and C6 (with $NH_2$ group attached) is 3 and 2, respectively. In the case of substitution at the nitrogen atom, the former conclusion that the lone electron pair causes a weaker substituent effect due to its possible direct interaction with the substituent can be repeated.

The ranges of cSAR(X) change for the examined substituents attached to the C8, C2 and N9 positions, and more generally the ranges for those attached to the carbon and nitrogen atom in adenine for the studied systems are shown in Figure S9 (Supplementary Materials). Changes in these ranges for a given substituent depend on the substitution position (C2, C8, and N9). In the extreme case (N9-Cl), the variation range is 0.042/0.160—i.e., 26% and slightly less for carbon substitution (NO, 0.066/0.272—i.e., 24%).

3.5. Interrelations between Hydrogen Bond Parameters

As shown above, substituents can significantly change the strength of intermolecular interactions. The presence of a substituent in the adenine molecule affects hydrogen bonds in base pairs, either strengthening or weakening them. The strength of individual H-bonds has been characterized by several descriptors, such as the energy of the H-bond (calculated according to the NBO approach),

$E_{HB}$; its length, $d_{HB}$; and the QTAIM parameters—i.e., the electron density at the bond critical point, $\rho_{BCP}$; its Laplacian, $\nabla^2 \rho_{BCP}$; and the delocalization index between hydrogen and H-bond acceptor atom A, $\delta(H,A)$. Thus, the results for the N···H and O···H hydrogen bonds in the studied dimers can be used to check the interrelationships of these descriptors of intermolecular interactions in the obtained range of their changes. The values of all applied characteristics are collected in Table S7 (Supplementary Materials).

In general, hydrogen bonds can be classified as weak, moderate, and strong. Jeffrey [61] distinguishes weak H-bonds as those for which the range of absolute energy value is 1–4 kcal/mol; for moderate H-bonds, these energies are 4–15 kcal/mol, and they are 15–40 kcal/mol for strong hydrogen bonds. However, it should be emphasized that there are no "sharp" borders between these hydrogen bonds [62]. The QTAIM theory is also a source of energetic parameters. Rozas, Alkorta, and Elguero [63] suggested that the Laplacian as well as the total electron energy density at hydrogen bond BCP, $H_{BCP}$, should both be used as criteria to characterize hydrogen bonding. They proposed for weak H-bonds that both $\nabla^2 \rho_{BCP}$ and $H_{BCP} > 0$; for medium H-bonds, they are $\nabla^2 \rho_{BCP} > 0$ and $H_{BCP} < 0$, while for strong ones, both $\nabla^2 \rho_{BCP}$ and $H_{BCP} < 0$. The topological QTAIM parameters also provide information on the nature of the interaction [64,65]. It has been shown that H-bonds shorter than 1.2 Å exhibit a covalent nature, bond lengths in the range 1.2–1.8 Å are associated with the partially covalent character, and H-bonds longer than 1.8 Å are noncovalent; this is also referred to as shared-shell, intermediate closed-shell, and closed-shell, respectively.

All the calculated energies and their corresponding H-bond lengths in the AU and AA pairs are shown in Figure S10 (Supplementary Materials). The linear equation of $\ln(|E_{HB}|)$ against the H-bond lengths (Figure S11 (Supplementary Materials)) indicates the exponential nature of the relationship shown in Figure S10 (Supplementary Materials). However, a deeper look at both figures indicates that three groups of H-bonds should be distinguished from all points, as shown in Figure 8. The first group (in red) contains O···H and HB2 (N···H) interactions in the WC pairs, the second (in green) contains HB2 in HG pairs and almost linear H-bonds (angle > 175°) in adenine dimers, while the third group (in blue) contains bonds with an H···N-H angle between 161° and 166°; this division is justified by the linear relations shown in Figure S12 (Supplementary Materials). For the same H-bond length, the strength of the interaction decreases as the group number increases. Interestingly, in the case of the strongest interactions, both the nitrogen and oxygen atoms act as acceptors of H-bond protons. Thus, the same relationship describes two types of hydrogen bonds. For comparison, a curve representing the EML equation [24] is added in Figure 8. This equation is based on the exponential fit for 83 experimentally observed in O···H hydrogen bonds.

Hydrogen bonds in AU base pairs are much stronger than in AA dimers (Table S7 (Supplementary Materials)). In the case of AU systems, N···H interactions (HB2) are stronger than O···H (HB1) H-bonds. For the same H-bond length, HB2 is slightly stronger in WC than the HG series. The HB1 bonds in WC pairs are much shorter than in HG dimers, and therefore significantly stronger. For AA dimers, the strength of interactions depends on the N···H-N angle, and, as expected, the linear hydrogen bonds are stronger (Figure 8). Additionally, the obtained values of $\nabla^2 \rho_{BCP}$ and $H_{BCP}$ for HB2 H-bonds (N···H) in the AU base pairs and the strongest AA dimers' H-bonds—AA4 C8-X and AA4 N9-X, X = NO, $NO_2$ ($\nabla^2 \rho_{BCP} > 0$ and $H_{BCP} < 0$, blue points in Figure S13 (Supplementary Materials))—reveal the partially covalent nature of these interactions. H-bonds in AA4 dimers are longer than 1.8 Å (in the range 1.82–1.88 Å). For the other hydrogen bonds, both QTAIM descriptors are positive, indicating the closed-shell nature of the interactions.

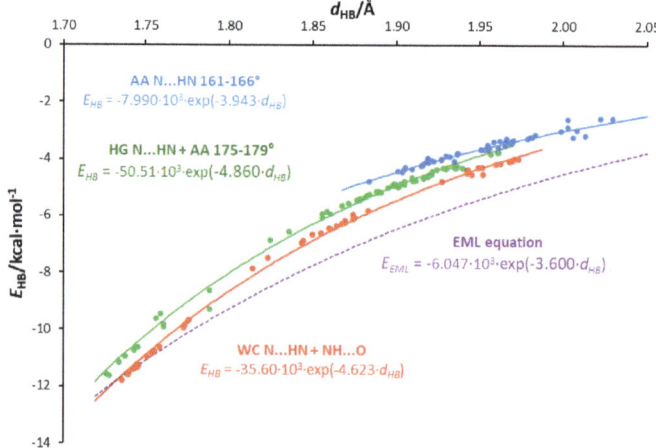

**Figure 8.** Dependences of the H-bond energies, $E_{HB}$, on their lengths, $d_{HB}$, in the studied adenine-uracil and adenine-adenine base pairs.

The relationships between the electron density at the H-bond critical point, $\rho_{BCP}$, or delocalization index, $\delta(H,A)$, and the H-bond length are shown in Figure S14 (Supplementary Materials) and Figure 9, respectively. The exponential nature of these relationships is confirmed by the linear equations $\ln(\rho_{BCP})$ and $\ln(\delta(H,A))$ in relation to the H-bond length, shown in Figures S15 and S16 (Supplementary Materials). The use of these electronic hydrogen bond descriptors allows us to distinguish three types of interaction:

- NH···O, where the oxygen atom of the uracil molecule is the H-bond acceptor and adenine amino group is the H-bond donor.
- N1···HN, where the N1 acceptor atom is a part of a six-membered adenine ring.
- N7···HN, where the N7 acceptor atom is a part of a five-membered adenine ring.

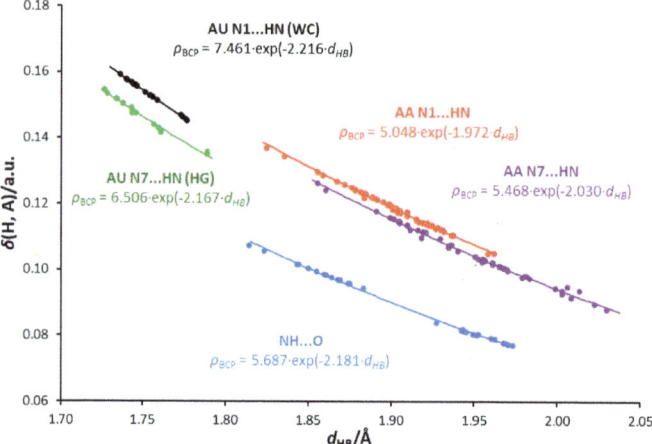

**Figure 9.** Dependences of the delocalization index, $\delta(H,A)$, on the H-bond lengths, $d_{HB}$, in the studied adenine-uracil and adenine-adenine base pairs.

For these interaction groups, the dependence of the H-bond energy on the electron density at its BCP is shown in Figure 10. In the case of N1···HN, energies corresponding to a given $\rho_{BCP}$ value are higher than for N7···HN; the same applies to the electron density at BCP for a given hydrogen bond length (Figure S14 (Supplementary Materials)).

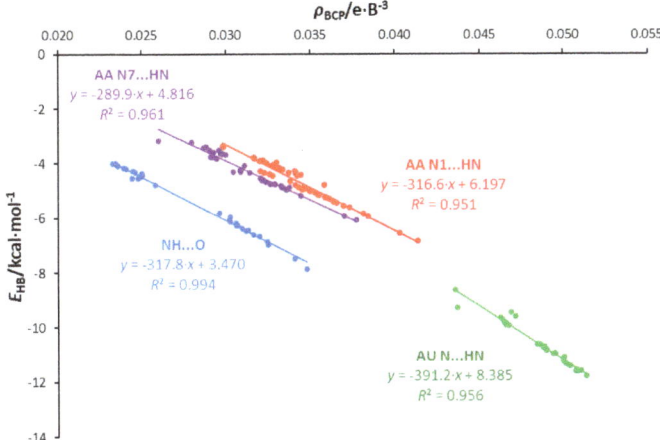

**Figure 10.** Adenine-adenine dimers. Energy of individual H-bonds as a function of the electron density at the H-bond critical point $\rho_{BCP}$.

Figure 9 shows an interesting relationship between the delocalization index $\delta(H,A)$ and the H-bond length $d_{HB}$. The values of $\delta(H,A)$ for a given $d_{HB}$ are higher for the N1···HN-type bonds than for N7···HN in the case of both AA and AU pairs. This shows that N1···HN bonds have a higher covalent character than N7···HN bonds, which is consistent with the values of $\nabla^2\rho_{BCP}$ and $H_{BCP}$ descriptors discussed earlier and presented in Figure S13 (Supplementary Materials). Even lower values of $\delta(H,A)$ are seen for the NH···O-type AU HB1 bonds (marked in blue). This may be interpreted by a stronger attraction of electrons by more electron-attracting oxygen than nitrogen atoms in these interactions, and, therefore, the less covalent character of the NH···O hydrogen bond compared to N···HN. Thus, the relationships for N···H bonds depend on both the type of the acceptor atom (N1 or N7) and the nature of the interactions. This also confirms the interrelationship between the delocalization index and the electron density at BCP of the H-bond (Figure S17 (Supplementary Materials)).

## 4. Conclusions

This theoretical study provides insight into the structural consequences of the substituent effect in biologically important adenine dimers. Four specific aspects were considered: classical SE on intra- and intermolecular interactions, effects on the transmitting moiety, and reverse SE. Additionally, the interrelations between hydrogen bond parameters are presented, revealing more information about the nature of H-bonding in adenine base pairs and other heterocyclic systems.

The classical SE on intramolecular interactions, described by the cSAR index, shows that the substituent has a diversified effect on the amino group, and this effect depends on its position. The transmission of the SE to the amino group from the C8 position is more effective than from the C2 or N9 positions. These differences can be explained by an analogy with benzene substituted in the para and meta positions. The C8 position in the adenine molecule is $n = 3$ bonds away (para-like effect), while C2 is $n = 2$ bonds away from the amino group (meta-like effect). N9 substitution has the weakest effect on the amino group due to an interaction of the substituent and N9 atom lone electron pair. These effects also explain the classical SE on intermolecular interactions—H-bonds. They are presented as relations between the cSAR(X) and H-bond energy, length, and H-bond critical point parameters. The H-bond formed by the substituted adenine acceptor atom shows more sensitivity to SE transmitted from N9, which is caused either by the close distance between the N9 and N7 atoms or by the effect on the lone electron pair of the N9 atom.

Changes in the HOMA index for adenine rings show that a substituent, depending on its position, may have either a strong or almost negligible effect on the aromaticity. The five-membered ring aromaticity is highly influenced by N9 because of an interaction between the substituent and the N9 lone electron pair, which greatly contributes to the ring's delocalized structure. The changes in the six-membered ring aromaticity caused by the substituent at the C8 position of the five-membered ring reveal an interesting long-distance SE.

The reverse SE is an effect of the amino group involved in varying intermolecular interactions on the attached substituent. The results presented as changes in the cSAR(X) values showed that the reverse SE is weaker when the amino group forms an H-bond (in base pairs) than in monomers. Moreover, its strength is consistent with para and meta-like effects in classical SE. Furthermore, the changes in the cSAR ranges for a given substituent may reach ca. 25% of the average variation in the *c*SAR(X) range for a particular substitution position (C-X or N-X).

A deeper analysis of the hydrogen bonds and interrelations between their parameters leads to the following observations:

- Hydrogen bond energies obtained by the NBO approach are smaller for a given length than those predicted by the EML equation for systems with an O···H single hydrogen bond. An exponential relation between the energy and H-bond length is proposed.
- The values of energy are dependent on the H-bond angle forced by the base pair geometry. The most favorable geometry is an anthracene-like system with two six-membered rings connected by two H-bonds. Base pairs of this geometry show the strongest H-bonding.
- Some of the H-bonds exhibit a partially covalent character, confirmed by positive values of the Laplacian and negative values of the total energy density at the bond critical point.
- Relations between the QTAIM parameters differentiate the type of acceptor atom—i.e., adenine N1, N7 atoms, and the uracil O carbonyl atom. The N1 acceptor atom of the six-membered ring forms H-bonds of higher covalent character than the N7 five-membered ring. Bonds formed by the carbonyl oxygen of uracil are less covalent in character, possibly due to the higher electronegativity and attraction of electrons.

**Supplementary Materials:** The following are available online at http://www.mdpi.com/1420-3049/25/16/3688/s1: Figure S1. Studied substituted adenine-uracil base pairs; Figure S2. Studied substituted adenine-adenine AA2 base pairs. Figure S3. Studied substituted adenine-adenine AA3 base pairs; Figure S4. Studied substituted adenine-adenine AA4 base pairs; Figure S5. Linear regressions between cSAR(X) values calculated by Hirshfeld (Hir) method and data from the VDD and NBO approaches for derivatives of the WC base pair substituted in positions C8-X and N9-X by X = NO, $NO_2$, Cl, F, H, Me, OH, $NH_2$; Figure S6. Relationships between cSAR($NH_2$) and cSAR(X) in substituted adenine-uracil and adenine-adenine systems; Figure S7. Dependence of HB1 and HB2 energies on cSAR(X) for adenine dimers substituted in C8, C2 and N9 positions of one adenine moiety; Figure S8. Energy of individual H-bonds as a function of cSAR(X) for adenine-adenine pairs with the same substituents attached to both adenines; Figure S9. Ranges of cSAR(X) changes for the examined X substituent attached to C2, C8, and the nitrogen (N9) and carbon atoms in the adenine monomer and its pairs (WC, HG, and adenine-adenine); Figure S10. Dependence of H-bond energies on their lengths for all hydrogen bonds in studied adenine-uracil and adenine-adenine base pairs; Figure S11. Relationship between $\ln(E_{HB})$ and H-bond lengths for all hydrogen bonds in studied adenine-uracil and adenine-adenine base pairs; Figure S12. Relationships between $\ln(|E_{HB}|)$ and hydrogen bond lengths for the groups shown in Figure 8; Figure S13. Dependence of Laplacian electron density on the length of the H-bond for all hydrogen bonds in studied adenine-uracil and adenine-adenine base pairs; Figure S14. Dependences of electron density at the H-bond critical points on their lengths in the studied adenine-uracil and adenine-adenine base pairs; Figure S15. Relationships between $\ln(\rho BCP)$ and H-bond lengths for all hydrogen bonds in studied adenine-uracil and adenine-adenine base pairs; Figure S16. Relationships between $\ln(\delta(H,A))$ and H-bond lengths, for all hydrogen bonds in studied adenine-uracil and adenine-adenine base pairs; Figure S17. Delocalization index $\delta(H,A)$ between H and A atoms, where A is the acceptor of H-bond as a function of electron density at H-bond critical point for AU and AA dimers; Table S1. Values of cSAR(X), cSAR($NH_2$) and HOMA for substituted adenine-uracil Hoogsteen (HG) and Watson-Crick (WC) pairs. HOMA values of five-membered adenine ring (HOMA AD5) and six-membered adenine ring (HOMA AD6); Table S2. Values of cSAR(X), cSAR(NH2) and HOMA for substituted adenine-adenine AA2 and AA3 pairs. HOMA values of five-membered adenine ring (HOMA AD5) and six-membered adenine ring (HOMA AD6); Table S3. Values of cSAR(X), cSAR(NH$_2$) and HOMA for substituted adenine-adenine AA4 pairs. HOMA values of five-membered adenine ring (HOMA AD5) and six-membered adenine ring (HOMA AD6); Table S4. Values of cSAR(X) and cSAR(NH$_2$). Symmetrically substituted adenine-adenine AA3 and AA4 pairs; Table S5. Values of cSAR(X) and cSAR(NH$_2$). Asymmetrically substituted adenine-adenine AA2 pairs; Table S6. HOMA values of uracil ring in adenine-uracil Hoogsteen (HG) and Watson-Crick (WC) pairs with substituents at C2, C8, N9 position of adenine moiety; Table S7. Calculated hydrogen bond parameters of all studied base pairs; Table S8. Changes in aromaticity expressed by the HOMA index for (a) AD6 and (b) AD5 rings due to substitution in WC and HG pairs; Table S9. Changes in aromaticity expressed by the HOMA index for AD6 and AD5 rings due to substitution in AA dimers.

**Author Contributions:** Conceptualization, H.S. and T.M.K.; Methodology, H.S. and T.M.K.; Investigation, P.A.W.; Formal analysis, P.A.W.; Validation, H.S. and P.A.W.; Funding acquisition, H.S.; Writing—original draft preparation, H.S. and P.A.W.; Writing—review and editing, T.M.K.; Visualization, P.A.W.; Supervision, H.S. All authors have read and agreed to the published version of the manuscript.

**Funding:** This research was funded by the National Science Centre of Poland (Grant no. UMO-2016/23/B/ST4/00082). APC was partly sponsored by MDPI.

**Acknowledgments:** The authors gratefully acknowledge the Interdisciplinary Center for Mathematical and Computational Modeling (Warsaw, Poland) and the Wrocław Centre for Networking and Supercomputing for providing computer time and facilities.

**Conflicts of Interest:** The authors declare no conflict of interest.

# References

1. Šponer, J.; Leszczynski, J.; Hobza, P. Structures and Energies of Hydrogen-Bonded DNA Base Pairs. A Nonempirical Study with Inclusion of Electron Correlation. *J. Phys. Chem.* **1996**, *100*, 1965–1974. [CrossRef]
2. Fonseca Guerra, C.; Bickelhaupt, F.M.; Snijders, J.G.; Baerends, E.J. Hydrogen Bonding in DNA Base Pairs: Reconciliation of Theory and Experiment. *J. Am. Chem. Soc.* **2000**, *122*, 4117–4128. [CrossRef]
3. Šponer, J.; Jurečka, P.; Hobza, P. Accurate Interaction Energies of Hydrogen-Bonded Nucleic Acid Base Pairs. *J. Am. Chem. Soc.* **2004**, *126*, 10142–10151. [CrossRef] [PubMed]
4. Poater, J.; Swart, M.; Bickelhaupt, F.M.; Fonseca Guerra, C. B-DNA Structure and Stability: The Role of Hydrogen Bonding, π–π Stacking Interactions, Twist-Angle, and Solvation. *Org. Biomol. Chem.* **2014**, *12*, 4691–4700. [CrossRef] [PubMed]
5. Šponer, J.; Leszczynski, J.; Hobza, P. Electronic Properties, Hydrogen Bonding, Stacking, and Cation Binding of DNA and RNA Bases. *Biopolymers* **2001**, *61*, 3–31. [CrossRef]
6. Fonseca Guerra, C.; Bickelhaupt, F.M.; Saha, S.; Wang, F. Adenine Tautomers: Relative Stabilities, Ionization Energies, and Mismatch with Cytosine. *J. Phys. Chem. A* **2006**, *110*, 4012–4020. [CrossRef] [PubMed]
7. Van der Wijst, T.; Fonseca Guerra, C.; Swart, M.; Bickelhaupt, F.M. Performance of Various Density Functionals for the Hydrogen Bonds in DNA Base Pairs. *Chem. Phys. Lett.* **2006**, *426*, 415–421. [CrossRef]
8. Grimme, S.; Antony, J.; Schwabe, T.; Mück-Lichtenfeld, C. Density Functional Theory with Dispersion Corrections for Supramolecular Structures, Aggregates, and Complexes of (Bio)Organic Molecules. *Org. Biomol. Chem.* **2007**, *5*, 741–758. [CrossRef]
9. Poater, J.; Swart, M.; Fonseca Guerra, F.; Bickelhaupt, F.M. Selectivity in DNA Replication. Interplay of Steric Shape, Hydrogen Bonds, π-Stacking and Solvent Effects. *Chem. Commun.* **2011**, *47*, 7326. [CrossRef]
10. Watson, J.D.; Crick, F.H.C. Molecular Structure of Nucleic Acids: A Structure for Deoxyribose Nucleic Acid. *Nature* **1953**, *171*, 737–738. [CrossRef]
11. Hoogsteen, K. The crystal and molecular structure of a hydrogen-bonded complex between 1-methylthymine and 9-methyladenine. *Acta Cryst.* **1963**, *16*, 907–916. [CrossRef]
12. Fonseca Guerra, C.; van der Wijst, T.; Bickelhaupt, F.M. Substituent Effects on Hydrogen Bonding in Watson-Crick Base Pairs. A Theoretical Study. *Struct. Chem.* **2005**, *16*, 211–221. [CrossRef]
13. Fonseca Guerra, C.; van der Wijst, T.; Bickelhaupt, F.M. Supramolecular Switches Based on the Guanine-Cytosine (GC) Watson-Crick Pair: Effect of Neutral and Ionic Substituents. *Chem. Eur. J.* **2006**, *12*, 3032–3042. [CrossRef] [PubMed]
14. Fonseca Guerra, C.; van der Wijst, T.; Bickelhaupt, F.M. Nanoswitches Based on DNA Base Pairs: Why Adenine-Thymine is Less Suitable than Guanine-Cytosine. *Chem. Phys. Chem.* **2006**, *7*, 1971–1979. [CrossRef] [PubMed]

15. Ebrahimi, A.; Habibi Khorassani, S.M.; Delarami, H.; Esmaeeli, H. The Effect of $CH_3$, F and $NO_2$ Substituents on the Individual Hydrogen Bond Energies in the Adenine–Thymine and Guanine–Cytosine Base Pairs. *J. Comput. Aided Mol. Des.* **2010**, *24*, 409–416. [CrossRef]
16. Nikolova, V.; Galabov, B. Effects of Structural Variations on the Hydrogen Bond Pairing between Adenine Derivatives and Thymine. *Maced. J. Chem. Chem. Eng.* **2015**, *34*, 159–167. [CrossRef]
17. Jana, K.; Ganguly, B. In Silico Studies with Substituted Adenines to Achieve a Remarkable Stability of Mispairs with Thymine Nucleobase. *New J. Chem.* **2016**, *40*, 1807–1816. [CrossRef]
18. Castro, A.C.; Swart, M. Fonseca Guerra, C. The Influence of Substituents and the Environment on the NMR Shielding Constants of Supramolecular Complexes Based on A–T and A–U Base Pairs. *Phys. Chem.* **2017**, *19*, 13496–13502.
19. Jezuita, A.; Szatylowicz, H.; Krygowski, T.M. Impact of the Substituents on the Electronic Structure of the Four Most Stable Tautomers of Purine and Their Adenine Analogues. *ACS Omega* **2020**, *5*, 11570–11577. [CrossRef]
20. Szatylowicz, H.; Marek, P.H.; Stasyuk, O.A.; Krygowski, T.M.; Solà, M. Substituted adenine quartets: Interplay between substituent effect, hydrogen bonding, and aromaticity. *RSC Adv.* **2020**, *10*, 23350–23358. [CrossRef]
21. Saenger, W. *Principles of Nucleic Acid Structure*; Neidle, S., Ed.; Springer: Berlin/Heidelberg, Germany, 1984.
22. Asensio, A.; Kobko, N.; Dannenberg, J.J. Cooperative Hydrogen-Bonding in Adenine–Thymine and Guanine–Cytosine Base Pairs. Density Functional Theory and Møller–Plesset Molecular Orbital Study. *J. Phys. Chem. A* **2003**, *107*, 6441–6443. [CrossRef]
23. Grunenberg, J. Direct Assessment of Interresidue Forces in Watson-Crick Base Pairs Using Theoretical Compliance Constants. *J. Am. Chem. Soc.* **2004**, *126*, 16310–16311. [CrossRef] [PubMed]
24. Espinosa, E.; Molins, E.; Lecomte, C. Hydrogen bond strengths revealed by topological analyses of experimentally observed electron densities. *Chem. Phys. Lett.* **1998**, *285*, 170–173. [CrossRef]
25. Matta, C.F.; Castillo, N.; Boyd, R.J. Extended Weak Bonding Interactions in DNA: π-Stacking (Base–Base), Base–Backbone, and Backbone-Backbone Interactions. *J. Phys. Chem. B* **2006**, *110*, 563–578. [CrossRef] [PubMed]
26. Dong, H.; Hua, W.; Li, S. Estimation on the Individual Hydrogen-Bond Strength in Molecules with Multiple Hydrogen Bonds. *J. Phys. Chem. A* **2007**, *111*, 2941–2945. [CrossRef]
27. Bader, R.F.W. *Atoms in Molecules: A Quantum Theory*; Oxford University Press: Oxford, UK, 1990.
28. Popelier, P.L.A. *Atoms in Molecules—An Introduction*; Pearson Education: London, UK, 2000.
29. Ebrahimi, A.; Habibi Khorassani, S.M.; Delarami, H. Estimation of individual binding energies in some dimers involving multiple hydrogen bonds using topological properties of electron charge density. *Chem. Phys.* **2009**, *365*, 18–23. [CrossRef]
30. Weinhold, F.; Landis, C.R. *Valency and Bonding. A Natural Bond. Orbital Donor-Acceptor Perspective*; Cambridge University Press: Cambridge, UK, 2005.
31. Szatylowicz, H.; Sadlej-Sosnowska, N. Characterizing the Strength of Individual Hydrogen Bonds in DNA Base Pairs. *J. Chem. Inf. Model.* **2010**, *50*, 2151–2161. [CrossRef]
32. Pandey, S.K.; Manogaran, D.; Manogaran, S.; Schaefer, H.F., III. Quantification of Hydrogen Bond Strength Based on Interaction Coordinates: A New Approach. *J. Phys. Chem. A* **2017**, *121*, 6090–6103. [CrossRef]
33. Poater, J.; Fradera, X.; Solà, M.; Duran, M.; Simon, S. On the Electron-Pair Nature of the Hydrogen Bond in the Framework of the Atoms in Molecules Theory. *Chem. Phys. Lett.* **2003**, *369*, 248–255. [CrossRef]
34. Hammett, L.P. The Effect of Structure upon the Reactions of Organic Compounds. Benzene Derivatives. *J. Am. Chem. Soc.* **1937**, *59*, 96–103. [CrossRef]
35. Hammett, L.P. *Physical Organic Chemistry*; McGraw-Hill Book Co. Inc.: New York, NY, USA, 1940.
36. Sadlej-Sosnowska, N. On the Way to Physical Interpretation of Hammett Constants: How Substituent Active Space Impacts on Acidity and Electron Distribution in p-Substituted Benzoic Acid Molecules. *Pol. J. Chem.* **2007**, *81*, 1123–1134.
37. Sadlej-Sosnowska, N. Substituent Active Region—A Gate for Communication of Substituent Charge with the Rest of a Molecule: Monosubstituted Benzenes. *Chem. Phys. Lett.* **2007**, *447*, 192–196. [CrossRef]
38. Sadlej-Sosnowska, N.; Krygowski, T.M. Substituent effect on geometry of the NO and $N(CH_3)_2$ groups in para substituted derivatives of nitrosobenzene and N,N-dimethylaniline. *Chem. Phys. Lett.* **2009**, *476*, 191–195. [CrossRef]

39. Stasyuk, O.A.; Szatylowicz, H.; Fonseca Guerra, C.; Krygowski, T.M. Theoretical Study of Electron-Attracting Ability of the Nitro Group: Classical and Reverse Substituent Effects. *Struct. Chem.* **2015**, *26*, 905–913. [CrossRef]
40. Bader, R.F.W. Atoms in molecules. *Acc. Chem. Res.* **1985**, *18*, 9–15. [CrossRef]
41. Frisch, M.J.; Trucks, G.W.; Schlegel, H.B.; Scuseria, G.E.; Robb, M.A.; Cheeseman, J.R.; Scalmani, G.; Barone, V.; Petersson, G.A.; Nakatsuji, H.; et al. *Gaussian 16, Revision C.01*; J. Gaussian Inc.: Wallingford, CT, USA, 2016.
42. Marek, P.H.; Szatylowicz, H.; Krygowski, T.M. Stacking of Nucleic Acid Bases: Optimization of the Computational Approach - the Case of Adenine Dimers. *Struct. Chem.* **2019**, *30*, 351–359. [CrossRef]
43. Becke, A.D. Density-Functional Thermochemistry. V. Systematic Optimization of Exchange-Correlation Functionals. *J. Chem. Phys.* **1997**, *107*, 8554–8560. [CrossRef]
44. Grimme, S.; Antony, J.; Ehrlich, S.; Krieg, H. A Consistent and Accurate Ab Initio Parametrization of Density Functional Dispersion Correction (DFT-D) for the 94 Elements H-Pu. *J. Chem. Phys.* **2010**, *132*, 154104. [CrossRef]
45. Kendall, R.A.; Dunning, T.H., Jr.; Harrison, R.J. Electron Affinities of the First-Row Atoms Revisited. Systematic Basis Sets and Wave Functions. *J. Chem. Phys.* **1992**, *96*, 6796–6806. [CrossRef]
46. Glendening, E.D.; Landis, C.R.; Weinhold, F. NBO 6.0: Natural Bond Orbital Analysis Program. *J. Comput. Chem.* **2013**, *34*, 1429–1437. [CrossRef]
47. Hirshfeld, F.L. Bonded-Atom Fragments for Describing Molecular Charge Densities. *Theor. Chim. Acta* **1977**, *44*, 129–138. [CrossRef]
48. Fonseca Guerra, C.; Handgraaf, J.W.; Baerends, E.J.; Bickelhaupt, F.M. Voronoi Deformation Density (VDD) Charges: Assessment of the Mulliken, Bader, Hirshfeld, Weinhold, and VDD Methods for Charge Analysis. *J. Comput. Chem.* **2004**, *25*, 189–210. [CrossRef] [PubMed]
49. Šponer, J.; Shukla, M.K.; Leszczynski, J. Computational modeling of DNA and RNA fragments. In *Handbook of Computational Chemistry*; Leszczynski, J., Ed.; Springer: Dordrecht, The Netherlands, 2012; pp. 1257–1275.
50. Boys, S.F.; Bernardi, F. The Calculation of Small Molecular Interactions by the Differences of Separate Total Energies. Some Procedures with Reduced Errors. *Mol. Phys.* **1970**, *19*, 553–566. [CrossRef]
51. Todd, A.; Keith, T.K. Gristmill Software, Overland Park KS, AIMAll (Version 19.10.12), USA, 2019. Available online: aim.tkgristmill.com (accessed on 11 August 2020).
52. Kruszewski, J.; Krygowski, T.M. Definition of Aromaticity Basing on the Harmonic Oscillator Model. *Tetrahedron Lett.* **1972**, *13*, 3839–3842. [CrossRef]
53. Lu, T.; Chen, F. Multiwfn: A Multifunctional Wavefunction Analyzer. *J. Comput. Chem.* **2012**, *33*, 580–592. [CrossRef]
54. Krygowski, T.M. Crystallographic Studies of Inter- and Intramolecular Interactions Reflected in Aromatic Character of Pi-Electron Systems. *J. Chem. Inf. Model.* **1993**, *33*, 70–78. [CrossRef]
55. Krygowski, T.M.; Palusiak, M.; Plonka, A.; Zachara–Horeglad, J.E. Relationship between substituent effect and aromaticity—Part III: Naphthalene as a transmitting moiety for substituent effect. *J. Phys. Org. Chem.* **2007**, *20*, 297–306. [CrossRef]
56. Shahamirian, M.; Cyrański, M.K.; Krygowski, T.M. Conjugation Paths in Monosubstituted 1,2- and 2,3-Naphthoquinones. *J. Phys. Chem. A* **2011**, *115*, 12688–12694. [CrossRef]
57. Koch, U.; Popelier, P.L.A. Characterization of C-H-O Hydrogen Bonds on the Basis of the Charge Density. *J. Phys. Chem.* **1995**, *99*, 9747–9754. [CrossRef]
58. Palusiak, M.; Krygowski, T.M. Application of AIM Parameters at Ring Critical Points for Estimation of π-Electron Delocalization in Six-Membered Aromatic and Quasi-Aromatic Rings. *Chem. Eur. J.* **2007**, *13*, 7996–8006. [CrossRef]
59. Hückel, E. Quantentheoretische Beiträge Zum Benzolproblem. *Z. Phys. Chem.* **1931**, *70*, 204–286.
60. Hansch, C.; Leo, A.; Taft, R.W. A Survey of Hammett Substituent Constants and Resonance and Field Parameters. *Chem. Rev.* **1991**, *91*, 165–195. [CrossRef]
61. Jeffrey, G.A. *An Introduction to Hydrogen Bonding*; Oxford University Press: New York, NY, USA, 1997.
62. Grabowski, S.J. Theoretical studies of strong hydrogen bonds. *Annu. Rep. Prog. Chem. Sect. C* **2006**, *102*, 131–165. [CrossRef]
63. Rozas, I.; Alkorta, I.; Elguero, J. Behavior of ylides containing N, O, and C atoms as hydrogen bond acceptors. *J. Am. Chem. Soc.* **2000**, *122*, 11154–11161. [CrossRef]

64. Grabowski, S.J.; Sokalski, W.A.; Dyguda, E.; Leszczynski, J. Quantitative Classification of Covalent and Noncovalent H-Bonds. *J. Phys. Chem. B* **2006**, *110*, 6444–6446. [CrossRef] [PubMed]
65. Grabowski, S.J. What Is the Covalency of Hydrogen Bonding? *Chem. Rev.* **2011**, *111*, 2597–2625. [CrossRef] [PubMed]

**Sample Availability:** Calculation output files for all studied dimers are available from the authors.

© 2020 by the authors. Licensee MDPI, Basel, Switzerland. This article is an open access article distributed under the terms and conditions of the Creative Commons Attribution (CC BY) license (http://creativecommons.org/licenses/by/4.0/).

Article

# Combined X-ray Crystallographic, IR/Raman Spectroscopic, and Periodic DFT Investigations of New Multicomponent Crystalline Forms of Anthelmintic Drugs: A Case Study of Carbendazim Maleate

Alexander P. Voronin [1], Artem O. Surov [1], Andrei V. Churakov [2], Olga D. Parashchuk [3], Alexey A. Rykounov [4] and Mikhail V. Vener [5],*

[1] Department of Physical Chemistry of Drugs, G.A. Krestov Institute of Solution Chemistry of RAS, 153045 Ivanovo, Russia; flox-av@yandex.ru (A.P.V.); aos@isc-ras.ru (A.O.S.)
[2] Department of Crystal Chemistry and X-ray Diffraction, N.S. Kurnakov Institute of General and Inorganic Chemistry of RAS, 119991 Moscow, Russia; churakov@igic.ras.ru
[3] Faculty of Physics, Lomonosov Moscow State University, 119991 Moscow, Russia; olga_par@rambler.ru
[4] Theoretical Department, FSUE "RFNC-VNIITF Named after Academ. E.I. Zababakhin", 456770 Snezhinsk, Russia; arykounov@gmail.com
[5] Department of Quantum Chemistry, D. Mendeleev University of Chemical Technology, 125047 Moscow, Russia
* Correspondence: mikhail.vener@gmail.com

Academic Editor: Ilya G. Shenderovich
Received: 6 May 2020; Accepted: 18 May 2020; Published: 21 May 2020

**Abstract:** Synthesis of multicomponent solid forms is an important method of modifying and fine-tuning the most critical physicochemical properties of drug compounds. The design of new multicomponent pharmaceutical materials requires reliable information about the supramolecular arrangement of molecules and detailed description of the intermolecular interactions in the crystal structure. It implies the use of a combination of different experimental and theoretical investigation methods. Organic salts present new challenges for those who develop theoretical approaches describing the structure, spectral properties, and lattice energy $E_{latt}$. These crystals consist of closed-shell organic ions interacting through relatively strong hydrogen bonds, which leads to $E_{latt} > 200$ kJ/mol. Some technical problems that a user of periodic (solid-state) density functional theory (DFT) programs encounters when calculating the properties of these crystals still remain unsolved, for example, the influence of cell parameter optimization on the $E_{latt}$ value, wave numbers, relative intensity of Raman-active vibrations in the low-frequency region, etc. In this work, various properties of a new two-component carbendazim maleate crystal were experimentally investigated, and the applicability of different DFT functionals and empirical Grimme corrections to the description of the obtained structural and spectroscopic properties was tested. Based on this, practical recommendations were developed for further theoretical studies of multicomponent organic pharmaceutical crystals.

**Keywords:** conventional and non-conventional H-bonds; empirical Grimme corrections; lattice energy of organic salts; computation of low-frequency Raman spectra

## 1. Introduction

Organic salts are crystalline ionic compounds that contain one or more organic ions in their structure. Organic salts have broad application in the pharmaceutical industry [1], non-linear optics [2], catalysis [3], green solvents for chemical production [4], etc. Rational design of organic salts and relative materials implies the development of computational methods capable of reliable prediction of industry-relevant properties such as spectroscopic features and crystal lattice energy.

There are several benchmark sets of single-component organic crystals consisting of small rigid molecules without organic fluorine, packed together by van der Waals forces and/or weak and moderate hydrogen bonds (H-bonds), whose lattice energy is accurately computed [5–9]. New and existing theoretical methods are developed and tested based on these sets, and then they are further applied to various compounds. However, most crystals with actual or potential practical application have little in common with the structures from the benchmark sets. Some examples include single-component crystals of larger, conformationally flexible molecules [10–14], fluoroorganic compounds [15,16], and multicomponent crystals [17–19], often with short (strong) [20,21] or ionic H-bonds [22]. The applicability of the methods tested against the benchmark sets for modeling the properties of non-model crystals (e.g., organic salts) is unclear.

In order to describe the properties of "real" crystals, semi-empirical methods based on additive schemes and/or parameterized force fields are often used [23–25]. Their area of application is often limited to a single property (usually to crystal lattice energy), and they are unable to describe a number of properties determined experimentally, including IR and Raman spectra, electron density distribution, etc. The semi-empirical methods provide accurate values of sublimation enthalpies of one-component crystals, consisting of molecules of an arbitrary size [26] and two-component crystals with non-conventional H-bonds [27]. However, these methods are very sensitive to force field parameterization.

For this reason, we chose periodic density functional theory (DFT) methods which allow describing a wide range of properties of the crystalline phase, and which have relatively low computational costs even when treating complex multicomponent crystals [28] of large flexible molecules [29,30] containing aromatic fluorine [31], as well as short (strong) or ionic H-bonds [32,33]. We believe that periodic (solid-state) DFT computations provide a grounded trade-off between the accuracy and the rate of calculations of experimentally observed properties of multi-component organic crystals. In the DFT methods, there are two main approaches based either on Gaussian-type orbitals (GTO) or on plane waves (PW), both with their advantages and disadvantages. Thus, GTO basis sets can better describe isolated molecules in the gas phase, which is essential for $E_{latt}$ calculation [33], while many solid-state properties such as solid-state infrared (IR) spectra are traditionally computed using PW [34]. Only a few articles provide a comparison of the results obtained with GTO and PW bases [35,36]. The choice of the functional is also important for the quality of the obtained data. For example, the B3LYP (Becke 3-parameter, Lee-Yang-Parr) functional is commonly applied in computations with GTOs [37,38], while PBE (Perdew-Burke-Ernzerhof) is used with PW basis sets [30,34,39]. It is well known that B3LYP describes systems with short (strong) or ionic H-bonds better than PBE, while the latter overestimates the stabilization energy in molecular complexes and crystals [40,41]. Non-directed dispersion interactions cause problems for DFT computations in both the GTO and PW versions, making it necessary to use the dispersion corrections of different nature. However, it is not yet investigated how the dispersion corrections [42–44], as well as other parameters such as optimization of cell parameters [7,9,30,34], type of the functional, and basis set, affect the observable properties (e.g., sublimation enthalpy [28,45], IR/Raman spectra [46–48], metric [18,22] and electron density features [42] at bond critical points of conventional and non-conventional hydrogen bonds) of molecular crystals with short (strong) or ionic H-bonds.

Some technical problems that a user of periodic (solid-state) DFT programs encounters when calculating the properties of multicomponent organic crystals remain unsolved. It is still unclear how full optimization (variation of cell parameters) influences the metric parameters of short/ionic H-bonds, $E_{latt}$ values, the wave numbers of normal vibrations in the low-frequency region <400 cm$^{-1}$, etc.

Soluble drug forms are one of the main areas of application of multicomponent crystals. For this reason, anthelmintic compounds with low aqueous solubility were selected as objects of the present study. Anthelmintic benzimidazole derivatives are basic compounds capable of forming a variety of two-component crystals with pharmaceutically acceptable acids [49–51]. Since the number of potential pharmaceutical crystal forms of target compounds is very high, accurate theoretical estimates of relevant properties are desired to avoid excessive experimental work.

In this paper, the new two-component crystalline form of the anthelmintic drug 1:1 carbendazim–maleic acid crystal [CRB + MLE] (1:1) is investigated using X-ray and IR/Raman spectroscopy in combination with periodic (solid-state) DFT calculations (Figure 1). The applicability of different DFT functionals and empirical Grimme corrections to reproducing the experimentally observed parameters was tested using the Crystal17 and QuantumEspresso DFT codes. As a result, "practical recipes" are proposed for computing multicomponent organic crystals for users of these programs.

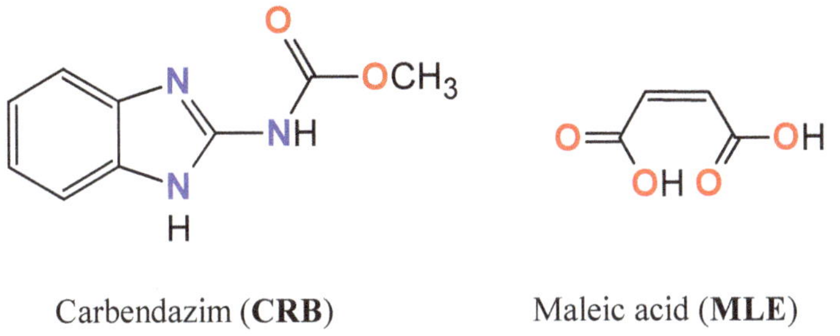

**Figure 1.** Molecular structures of carbendazim (**CRB**) and maleic acid (**MLE**).

## 2. Results and Discussion

The crystallographic data of the [CRB + MLE] (1:1) salt were recorded at 120 K and at room temperature to study the temperature effect on the metric parameters of the unit cell and most notable H-bonds. The crystallographic information is collected in Table S1 (Supplementary Materials). We see that the thermal expansion in the interval between 120 K and 296 K is almost negligible, as the cell volume increases only by 3% (44 Å$^3$).

The asymmetric unit contains one CRB cation and one MLE anion. The crystal has H-bonds of different types and strengths: conventional intra- and intermolecular H-bonds and non-conventional C–H···O contacts. A number of equations were proposed to assess the dependence of the H-bond stabilization energy from the distances between the heavy atoms [52,53], H···O/N distance [54], and electron density descriptors [41,42]. According to these approaches, the intramolecular O24–H24···O21 bond can be considered strong, while the two intermolecular N–H···O bonds can be classified as medium or weak hydrogen bonds (Table S2, Supplementary Materials).

### 2.1. Hydrogen Bond Patterns

The molecules in the asymmetric unit are combined into a heterodimer formed by ionic N$^+$1–H1···O21 and medium or weak N3–H3···O22 bonds, which build the eight-membered cyclic motif with the $R_2^2(8)$ graph set notation [55]. Another conventional N2–H2···O23 hydrogen bond connects the N–H group of the CRB cation with the adjacent MLE anion and is assisted by two C–H···O contacts (Figure S1, Supplementary Materials). The O21 atom acts as an acceptor of two H-bonds, short (strong) intramolecular and ionic intermolecular ones (Figure 2).

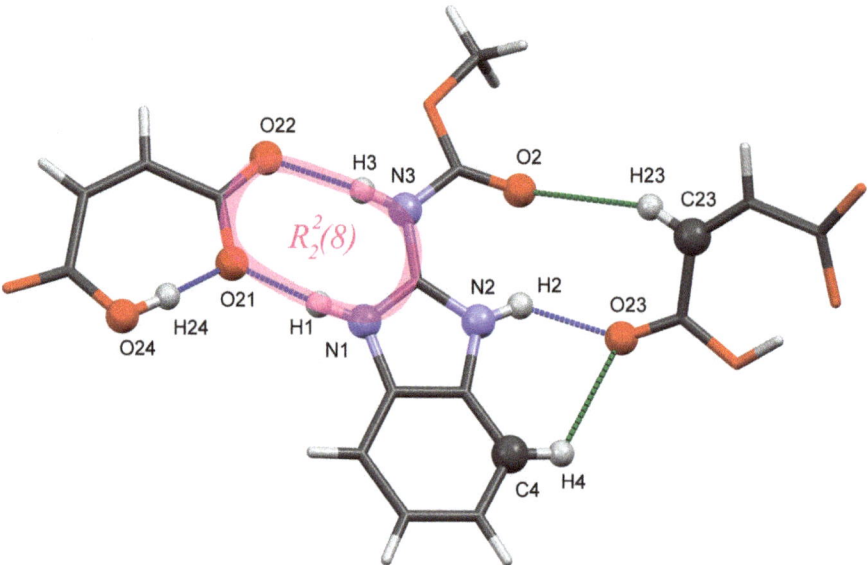

**Figure 2.** Part of the hydrogen bond network in [CRB + MLE] (1:1). The H-bonds and C–H⋯O contacts are colored blue and green, respectively.

### 2.2. Effect of Optimization on Cell Parameters

The volume of the crystallographic cell of the considered two-component crystal increases from 1400.5 to 1444.6 Å$^3$ as the temperature rises from 120 to 296 K. This means that the cell parameters change slightly when the temperature increases. This result is consistent with the published data [56], according to which the thermal expansion from 120 to 296 K for organic crystals is estimated to range between ~1% and ~3%. The sign and absolute value of the relative change in the volume of the crystallographic cell depend on the functional and type of Grimme corrections (Table 1). The data given in this table indicate that none of the approximations reproduce the experimental value of the thermal expansion of the considered crystal. Some approximations give negative values of this coefficient. Note that such a result was already obtained in a number of articles (see Tables 2 and 3 in Reference [57], Table 2 in Reference [58], Figure 4 in Reference [59], and Table S19 in Reference [34]). These results demonstrate that the change in the unit cell volume during the lattice optimization does not always correspond to the experimental data.

### 2.3. Metric Parameters of Conventional and Non-Conventional H-Bonds

The experimental values of the distances between the heavy atoms involved in the formation of conventional H-bonds are compared in Table 1 with the theoretical values computed using different levels of approximation with fixed unit cell parameters and full unit cell relaxation. We assume that the calculations are in agreement with the X-ray data if the theoretical values of the O⋯O and O⋯N distances differ from the experimental ones by no more than 0.01 Å. The data presented in Table 1 suggest that (i) the considered distances are very sensitive to the choice of the functional (B3LYP or PBE), the inclusion of the dispersion correction, and its type (D2, D3, or none), (ii) the variation of the cell parameters greatly changes the distances if the relative change in the volume of the crystallographic cell is more than a few percent, and (iii) the results we get also depend on the basis set type (PW or GTO).

The metric parameters of the non-conventional H-bonds extracted from the experimental crystal structure are poorly reproduced by all the approximations used (Table S3, Supplementary Materials). We can draw two conclusions. Firstly, the B3LYP approximation with the fixed cell parameters gives the best description of the metric parameters of the conventional H-bonds in the considered crystal. Secondly, it is almost impossible to describe the distances between the heavy atoms involved in the formation of the non-conventional H bonds with an accuracy of ~0.01 Å in the framework of the approximations used.

## 2.4. IR Spectrum in the Low-Frequency Region

The IR spectrum of [CRB + MLE] (1:1) can be divided into high-frequency (>1800 cm$^{-1}$), low-frequency (<400 cm$^{-1}$), and mid-frequency spectral ranges (Figures S2 and S3, Supplementary Materials). For a correct description of the IR frequencies of the asymmetric vibrations of the O–H$\cdots$O and O–H$\cdots$N/$^-$O$\cdots$H–N$^+$ fragments in the high-frequency range, it is necessary to go beyond the double harmonic approximation framework [60]. Explicit accounting of mechanical and electric anharmonicity is very cumbersome and time-consuming [61], especially in the case of organic crystals with intermolecular H-bonds [62,63]. The mid-frequency spectral range is usually described well in the cluster approximation; in some cases, the cluster can consist of one molecule [64].

We focus on reproducing the frequencies, as well as IR and Raman activities, in the low-frequency spectral range. It is currently being intensively studied [29,65,66], since various intermolecular vibrations can be observed in it, in particular, because of the presence of intermolecular H-bonds [36,67–69]. The double harmonic approximation provides a reasonable description of the IR/Raman spectra of organic crystals in the low-frequency spectral range [29,30,70,71].

The experimental IR frequencies of [CRB + MLE] (1:1) in the low-frequency spectral range are compared with the theoretical values computed at different levels of approximation (PBE-D3/6-31G(d,p), B3LYP/6-31G(d,p), B3LYP-D2/6-31G(d,p), B3LYP-D3/6-31G(d,p), and PBE-D3/PW) with fixed unit cell parameters (AtomOnly) and full unit cell relaxation (FullOpt) and the values are collected in Table 2. The periodic DFT calculations in all the approximations used produce reasonable values of vibration frequencies. The IR intensities are reproduced by all the approximations only semi-quantitatively. The use of B3LYP-D3 approximation in the periodic DFT calculations leads to the termination of the IR and Raman activity computations. For this reason, B3LYP-D2 was used to calculate the IR and Raman activities.

The frequencies of the IR-active vibrations of the parent CRB crystal in the range of 400–150 cm$^{-1}$ are reproduced well by all the approximations (Table S4, Supplementary Materials). This is due to the absence of short (strong) or ionic H-bonds in this crystal. The IR intensities are reproduced by all the approximations only semi-quantitatively.

Periodic DFT computations of molecular crystals sometimes lead to the appearance of imaginary frequencies [6,72,73]. We encountered this problem when calculating the IR/Raman spectra of [CRB + MLE] (1:1) using the PBE-D3/PW (FullOpt) approximation (see Table S5, Supplementary Materials). Unlike calculations of non-periodic systems, there is no universal recipe for solving the problem of imaginary frequencies appearing in periodic calculations. This problem is usually solved by reducing the space symmetry of a crystal [72,74]. Other methods include (i) the use of extended basis sets [73], (ii) variation of the cell parameters [74], and (iii) increasing the atomic displacement value in numerical second derivative calculations [41]. However, in some cases, these tricks fail to result in a stable structure.

**Table 1.** Distances between the heavy atoms involved in the formation of conventional hydrogen bonds in [CRB + MLE] (1:1). Experiment (Exp.) vs. theoretical values. Computations at different levels of approximation with fixed unit cell parameters (AtomOnly) and full unit cell relaxation (FullOpt). The FullOpt values are given in parentheses. The units are Å. The relative change in the volume of the crystallographic cell $\Delta V$ is given in the last line.

| Fragment [a] | Exp. | Computations | | | |
|---|---|---|---|---|---|
| | | PBE-D3/6-31G(d,p) | B3LYP/6-31G(d,p) | B3LYP-D2/6-31G(d,p) | B3LYP-D3/6-31G(d,p) | PBE-D3/PW [b] |
| O24–H24⋯O21 (intra-) | 2.442 | 2.462 (2.481) | 2.446 (2.399) | 2.468 (2.433) | 2.457 (2.443) | 2.453 (2.456) |
| N1–H1⋯O21 | 2.685 | 2.662 (2.700) | 2.665 (2.622) | 2.700 (2.640) | 2.680 (2.641) | 2.675 (2.670) |
| N3–H3⋯O22 | 2.761 | 2.735 (2.798) | 2.768 (2.718) | 2.762 (2.746) | 2.757 (2.745) | 2.741 (2.744) |
| N2–H2⋯O23 | 2.756 | 2.701 (2.665) | 2.714 (2.684) | 2.750 (2.681) | 2.701 (2.677) | |
| $\Delta V = (V_{exp} - V_{theor})/V_{exp}$ (%) | 3.1 | 6.8 | −9.2 | 13.3 | 9.4 | <−0.1 |

[a] Atomic numbering is given in Figure 2; [b] PW stands for the plane-wave basis set with a cut-off energy of 100 Ry and PAW pseudopotentials.

**Table 2.** Frequencies and infrared (IR) intensities of the low-frequency vibrations of [CRB + MLE] (1:1). Experiment vs. computations at different levels of approximation. The values obtained using the FullOpt option are shown in italics. The units are cm$^{-1}$ (wave numbers) and kM/mol (intensities).

| Exp. [a] | Normal Mode Symmetry, Assignment [b] | Computations [c] | | | | |
|---|---|---|---|---|---|---|
| | | PBE-D3/6-31G(d,p) | B3LYP/6-31G(d,p) | B3LYP-D2/6-31G(d,p) | B3LYP-D3/6-31G(d,p) | PBE-D3/PW [d] |
| 180–167 vs, broad | B$_u$, ν(N3⋯O22) + ν(N1⋯O21) | 179 (453) 186 (336) [e] | 180 (353) 171 (390) | | 150 (66) [e] | 177 |
| 215 s | B$_u$, CH$_3$ twist | 219 (62) 228 (80) | 222 (63) 219 (51) | | 202 (313) | |
| 266 s | B$_u$, ν(O24⋯O21) | 257 (172) 260 (194) | 264 (165) 263 (153) | | 278 (152) | 258 |
| 302 s | B$_u$, CNC(=O) bending | 299 (58) 302 (43) | 303 (68) 302 (65) | | 312 (95) | |
| 330–346 vs, broad | B$_u$, ν(N1⋯O21) | 362 (132) 365 (46) | 358 (223) 334 (122) | | 368 (156) | 349 |
| 380 s | B$_u$, ν(N3⋯O22) | 368 (160) 376 (206) | 374 (76) 351 (183) | | 383 (88) [e] | 364 |

[a] the abbreviations used for relative intensities are vs, very strong; s, strong; [b] Atomic numbering is given in Figure 2; [c] the IR intensities are given in parenthesis; [d] PW stands for the plane-wave basis set with a cut-off energy of 100 Ry and PAW pseudopotentials; [e] in the calculations, this is a doublet of bands with almost identical wave numbers and IR intensities.

## 2.5. Raman Spectrum in the Low-Frequency Region

The wavenumber of the lowest Raman-active vibration of [CRB + MLE] (1:1) is ~25 cm$^{-1}$ (Figure 3). Its theoretical wavenumber is very sensitive to the level of approximation (Table S5, Supplementary Materials). The optimization of the cell parameters greatly affects this value, as well as the number of IR/Raman active vibrations below 100 cm$^{-1}$. A significant decrease in the cell volume (about 10%) as a result of optimization leads to a blue shift in the wave number of the lowest IR/Raman active vibration by ~10 cm$^{-1}$, in accordance with References [75,76].

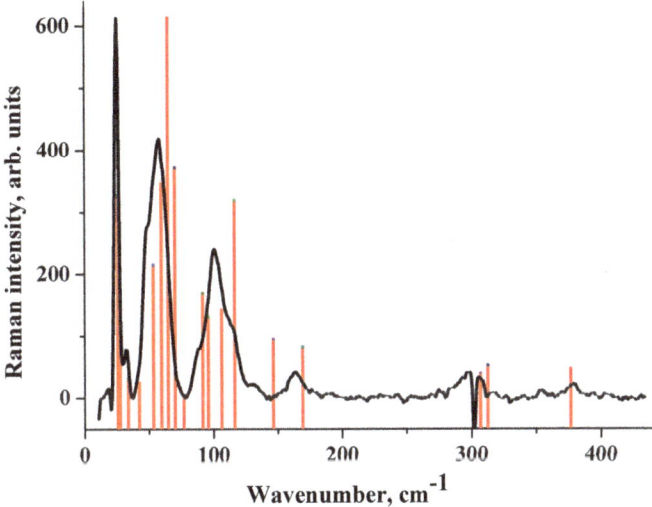

**Figure 3.** Raman spectrum of [CRB + MLE] (1:1). Experiment (black line) vs. B3LYP(AtomOnly) computations (red bars). The height of the bars is proportional to the relative Raman intensity of the corresponding transition.

The experimental Raman spectrum of [CRB + MLE] (1:1) in the low-frequency region is shown in Figure 3. The dips in the spectrum at 20.2 cm$^{-1}$ and at 302 cm$^{-1}$ are the artefacts of the measurements associated with the presence of dust particles on the mirrors. B3LYP with the fixed cell parameters provides a reasonable description of the Raman spectrum of [CRB + MLE] (1:1) (Figure 3). This applies to both the wave numbers and the Raman intensities. In contrast to B3LYP(AtomOnly), B3LYP-D2(FullOpt) does not provide an adequate description of the Raman spectrum (Figure S4, Supplementary Materials). PBE-D3(FullOpt) reproduces the Raman spectrum of the salt somewhat better than B3LYP-D2(FullOpt) (see Figures S4 and S5, Supplementary Materials). However, the calculated wavenumbers of the most intense bands in the region below 100 cm$^{-1}$ are blue-shifted compared with the experiment, and the Raman intensity of the vibrations in the region of 100–400 cm$^{-1}$ turns out to be very high. This result can be explained by a significant reduction in the cell volume of [CRB + MLE] (1:1) as a result of full optimization (see Table 2).

In the Raman spectrum of crystalline maleic acid (Figure 4), the most intense band lies in the region of 100 cm$^{-1}$, while the lowest Raman-active vibration is most intense in [CRB + MLE] (1:1). The B3LYP(AtomOnly) approximation reproduces these differences. This approximation provides a reasonable description of the acid Raman spectrum (Figure 4). PBE-D3(FullOpt) and B3LYP-D2(FullOpt) do not reproduce the acid Raman spectrum (Figures S6 and S7, Supplementary Materials).

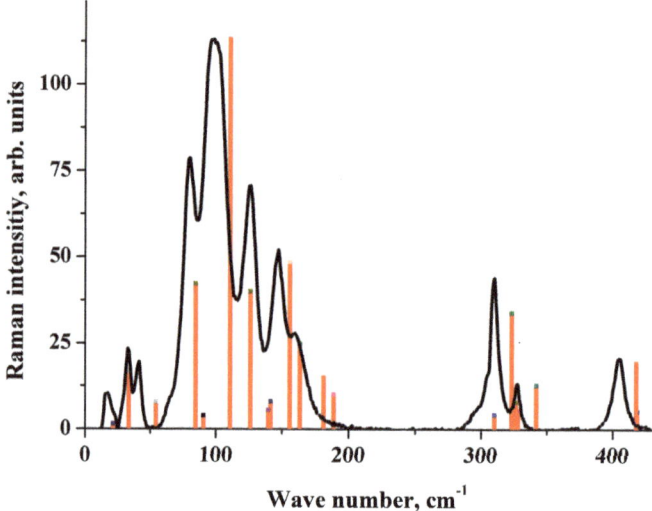

**Figure 4.** Raman spectrum of crystalline maleic acid. Experiment (black line) vs. B3LYP(AtomOnly) computations (red bars). The height of the bars is proportional to the relative Raman intensity of the corresponding transition.

The signal from the CRB crystal contains a strong (apparently luminescent) background, and its Raman spectrum is very noisy (Figure S8, Supplementary Materials). Therefore, we focus on reproducing the spectrum below 100 cm$^{-1}$, i.e., in the THz region. The B3LYP(AtomOnly) approximation reproduces the position of the most intense low-lying vibration and provides a reasonable description of the spectrum in the considered frequency region (Figure 5). B3LYP-D2(FullOpt) and PBE-D3(FullOpt) do not reproduce the position of the most intense low-lying vibration (Figures S9 and S10, Supplementary Materials).

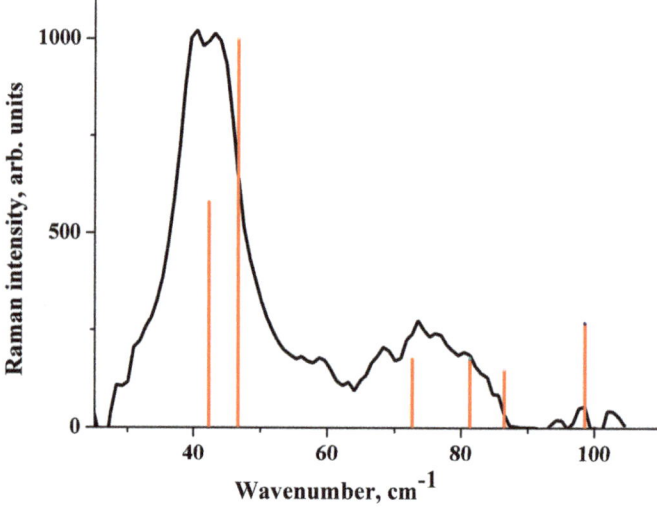

**Figure 5.** Raman spectrum of the CRB crystal in the region of 25–100 cm$^{-1}$ (see text). Experiment (black line) vs. B3LYP(AtomOnly) computations (red bars). The height of the bars is proportional to the relative Raman intensity of the corresponding transition.

To clarify the effect of cell parameter optimization on the Raman spectrum in the low-frequency region, calculations were performed in the PBE-D3(AtomOnly) approximation. Note that this approximation is used in periodic DFT computations with both GTO [77] and PW [78] basis sets. According to Figure 6, Figures S6 and S10 (Supplementary Materials), PBE-D3(AtomOnly) provides a reasonable description of the Raman spectrum of [CRB + MLE] (1:1) and crystals of pure CRB and MLE.

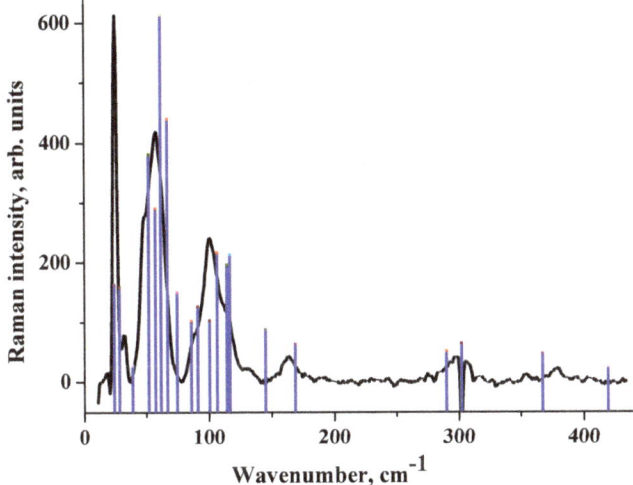

**Figure 6.** Raman spectrum of [CRB + MLE] (1:1). Experiment (black line) vs. PBE-D3(AtomOnly) computations (blue bars). The height of the bars is proportional to the relative Raman intensity of the corresponding transition.

We conclude that the approximations using cell parameter optimization cannot satisfactorily describe the low-frequency Raman spectrum of the crystals with intra- and intermolecular H bonds of different strengths. This is due to the overestimation of the thermal expansion of the crystals by PBE-D3 and B3LYP-D2 (Table 2 and Table S6, Supplementary Materials). The relative changes in the cell volume for B3LYP-D2(FullOpt) are more than 10%; therefore, this approximation provides a poor description of the low-frequency Raman spectra.

The results obtained in this work show that the low-frequency Raman spectra of organic crystals with intramolecular O–H···O, intermolecular O–H···N and ⁻O···H–N⁺ bonds are reproduced in the approximations B3LYP(AtomOnly) and PBE-D3(AtomOnly). According to References [31,77], the structure and IR/Raman spectra of crystals containing organic fluorine and non-conventional C–H···F/C–H···O bonds are adequately described by PBE-D3 and modified PBE functionals with modest basis sets in terms of the AtomOnly approximation.

*2.6. Lattice Energy Evaluation*

A number of computational approaches to $E_{latt}$ assessment are reported in the literature. They mostly concern single-component crystals and two-component crystals without an intermolecular proton transfer (cocrystals), and they use either careful quantum chemical modeling [6,79–82] or semi-empirical schemes [23,24,83–86]. Molecular salts present new challenges for those developing theoretical approaches describing the lattice energy $E_{latt}$. These crystals consist of closed-shell organic ions interacting through ionic H-bonds, which may be partially covalent [21,87]. This is one of the reasons for high $E_{latt}$ values in two-component organic crystals, e.g., 246 kJ·mol$^{-1}$ (Reference [88]), 259–286 kJ·mol$^{-1}$ (Reference [89]), 272 kJ·mol$^{-1}$ (Reference [18]), 253–295 kJ·mol$^{-1}$ (Reference [90]), 299 kJ·mol$^{-1}$ (Reference [91]), etc. It should be pointed out that the $E_{latt}$ values for one-component crystals included in the benchmark sets vary from 25 kJ·mol$^{-1}$ for $CO_2$ to 163 kJ·mol$^{-1}$ for cytosine [5]. The intermolecular proton transfer occurs only in the condensed phase [92], which means that there are closed-shell ions in a molecular crystal (BH$^+$ cation and A$^-$ anion), and neutral organic molecules B and HA in the gas phase. Gas-phase energies of these species ($E_{mol}$) are different for neutral molecules and closed-shell ions. This leads to some ambiguity in the choice of gas-phase structures during the $E_{latt}$ calculation (see Equations (1) and (2) in Section 3.6). In our case, these structures can be either a CRBH$^+$ cation and a maleate ion, or molecules of carbendazim and maleic acid (see the Supplementary Materials for the calculation details).

GTO basis sets require evaluation of the correction of the basic set superposition error (BSSE) (Equation (2)). The Crystal17 evaluation scheme for this correction involves only neutral molecules. It was assumed that BSSEs for the neutral molecules are equal to the values of BSSE for the ionic species in [CRB + MLE] (1:1). The $E_{latt}$ value for the neutral molecules in the gas phase was found to be around 250 kJ·mol$^{-1}$ (Table 3), which is comparable to the known $E_{latt}$ values for multicomponent pharmaceutical crystals estimated using other schemes [45,93,94]. The $E_{latt}$ values obtained by PBE-D3 with and without variation of cell parameters agree well with each other.

**Table 3.** Crystal lattice energy of [CRB + MLE] (1:1) derived from the periodic DFT computations with plane wave and Gaussian-type orbitals. $^a$ The units are kJ·mol$^{-1}$.

|  | B3LYP-D3/6-31G(d,p) (AtomOnly) | PBE-D3/6-31G(d,p) (AtomOnly) | PBE-D3/6-31G(d,p) (FullOpt) | PBE-D3/PW (AtomOnly) |
|---|---|---|---|---|
| Neutral molecules in the gas phase | 277.8 | 266.8 | 258.4 | 278.3 |
| Charged ions in the gas phase $^a$ | 817.2 | 655.6 | 647.2 | 625.0 |

$^a$ See the Supplementary Materials.

In addition, they corresponded to $E_{latt}$ computed by PBE-D3/PW. The lattice energies obtained with the CRBH$^+$ cation and maleate ion in the gas phase were found to be above 600 kJ·mol$^{-1}$ (Table 3). Moreover, this value depended on the DFT functional and the applied BSSE correction. Such large $E_{latt}$ values for multicomponent crystals of organic salts are known in the literature [95–99]. It should be noted that there is a special class of one-component crystals consisting of zwitterionic molecules, which are also characterized by large $E_{latt}$ values [33]. We have to admit that the scheme we used for $E_{latt}$ evaluation of crystals of organic salts requires further development. The raised problem of accounting for BSSE for organic salts and uniform description of the species in crystals and gas phase originates from two independent sources: (1) an assumption of additivity of BSSE corrections for multicomponent crystals; (2) limited ability of the existing approaches to treat crystals of organic salts, e.g., BH$^+$ and A$^-$ in a crystal and B and HA in the gas phase.

In summary, the optimization of cell parameters does not lead to a noticeable change in the $E_{latt}$ value despite the significant variation of the cell volume. The use of the GTO and PW basis sets in the PBE-D3 approximation leads to close $E_{latt}$ values.

## 3. Materials and Methods

### 3.1. Compounds and Solvents

Carbendazim ($C_9H_9N_3O_2$, 98%) was purchased from Acros Organics (Geel, Belgium), and maleic acid ($C_4H_4O_4$, 98%) was purchased from Merck (Darmstadt, Germany). The solvents were purchased from different suppliers and were used as received without further purification.

### 3.2. Cocrystal Preparation

The grinding experiments were performed using a Fritsch planetary micro mill (Fritsch, Idar-Oberstein, Germany), model Pulverisette 7, in 12-mL agate grinding jars with ten 5-mm agate balls at a rate of 500 rpm for 40 min. In a typical experiment, 80–100 mg of the carbendazim/maleic acid mixture in the 1:1 stoichiometric ratio was placed into a grinding jar, and

40–50 µL of methanol was added with a micropipette. In the other method, 200 mg of the 1:1 mixture of carbendazim and maleic acid was suspended in 2 mL of methanol and left to be stirred overnight on a magnetic stirrer at room temperature. The precipitate was filtered from the solution and dried at room temperature.

The diffraction-quality single crystals of [CRB + MLE] (1:1) were obtained by dissolving 90 mg of the stoichiometric 1:1 mixture of the components in 5 mL of methanol at 40 °C. After complete dissolution, the solution was gently cooled to room temperature; then, it was covered with Parafilm with a few small holes pierced in it and left for the solvent to evaporate. After a week, small colorless crystals appeared in the solution.

### 3.3. Thermal Analysis

The thermal analysis was carried out using a differential scanning calorimeter (DSC) with a refrigerated cooling system (Perkin Elmer DSC 4000, Perkin Elmer Inc., Waltham, MS, USA). The sample was heated in a standard sealed aluminum pan (40 µL volume) at a rate of 10 °C·min$^{-1}$ in a nitrogen atmosphere. The unit was calibrated with indium and zinc standards. The accuracy of the weighing procedure was ±0.01 mg. The results of the DSC analysis for [CRB + MLE] (1:1) and pure components are presented in Figure S11 (Supplementary Materials).

### 3.4. Single-Crystal and Powder X-ray Diffraction (XRD) Experiments

Single-crystal XRD data were collected on a SMART APEX II diffractometer (Bruker AXS, Karlsruhe, Germany) using graphite-monochromated MoK$_\alpha$ radiation ($\lambda$ = 0.71073 Å) at 120 and 296 K. Absorption corrections based on the measurements of equivalent reflections were applied [100]. The structures were solved by direct methods and refined by full matrix least-squares on $F^2$ with anisotropic thermal parameters for all the non-hydrogen atoms [101]. All the hydrogen atoms were found from a difference Fourier map and refined isotropically. The crystallographic data for [CRB + MLE] (1:1) were deposited by the Cambridge Crystallographic Data Center (Cambridge, UK) as supplementary publications under the CCDC numbers 1,994,877 and 1,994,878 for 120 and 296 K, respectively. This information can be obtained free of charge from the Cambridge Crystallographic Data Center via www.ccdc.cam.ac.uk/data_request/cif.

The X-ray powder diffraction (PXRD) data of the bulk materials were recorded under ambient conditions in Bragg–Brentano geometry on a Bruker D2 Phaser diffractometer equipped with a second-generation LynxEye detector (Bruker AXS, Karlsruhe, Germany) with CuK$\alpha$ radiation ($\lambda$ = 1.5406 Å). PXRD patterns of salt and parent solids are given in Figure S12 (Supplementary Materials).

### 3.5. IR and Raman Spectroscopy

The Fourier-transform infrared (FT-IR) spectra of the compounds were recorded in the spectral range of 400–150 cm$^{-1}$ from CsBr pellets on a Bruker Vertex 80 V device (Bruker Optik, Ettlingen, Germany) equipped with a Mylar multilayer beamsplitter. The high-quality spectra were obtained and analyzed using the OPUS 6.5.83 software (Bruker Optik, Ettlingen, Germany).

The Raman measurements in the spectral range of 10–440 cm$^{-1}$ were performed using a Raman microscope (inVia, Renishaw plc, Spectroscopy Product Division, Old Town Wotton-Under-Edge, Gloucestershire, GL12 7DW, UK) with a 50× objective lens (Leica DM 2500 M, NA = 0.75, Leica Mikrosysteme Vertrieb GmbH, Mikroskopie und Histologie Ernst-Leitz-Strasse 17-37Wetzlar, 35578 Germany). The measurements were made with a NExT monochromator (Renishaw plc, Spectroscopy Product Division, Old Town Wotton-Under-Edge, Gloucestershire, GL12 7DW, UK). The excitation wavelength was 633 nm, as provided by an He–Ne laser (RL633, Renishaw plc, Spectroscopy Product Division, Old Town Wotton-Under-Edge, Gloucestershire, GL12 7DW, UK) with the maximum power of 17 mW. The acquisition time and number of accumulations were adjusted to maximize the signal-to-noise ratio with the minimal sample degradation. All the spectra for the powder samples were measured at several points and then averaged to reduce the anisotropy effect on the Raman spectra and to increase the single-to-noise ratio. The background from the Raman spectra was subtracted by the cubic spline interpolation method. All the spectra were divided by the number of accumulations and acquisition time. The dips in the spectra at wavenumbers of 20.2 cm$^{-1}$ and 302 cm$^{-1}$ are the artefacts of the measurements associated with the presence of dust particles on the NExT monochromator mirrors.

## 3.6. Periodic (Solid-State) DFT Computations

In the CRYSTAL17 calculations [102], we employed the B3LYP [103,104] and PBE [105] functionals with the all-electron Gaussian-type localized orbital basis sets 6-31G(d,p). The London dispersion interactions were taken into account by introducing the D3 correction with Becke–Jones damping (B3LYP-D3 and PBE-D3) and the D2 correction (B3LYP-D2) developed by Grimme et al. [106,107]. In the QuantumExpresso calculations [108,109], we employed PBE with a plane-wave basis set. PAW pseudopotentials with a cut-off energy of 100 Ry were used [110]. The London dispersion interactions were taken into account by introducing the D3 correction with Becke–Jones damping (PBE-D3). In one series of calculations, the space groups and the unit cell parameters of the crystals obtained from the X-ray diffraction experiment were fixed, and the structural relaxations were limited to the positional parameters of the atoms (AtomOnly). In the other series, the optimization was also performed by the cell parameters without cell volume restrictions (FullOpt). The symmetry of crystals was kept during all computations.

The crystal lattice energy $E_{latt}$ of the $n$-component crystal was estimated from periodic DFT as the difference between the sum of total electronic energies of relaxed isolated species $E_{mol}$ and the total energy of the crystal $E_{cry}$ calculated per asymmetric unit [111]

$$E_{latt} = \sum_{i=1}^{n} E_{mol,i} - \frac{E_{cry}}{Z} \qquad (1)$$

Equation (1) was used in PBE/PW calculations. In the case of GTO basis sets, the basis set superposition error (BSSE) [112] was taken into account.

$$E_{latt} = \sum_{i=1}^{n} (E_{mol,i} + BSSE_i) - \frac{E_{cry}}{Z} \qquad (2)$$

Further details of the calculations are given in Section S1 of the Supplementary Materials.

## 4. Conclusions

In this work, we investigated the influence of cell parameter optimization on the $E_{latt}$ value, as well as the structural and spectroscopic properties of the new two-component carbendazim maleate crystal. The sign and absolute value of the relative change in the cell volume of the crystal depends on the functional and type of Grimme correction. Some properties of the considered crystal (metric parameters of short/ionic H-bonds, low-frequency Raman spectra) strongly depend on the changes in the cell volume, while other properties (lattice energy $E_{latt}$, infrared spectra in the 400–150 cm$^{-1}$ frequency region) are weakly related to these variations.

Optimization of the cell parameters of [CRB + MLE] (1:1) and crystals made up of its constituents greatly affects the wavenumber of the lowest Raman-active vibration, the number of Raman active vibrations below 100 cm$^{-1}$, and the relative intensity of these vibrations. B3LYP and PBE-D3 with fixed cell parameters provide a reasonable description of the low-frequency Raman spectra of the considered molecular crystals. This applies both to the wave numbers and Raman intensities. B3LYP and PBE-D3 with modest basis sets and fixed cell parameters can be recommended for evaluation of the structure, H-bond pattern, and infrared/Raman spectra of multicomponent pharmaceutical crystals.

The applicability of different DFT approximations to the $E_{latt}$ calculation of the pharmaceutical salt of carbendazim and maleic acid was examined. It is shown that the existing methods for the calculation of $E_{latt}$ require further developments for solving the problem of accounting for BSSE for organic salts and uniform description of the species in crystals (closed-shell organic ions) and gas phase (neutral organic molecules). It is shown that optimization of the cell parameters does not lead to a noticeable change in the $E_{latt}$ value despite the significant variation of the cell volume. The use of the GTO and PW basis sets in the PBE-D3 approximation leads to close $E_{latt}$ values.

The PBE-D3 method with modest basis sets and fixed cell parameters provides a reasonable trade-off between the accuracy and the computational cost in evaluation of a number of relevant properties of multicomponent pharmaceutical crystals.

**Supplementary Materials:** The following are available online: Section S1. Computational details; Table S1. Crystallographic data for [CRB + MLE] (1:1); Table S2. Experimental metric parameters and interaction energies of conventional intermolecular hydrogen bonds in [CRB + MLE] (1:1); Table S3. Experimental vs. theoretical distances between heavy atoms involved in the formation of nonconventional hydrogen bonds and C-H-O bond angles in [CRB + MLE] (1:1); Table S4. Tentative assignment of the bands in the IR spectrum of the pure CRB crystal below 400 cm$^{-1}$; Table S5. Calculated values of low frequency IR or Raman active phonons in [CRB + MLE] (1:1) and the volume of the crystallographic cell; Table S6. Volume of the crystallographic cell of the CRB and MLE crystals after optimization; Figure S1. Part of the crystal structure with notable non-conventional hydrogen bonds; Figure S2. Experimental and theoretical IR spectra in the low-frequency range; Figure S3. Experimental and theoretical IR spectra of [CRB + MLE] (1:1) in high-frequency and mid-frequency range; Figure S4. Experimental vs. theoretical [B3LYP-D2(FullOpt)] Raman spectrum of [CRB + MLE] (1:1); Figure S5. Experimental vs. theoretical [PBE-D3(FullOpt)] Raman spectrum of [CRB + MLE] (1:1); Figure S6. Experimental vs. theoretical [PBE-D3(AtomOnly) and PBE-D3(FullOpt)] Raman spectrum of MLE; Figure S7. Experimental vs. theoretical [B3LYP-D2(FullOpt)] Raman spectrum of MLE; Figure S8. Experimental Raman spectrum of CRB; Figure S9. Experimental vs. theoretical [B3LYP-D2(FullOpt)] Raman spectrum of CRB in the THz region; Figure S10. Experimental vs. theoretical [PBE-D3(AtomOnly) and PBE-D3(FullOpt)] Raman spectrum of CRB in the THz region; Figure S11. Results of DSC analysis of [CRB + MLE] (1:1) and its pure components; Figure S12. Powder X-Ray diffraction patterns for pure CRB, MLE, and [CRB + MLE] (1:1) (experimental and theoretical).

**Author Contributions:** Conceptualization, A.P.V., A.O.S., and M.V.V.; experimental methodology, A.O.S. and A.P.V.; theoretical methodology, M.V.V., A.P.V., and A.A.R.; investigation, A.P.V., A.O.S., A.V.C., O.D.P., A.A.R., and M.V.V.; single-crystal XRD experiment, A.V.C.; Raman spectroscopy, O.D.P.; PW computations, A.A.R.; writing and visualization, A.P.V., A.O.S., and M.V.V.; supervision, M.V.V. and A.O.S.; project administration, A.O.S.; funding acquisition, A.O.S. All authors have read and agreed to the published version of the manuscript.

**Funding:** The work was supported by the Russian Scientific Foundation (project No. 19-73-10005).

**Acknowledgments:** We thank Matvey S. Gruzdev (Institute of Solution Chemistry of Russian Academy of Sciences) for performing the FT-IR spectroscopic experiment in the low-frequency range. M.V.V. thanks Oleg G. Kharlanov (Faculty of Physics, Lomonosov Moscow State University) for drawing his attention to the relationship between the variation of the cell parameters and the low-frequency Raman spectrum of the crystal. We thank "the Upper Volga Region Centre of Physicochemical Research" for technical assistance with PXRD and FT-IR experiments.

**Conflicts of Interest:** The authors declare no conflict of interest.

### References

1. Paulekuhn, G.S.; Dressman, J.B.; Saal, C. Trends in Active Pharmaceutical Ingredient Salt Selection based on Analysis of the Orange Book Database. *J. Med. Chem.* **2007**, *50*, 6665–6672. [CrossRef] [PubMed]
2. Marder, S.R.; Perry, J.W.; Schaefer, W.P. Synthesis of Organic Salts with Large Second-Order Optical Nonlinearities. *Science* **1989**, *245*, 626–628. [CrossRef] [PubMed]
3. Zhao, D.; Wu, M.; Kou, Y.; Min, E. Ionic liquids: Applications in catalysis. *Catal. Today* **2002**, *74*, 157–189. [CrossRef]
4. Kubisa, P. Application of ionic liquids as solvents for polymerization processes. *Prog. Polym. Sci.* **2004**, *29*, 3–12. [CrossRef]
5. Otero-de-la-Roza, A.; Johnson, E.R. A benchmark for non-covalent interactions in solids. *J. Chem. Phys.* **2012**, *137*, 054103. [CrossRef]
6. Reilly, A.M.; Tkatchenko, A. Understanding the role of vibrations, exact exchange, and many-body van der Waals interactions in the cohesive properties of molecular crystals. *J. Chem. Phys.* **2013**, *139*, 024705. [CrossRef]

7. Brandenburg, J.G.; Alessio, M.; Civalleri, B.; Peintinger, M.F.; Bredow, T.; Grimme, S. Geometrical Correction for the Inter- and Intramolecular Basis Set Superposition Error in Periodic Density Functional Theory Calculations. *J. Phys. Chem. A* **2013**, *117*, 9282–9292. [CrossRef]
8. Červinka, C.; Fulem, M.; Růžička, K. CCSD(T)/CBS fragment-based calculations of lattice energy of molecular crystals. *J. Chem. Phys.* **2016**, *144*, 064505. [CrossRef]
9. Cutini, M.; Civalleri, B.; Corno, M.; Orlando, R.; Brandenburg, J.G.; Maschio, L.; Ugliengo, P. Assessment of Different Quantum Mechanical Methods for the Prediction of Structure and Cohesive Energy of Molecular Crystals. *J. Chem. Theory Comput.* **2016**, *12*, 3340–3352. [CrossRef]
10. Wang, K.; Mishra, M.K.; Sun, C.C. Exceptionally Elastic Single-Component Pharmaceutical Crystals. *Chem. Mater.* **2019**, *31*, 1794–1799. [CrossRef]
11. López-Mejías, V.; Kampf, J.W.; Matzger, A.J. Nonamorphism in Flufenamic Acid and a New Record for a Polymorphic Compound with Solved Structures. *J. Am. Chem. Soc.* **2012**, *134*, 9872–9875. [CrossRef] [PubMed]
12. Greenwell, C.; McKinley, J.L.; Zhang, P.; Zeng, Q.; Sun, G.; Li, B.; Wen, S.; Beran, G.J.O. Overcoming the difficulties of predicting conformational polymorph energetics in molecular crystals via correlated wavefunction methods. *Chem. Sci.* **2020**, *11*, 2200–2214. [CrossRef] [PubMed]
13. Geng, W.-J.; Ma, Q.; Chen, Y.; Yang, W.; Jia, Y.-F.; Li, J.-S.; Zhang, Z.-Q.; Fan, G.-J.; Wang, S.-M. Structure–Performance Relationship in Thermally Stable Energetic Materials: Tunable Physical Properties of Benzopyridotetraazapentalene by Incorporating Amino Groups, Hydrogen Bonding, and π–π Interactions. *Cryst. Growth Des.* **2020**, *20*, 2106–2114. [CrossRef]
14. Srirambhatla, V.K.; Guo, R.; Dawson, D.M.; Price, S.L.; Florence, A.J. Reversible, Two-Step Single-Crystal to Single-Crystal Phase Transitions between Desloratadine Forms I, II, and III. *Cryst. Growth Des.* **2020**, *20*, 1800–1810. [CrossRef]
15. Usta, H.; Kim, D.; Ozdemir, R.; Zorlu, Y.; Kim, S.; Ruiz Delgado, M.C.; Harbuzaru, A.; Kim, S.; Demirel, G.; Hong, J.; et al. High Electron Mobility in [1]Benzothieno [3,2-b][1]benzothiophene-Based Field-Effect Transistors: Toward n-Type BTBTs. *Chem. Mater.* **2019**, *31*, 5254–5263. [CrossRef]
16. Komissarova, E.A.; Dominskiy, D.I.; Zhulanov, V.E.; Abashev, G.G.; Siddiqui, A.; Singh, S.P.; Sosorev, A.Y.; Paraschuk, D.Y. Unraveling the unusual effect of fluorination on crystal packing in an organic semiconductor. *Phys. Chem. Chem. Phys.* **2020**, *22*, 1665–1673. [CrossRef]
17. Zhu, L.; Yi, Y.; Li, Y.; Kim, E.-G.; Coropceanu, V.; Brédas, J.-L. Prediction of Remarkable Ambipolar Charge-Transport Characteristics in Organic Mixed-Stack Charge-Transfer Crystals. *J. Am. Chem. Soc.* **2012**, *134*, 2340–2347. [CrossRef]
18. Manin, A.N.; Voronin, A.P.; Shishkina, A.V.; Vener, M.V.; Churakov, A.V.; Perlovich, G.L. Influence of Secondary Interactions on the Structure, Sublimation Thermodynamics, and Solubility of Salicylate:4-Hydroxybenzamide Cocrystals. Combined Experimental and Theoretical Study. *J. Phys. Chem. B* **2015**, *119*, 10466–10477. [CrossRef]
19. Sharada, D.; Saha, A.; Saha, B.K. Charge transfer complexes as colour changing and disappearing–reappearing colour materials. *New J. Chem.* **2019**, *43*, 7562–7566. [CrossRef]
20. Vishweshwar, P.; Jagadeesh Babu, N.; Nangia, A.; Mason, S.A.; Puschmann, H.; Mondal, R.; Howard, J.A.K. Variable Temperature Neutron Diffraction Analysis of a Very Short O–H···O Hydrogen Bond in 2,3,5,6-Pyrazinetetracarboxylic Acid Dihydrate: Synthon-Assisted Short Oacid–H···Owater Hydrogen Bonds in a Multicenter Array. *J. Phys. Chem. A* **2004**, *108*, 9406–9416. [CrossRef]
21. Vener, M.V.; Manaev, A.V.; Egorova, A.N.; Tsirelson, V.G. QTAIM Study of Strong H-Bonds with the O−H···A Fragment (A = O, N) in Three-Dimensional Periodical Crystals. *J. Phys. Chem. A* **2007**, *111*, 1155–1162. [CrossRef] [PubMed]
22. Surov, A.O.; Voronin, A.P.; Vener, M.V.; Churakov, A.V.; Perlovich, G.L. Specific features of supramolecular organisation and hydrogen bonding in proline cocrystals: A case study of fenamates and diclofenac. *CrystEngComm* **2018**, *20*, 6970–6981. [CrossRef]
23. Gavezzotti, A. Calculation of Intermolecular Interaction Energies by Direct Numerical Integration over Electron Densities. 2. An Improved Polarization Model and the Evaluation of Dispersion and Repulsion Energies. *J. Phys. Chem. B* **2003**, *107*, 2344–2353. [CrossRef]

24. Mackenzie, C.F.; Spackman, P.R.; Jayatilaka, D.; Spackman, M.A. CrystalExplorer model energies and energy frameworks: Extension to metal coordination compounds, organic salts, solvates and open-shell systems. *IUCrJ* **2017**, *4*, 575–587. [CrossRef] [PubMed]
25. Day, G.M. Current approaches to predicting molecular organic crystal structures. *Crystallogr. Rev.* **2011**, *17*, 3–52. [CrossRef]
26. Chickos, J.S.; Gavezzotti, A. Sublimation Enthalpies of Organic Compounds: A Very Large Database with a Match to Crystal Structure Determinations and a Comparison with Lattice Energies. *Cryst. Growth Des.* **2019**, *19*, 6566–6576. [CrossRef]
27. Colombo, V.; Presti, L.L.; Gavezzotti, A. Two-component organic crystals without hydrogen bonding: Structure and intermolecular interactions in bimolecular stacking. *CrystEngComm* **2017**, *19*, 2413–2423. [CrossRef]
28. Manin, A.N.; Voronin, A.P.; Manin, N.G.; Vener, M.V.; Shishkina, A.V.; Lermontov, A.S.; Perlovich, G.L. Salicylamide Cocrystals: Screening, Crystal Structure, Sublimation Thermodynamics, Dissolution, and Solid-State DFT Calculations. *J. Phys. Chem. B* **2014**, *118*, 6803–6814. [CrossRef]
29. King, M.D.; Buchanan, W.D.; Korter, T.M. Identification and Quantification of Polymorphism in the Pharmaceutical Compound Diclofenac Acid by Terahertz Spectroscopy and Solid-State Density Functional Theory. *Anal. Chem.* **2011**, *83*, 3786–3792. [CrossRef]
30. Bedoya-Martínez, N.; Schrode, B.; Jones, A.O.F.; Salzillo, T.; Ruzié, C.; Demitri, N.; Geerts, Y.H.; Venuti, E.; Della Valle, R.G.; Zojer, E.; et al. DFT-Assisted Polymorph Identification from Lattice Raman Fingerprinting. *J. Phys. Chem. Lett.* **2017**, *8*, 3690–3695. [CrossRef]
31. Levina, E.O.; Chernyshov, I.Y.; Voronin, A.P.; Alekseiko, L.N.; Stash, A.I.; Vener, M.V. Solving the enigma of weak fluorine contacts in the solid state: A periodic DFT study of fluorinated organic crystals. *Rsc. Adv.* **2019**, *9*, 12520–12537. [CrossRef]
32. Vener, M.V.; Medvedev, A.G.; Churakov, A.V.; Prikhodchenko, P.V.; Tripol'skaya, T.A.; Lev, O. H-Bond Network in Amino Acid Cocrystals with $H_2O$ or $H_2O_2$. The DFT Study of Serine–H2O and Serine–$H_2O_2$. *J. Phys. Chem. A* **2011**, *115*, 13657–13663. [CrossRef] [PubMed]
33. Červinka, C.; Fulem, M. Cohesive properties of the crystalline phases of twenty proteinogenic α-aminoacids from first-principles calculations. *Phys. Chem. Chem. Phys.* **2019**, *21*, 18501–18515. [CrossRef] [PubMed]
34. Tukachev, N.V.; Maslennikov, D.R.; Sosorev, A.Y.; Tretiak, S.; Zhugayevych, A. Ground-State Geometry and Vibrations of Polyphenylenevinylene Oligomers. *J. Phys. Chem. Lett.* **2019**, *10*, 3232–3239. [CrossRef]
35. Tosoni, S.; Tuma, C.; Sauer, J.; Civalleri, B.; Ugliengo, P. A comparison between plane wave and Gaussian-type orbital basis sets for hydrogen bonded systems: Formic acid as a test case. *J. Chem. Phys.* **2007**, *127*, 154102. [CrossRef]
36. Drużbicki, K.; Mikuli, E.; Pałka, N.; Zalewski, S.; Ossowska-Chruściel, M.D. Polymorphism of Resorcinol Explored by Complementary Vibrational Spectroscopy (FT-RS, THz-TDS, INS) and First-Principles Solid-State Computations (Plane-Wave DFT). *J. Phys. Chem. B* **2015**, *119*, 1681–1695. [CrossRef]
37. Zvereva, E.E.; Shagidullin, A.R.; Katsyuba, S.A. Ab Initio and DFT Predictions of Infrared Intensities and Raman Activities. *J. Phys. Chem. A* **2011**, *115*, 63–69. [CrossRef]
38. Katsyuba, S.A.; Zvereva, E.E.; Burganov, T.I. Is There a Simple Way to Reliable Simulations of Infrared Spectra of Organic Compounds? *J. Phys. Chem. A* **2013**, *117*, 6664–6670. [CrossRef]
39. Churakov, A.V.; Prikhodchenko, P.V.; Lev, O.; Medvedev, A.G.; Tripol'skaya, T.A.; Vener, M.V. A model proton-transfer system in the condensed phase: $NH_4^+ OOH^-$, a crystal with short intermolecular H-bonds. *J. Chem. Phys.* **2010**, *133*, 164506. [CrossRef]
40. Tuma, C.; Daniel Boese, A.; Handy, N.C. Predicting the binding energies of H-bonded complexes: A comparative DFT study. *Phys. Chem. Chem. Phys.* **1999**, *1*, 3939–3947. [CrossRef]
41. Vener, M.V.; Levina, E.O.; Koloskov, O.A.; Rykounov, A.A.; Voronin, A.P.; Tsirelson, V.G. Evaluation of the Lattice Energy of the Two-Component Molecular Crystals Using Solid-State Density Functional Theory. *Cryst. Growth Des.* **2014**, *14*, 4997–5003. [CrossRef]
42. Shishkina, A.V.; Zhurov, V.V.; Stash, A.I.; Vener, M.V.; Pinkerton, A.A.; Tsirelson, V.G. Noncovalent Interactions in Crystalline Picolinic Acid N-Oxide: Insights from Experimental and Theoretical Charge Density Analysis. *Cryst. Growth Des.* **2013**, *13*, 816–828. [CrossRef]
43. Akin-Ojo, O.; Wang, F. Effects of the dispersion interaction in liquid water. *Chem. Phys. Lett.* **2011**, *513*, 59–62. [CrossRef]

44. Katsyuba, S.A.; Vener, M.V.; Zvereva, E.E.; Brandenburg, J.G. The role of London dispersion interactions in strong and moderate intermolecular hydrogen bonds in the crystal and in the gas phase. *Chem. Phys. Lett.* **2017**, *672*, 124–127. [CrossRef]
45. Voronin, A.P.; Perlovich, G.L.; Vener, M.V. Effects of the crystal structure and thermodynamic stability on solubility of bioactive compounds: DFT study of isoniazid cocrystals. *Comput. Theor. Chem.* **2016**, *1092*, 1–11. [CrossRef]
46. Mohaček-Grošev, V.; Grdadolnik, J.; Stare, J.; Hadži, D. Identification of hydrogen bond modes in polarized Raman spectra of single crystals of α-oxalic acid dihydrate. *J. Raman Spectrosc.* **2009**, *40*, 1605–1614. [CrossRef]
47. Yushina, I.D.; Kolesov, B.A.; Bartashevich, E.V. Raman spectroscopy study of new thia- and oxazinoquinolinium triodides. *New J. Chem.* **2015**, *39*, 6163–6170. [CrossRef]
48. Katsyuba, S.A.; Vener, M.V.; Zvereva, E.E.; Fei, Z.; Scopelliti, R.; Laurenczy, G.; Yan, N.; Paunescu, E.; Dyson, P.J. How Strong Is Hydrogen Bonding in Ionic Liquids? Combined X-ray Crystallographic, Infrared/Raman Spectroscopic, and Density Functional Theory Study. *J. Phys. Chem. B* **2013**, *117*, 9094–9105. [CrossRef]
49. Bolla, G.; Nangia, A. Novel pharmaceutical salts of albendazole. *CrystEngComm* **2018**, *20*, 6394–6405. [CrossRef]
50. Chen, J.-M.; Wang, Z.-Z.; Wu, C.-B.; Li, S.; Lu, T.-B. Crystal engineering approach to improve the solubility of mebendazole. *CrystEngComm* **2012**, *14*, 6221–6229. [CrossRef]
51. Chen, J.; Lu, T. New Crystalline Forms of Mebendazole with n-Alkyl Carboxylic Acids: Neutral and Ionic Status. *Chin. J. Chem.* **2013**, *31*, 635–640. [CrossRef]
52. Musin, R.N.; Mariam, Y.H. An integrated approach to the study of intramolecular hydrogen bonds in malonaldehyde enol derivatives and naphthazarin: Trend in energetic versus geometrical consequences. *J. Phys. Org. Chem.* **2006**, *19*, 425–444. [CrossRef]
53. Filarowski, A.; Koll, A.; Sobczyk, L. Intramolecular Hydrogen Bonding in o-hydroxy Aryl Schiff Bases. *Curr. Org. Chem.* **2009**, *13*, 172–193. [CrossRef]
54. Rozenberg, M.; Loewenschuss, A.; Marcus, Y. An empirical correlation between stretching vibration redshift and hydrogen bond length. *Phys. Chem. Chem. Phys.* **2000**, *2*, 2699–2702. [CrossRef]
55. Etter, M.C. Encoding and decoding hydrogen-bond patterns of organic compounds. *Acc. Chem. Res.* **1990**, *23*, 120–126. [CrossRef]
56. Nyman, J.; Day, G.M. Static and lattice vibrational energy differences between polymorphs. *CrystEngComm* **2015**, *17*, 5154–5165. [CrossRef]
57. Karamertzanis, P.G.; Day, G.M.; Welch, G.W.A.; Kendrick, J.; Leusen, F.J.J.; Neumann, M.A.; Price, S.L. Modeling the interplay of inter- and intramolecular hydrogen bonding in conformational polymorphs. *J. Chem. Phys.* **2008**, *128*, 244708. [CrossRef]
58. Juliano, T.R.; Korter, T.M. Terahertz Vibrations of Crystalline Acyclic and Cyclic Diglycine: Benchmarks for London Force Correction Models. *J. Phys. Chem. A* **2013**, *117*, 10504–10512. [CrossRef]
59. Stein, M.; Heimsaat, M. Intermolecular Interactions in Molecular Organic Crystals upon Relaxation of Lattice Parameters. *Crystals* **2019**, *9*, 665. [CrossRef]
60. Barone, V.; Biczysko, M.; Bloino, J. Fully anharmonic IR and Raman spectra of medium-size molecular systems: Accuracy and interpretation. *PCCP* **2014**, *16*, 1759–1787. [CrossRef]
61. Vener, M.V.; Sauer, J. Quantum anharmonic frequencies of the O$\cdots$H$\cdots$O fragment of the H5O2$^+$ ion: A model three-dimensional study. *Chem. Phys. Lett.* **1999**, *312*, 591–597. [CrossRef]
62. Brela, M.; Stare, J.; Pirc, G.; Sollner-Dolenc, M.; Boczar, M.; Wójcik, M.J.; Mavri, J. Car–Parrinello Simulation of the Vibrational Spectrum of a Medium Strong Hydrogen Bond by Two-Dimensional Quantization of the Nuclear Motion: Application to 2-Hydroxy-5-nitrobenzamide. *J. Phys. Chem. B* **2012**, *116*, 4510–4518. [CrossRef] [PubMed]
63. Brela, M.Z.; Wójcik, M.J.; Witek, Ł.J.; Boczar, M.; Wrona, E.; Hashim, R.; Ozaki, Y. Born–Oppenheimer Molecular Dynamics Study on Proton Dynamics of Strong Hydrogen Bonds in Aspirin Crystals, with Emphasis on Differences between Two Crystal Forms. *J. Phys. Chem. B* **2016**, *120*, 3854–3862. [CrossRef] [PubMed]
64. Sushchinskiĭ, M.M. *Raman Spectra of Molecules and Crystals*; Wiley: New York, NY, USA, 1972.
65. Takahashi, M. Terahertz Vibrations and Hydrogen-Bonded Networks in Crystals. *Crystals* **2014**, *4*, 74–103. [CrossRef]

66. Zhang, F.; Wang, H.-W.; Tominaga, K.; Hayashi, M. Mixing of intermolecular and intramolecular vibrations in optical phonon modes: Terahertz spectroscopy and solid-state density functional theory. *Wiley Interdiscip. Rev. Comput. Mol. Sci.* **2016**, *6*, 386–409. [CrossRef]
67. Sakun, V.P.; Vener, M.V.; Sokolov, N.D. Proton tunneling assisted by the intermolecular vibration excitation. Temperature dependence of the proton spin-lattice relaxation time in benzoic acid powder. *J. Chem. Phys.* **1996**, *105*, 379–387. [CrossRef]
68. Yang, J.; Li, S.; Zhao, H.; Song, B.; Zhang, G.; Zhang, J.; Zhu, Y.; Han, J. Molecular Recognition and Interaction between Uracil and Urea in Solid-State Studied by Terahertz Time-Domain Spectroscopy. *J. Phys. Chem. A* **2014**, *118*, 10927–10933. [CrossRef]
69. Takahashi, M.; Okamura, N.; Ding, X.; Shirakawa, H.; Minamide, H. Intermolecular hydrogen bond stretching vibrations observed in terahertz spectra of crystalline vitamins. *CrystEngComm* **2018**, *20*, 1960–1969. [CrossRef]
70. Jepsen, P.U.; Clark, S.J. Precise ab-initio prediction of terahertz vibrational modes in crystalline systems. *Chem. Phys. Lett.* **2007**, *442*, 275–280. [CrossRef]
71. Ruggiero, M.T. Invited Review: Modern Methods for Accurately Simulating the Terahertz Spectra of Solids. *J. InfraredMillim. Terahertz Waves* **2020**. [CrossRef]
72. Vener, M.V.; Sauer, J. Environmental effects on vibrational proton dynamics in $H_5O_2^+$: DFT study on crystalline H5O2$^+$ ClO$_4^-$. *Phys. Chem. Chem. Phys.* **2005**, *7*, 258–263. [CrossRef] [PubMed]
73. Shishkina, A.; Stash, A.; Civalleri, B.; Ellern, A.; Tsirelson, V. Electron-density and electrostatic-potential features of orthorhombic chlorine trifluoride. *Mendeleev Commun.* **2010**, *20*, 161–164. [CrossRef]
74. Sen, A.; Mitev, P.D.; Eriksson, A.; Hermansson, K. H-bond and electric field correlations for water in highly hydrated crystals. *Int. J. Quantum Chem.* **2016**, *116*, 67–80. [CrossRef]
75. King, M.D.; Buchanan, W.D.; Korter, T.M. Application of London-type dispersion corrections to the solid-state density functional theory simulation of the terahertz spectra of crystalline pharmaceuticals. *Phys. Chem. Chem. Phys.* **2011**, *13*, 4250–4259. [CrossRef]
76. Kim, J.; Kwon, O.P.; Jazbinsek, M.; Park, Y.C.; Lee, Y.S. First Principles Calculation of Terahertz Absorption with Dispersion Correction of 2,2′-Bithiophene as Model Compound. *J. Phys. Chem. C* **2015**, *119*, 12598–12607. [CrossRef]
77. Fonari, A.; Corbin, N.S.; Vermeulen, D.; Goetz, K.P.; Jurchescu, O.D.; McNeil, L.E.; Bredas, J.L.; Coropceanu, V. Vibrational properties of organic donor-acceptor molecular crystals: Anthracene-pyromellitic-dianhydride (PMDA) as a case study. *J. Chem. Phys.* **2015**, *143*, 224503. [CrossRef]
78. Deringer, V.L.; George, J.; Dronskowski, R.; Englert, U. Plane-Wave Density Functional Theory Meets Molecular Crystals: Thermal Ellipsoids and Intermolecular Interactions. *Acc. Chem. Res.* **2017**, *50*, 1231–1239. [CrossRef]
79. Beran, G.J.O.; Nanda, K. Predicting Organic Crystal Lattice Energies with Chemical Accuracy. *J. Phys. Chem. Lett.* **2010**, *1*, 3480–3487. [CrossRef]
80. Moellmann, J.; Grimme, S. DFT-D3 Study of Some Molecular Crystals. *J. Phys. Chem. C* **2014**, *118*, 7615–7621. [CrossRef]
81. Yang, J.; Hu, W.; Usvyat, D.; Matthews, D.; Schütz, M.; Chan, G.K.-L. Ab initio determination of the crystalline benzene lattice energy to sub-kilojoule/mole accuracy. *Science* **2014**, *345*, 640–643. [CrossRef]
82. Červinka, C.; Fulem, M. State-of-the-Art Calculations of Sublimation Enthalpies for Selected Molecular Crystals and Their Computational Uncertainty. *J. Chem. Theory Comput.* **2017**, *13*, 2840–2850. [CrossRef] [PubMed]
83. Gharagheizi, F.; Ilani-Kashkouli, P.; Acree, W.E.; Mohammadi, A.H.; Ramjugernath, D. A group contribution model for determining the sublimation enthalpy of organic compounds at the standard reference temperature of 298K. *Fluid Phase Equilibria* **2013**, *354*, 265–285. [CrossRef]
84. Bygrave, P.J.; Allan, N.L.; Manby, F.R. The embedded many-body expansion for energetics of molecular crystals. *J. Chem. Phys.* **2012**, *137*, 164102. [CrossRef] [PubMed]
85. McDonagh, J.L.; Palmer, D.S.; Mourik, T.V.; Mitchell, J.B.O. Are the Sublimation Thermodynamics of Organic Molecules Predictable? *J. Chem. Inf. Modeling* **2016**, *56*, 2162–2179. [CrossRef]
86. Perlovich, G.L.; Raevsky, O.A. Sublimation of Molecular Crystals: Prediction of Sublimation Functions on the Basis of HYBOT Physicochemical Descriptors and Structural Clusterization. *Cryst. Growth Des.* **2010**, *10*, 2707–2712. [CrossRef]

87. Grabowski, S.J. What Is the Covalency of Hydrogen Bonding? *Chem. Rev.* **2011**, *111*, 2597–2625. [CrossRef]
88. Landeros-Rivera, B.; Moreno-Esparza, R.; Hernández-Trujillo, J. Theoretical study of intermolecular interactions in crystalline arene–perhaloarene adducts in terms of the electron density. *Rsc. Adv.* **2016**, *6*, 77301–77309. [CrossRef]
89. Surov, A.O.; Churakov, A.V.; Perlovich, G.L. Three Polymorphic Forms of Ciprofloxacin Maleate: Formation Pathways, Crystal Structures, Calculations, and Thermodynamic Stability Aspects. *Cryst. Growth Des.* **2016**, *16*, 6556–6567. [CrossRef]
90. Jarzembska, K.N.; Hoser, A.A.; Varughese, S.; Kamiński, R.; Malinska, M.; Stachowicz, M.; Pedireddi, V.R.; Woźniak, K. Structural and Energetic Analysis of Molecular Assemblies in a Series of Nicotinamide and Pyrazinamide Cocrystals with Dihydroxybenzoic Acids. *Cryst. Growth Des.* **2017**, *17*, 4918–4931. [CrossRef]
91. Tao, Q.; Hao, Q.-Q.; Voronin, A.P.; Dai, X.-L.; Huang, Y.; Perlovich, G.L.; Lu, T.-B.; Chen, J.-M. Polymorphic Forms of a Molecular Salt of Phenazopyridine with 3,5-Dihydroxybenzoic Acid: Crystal Structures, Theoretical Calculations, Thermodynamic Stability, and Solubility Aspects. *Cryst. Growth Des.* **2019**, *19*, 5636–5647. [CrossRef]
92. Kong, S.; Shenderovich, I.G.; Vener, M.V. Density Functional Study of the Proton Transfer Effect on Vibrations of Strong (Short) Intermolecular O–H···N/O–····H–N$^+$ Hydrogen Bonds in Aprotic Solvents. *J. Phys. Chem. A* **2010**, *114*, 2393–2399. [CrossRef] [PubMed]
93. Dai, X.-L.; Voronin, A.P.; Huang, Y.-L.; Perlovich, G.L.; Zhao, X.-H.; Lu, T.-B.; Chen, J.-M. 5-Fluorouracil Cocrystals with Lipophilic Hydroxy-2-Naphthoic Acids: Crystal Structures, Theoretical Computations, and Permeation Studies. *Cryst. Growth Des.* **2020**, *20*, 923–933. [CrossRef]
94. Manin, A.N.; Voronin, A.P.; Drozd, K.V.; Churakov, A.V.; Perlovich, G.L. Pharmaceutical salts of emoxypine with dicarboxylic acids. *Acta Crystallogr. Sect. C* **2018**, *74*, 797–806. [CrossRef] [PubMed]
95. Preiss, U.P.; Zaitsau, D.H.; Beichel, W.; Himmel, D.; Higelin, A.; Merz, K.; Caesar, N.; Verevkin, S.P. Estimation of Lattice Enthalpies of Ionic Liquids Supported by Hirshfeld Analysis. *Chem. Phys. Chem.* **2015**, *16*, 2890–2898. [CrossRef]
96. Mohamed, S.; Tocher, D.A.; Price, S.L. Computational prediction of salt and cocrystal structures—Does a proton position matter? *Int. J. Pharm.* **2011**, *418*, 187–198. [CrossRef]
97. Chen, L.; Bryantsev, V.S. A density functional theory based approach for predicting melting points of ionic liquids. *Phys. Chem. Chem. Phys.* **2017**, *19*, 4114–4124. [CrossRef]
98. Aakeröy, C.B.; Seddon, K.R.; Leslie, M. Hydrogen-bonding contributions to the lattice energy of salts for second harmonic generation. *Struct. Chem.* **1992**, *3*, 63–65. [CrossRef]
99. Blazejowski, J.; Lubkowski, J. Lattice energy of organic ionic crystals and its importance in analysis of features, behaviour and reactivity of solid-state systems. *J. Therm. Anal.* **1992**, *38*, 2195–2210. [CrossRef]
100. Sheldrick, G. *SADABS, Program for Scaling and Correction of Area Detector Data*; University of Göttingen: Gottingen, Germany, 1997.
101. Sheldrick, G. A short history of SHELX. *Acta Crystallogr. Sect. A: Found. Crystallogr.* **2008**, *64*, 112–122. [CrossRef]
102. Dovesi, R.; Erba, A.; Orlando, R.; Zicovich-Wilson, C.M.; Civalleri, B.; Maschio, L.; Rérat, M.; Casassa, S.; Baima, J.; Salustro, S.; et al. Quantum-mechanical condensed matter simulations with CRYSTAL. *Wiley Interdiscip. Rev. Comput. Mol. Sci.* **2018**, *8*, e1360. [CrossRef]
103. Becke, A.D. Density-functional thermochemistry. III. The role of exact exchange. *J. Chem. Phys.* **1993**, *98*, 5648–5652. [CrossRef]
104. Vosko, S.H.; Wilk, L.; Nusair, M. Accurate spin-dependent electron liquid correlation energies for local spin density calculations: A critical analysis. *Can. J. Phys.* **1980**, *58*, 1200–1211. [CrossRef]
105. Perdew, J.P.; Burke, K.; Ernzerhof, M. Generalized Gradient Approximation Made Simple. *Phys. Rev. Lett.* **1996**, *77*, 3865–3868. [CrossRef] [PubMed]
106. Grimme, S.; Antony, J.; Ehrlich, S.; Krieg, H. A consistent and accurate ab initio parametrization of density functional dispersion correction (DFT-D) for the 94 elements H-Pu. *J. Chem. Phys.* **2010**, *132*, 154104. [CrossRef] [PubMed]
107. Grimme, S.; Ehrlich, S.; Goerigk, L. Effect of the damping function in dispersion corrected density functional theory. *J. Comput. Chem.* **2011**, *32*, 1456–1465. [CrossRef]

108. Giannozzi, P.; Baroni, S.; Bonini, N.; Calandra, M.; Car, R.; Cavazzoni, C.; Ceresoli, D.; Chiarotti, G.L.; Cococcioni, M.; Dabo, I.; et al. QUANTUM ESPRESSO: A modular and open-source software project for quantum simulations of materials. *J. Phys. Condens. Matter* **2009**, *21*, 395502. [CrossRef]
109. Giannozzi, P.; Andreussi, O.; Brumme, T.; Bunau, O.; Buongiorno Nardelli, M.; Calandra, M.; Car, R.; Cavazzoni, C.; Ceresoli, D.; Cococcioni, M.; et al. Advanced capabilities for materials modelling with Quantum ESPRESSO. *J. Phys. Condens. Matter* **2017**, *29*, 465901. [CrossRef]
110. Blöchl, P.E. Projector augmented-wave method. *Phys. Rev. B* **1994**, *50*, 17953–17979. [CrossRef]
111. Seaton, C.C.; Parkin, A. Making Benzamide Cocrystals with Benzoic Acids: The Influence of Chemical Structure. *Cryst. Growth Des.* **2011**, *11*, 1502–1511. [CrossRef]
112. Boys, S.F.; Bernardi, F. The calculation of small molecular interactions by the differences of separate total energies. Some procedures with reduced errors. *Mol. Phys.* **1970**, *19*, 553–566. [CrossRef]

**Sample Availability:** Samples of crystalline [CRB + MLE] (1:1), input and output files of the computations are available from the authors.

© 2020 by the authors. Licensee MDPI, Basel, Switzerland. This article is an open access article distributed under the terms and conditions of the Creative Commons Attribution (CC BY) license (http://creativecommons.org/licenses/by/4.0/).

Article

# Catalytic Effect of Hydrogen Bond on Oxhydryl Dehydrogenation in Methanol Steam Reforming on Ni(111)

Changming Ke [1] and Zijing Lin [2,*]

[1] Hefei National Laboratory for Physical Sciences at Microscales, Department of Physics, School of Physical Sciences, University of Science and Technology of China, Hefei 230052, Anhui, China; cmk@mail.ustc.edu.cn

[2] CAS Key Laboratory of Strongly-Coupled Quantum Matter Physics, Department of Physics, School of Physical Sciences, University of Science and Technology of China, Hefei 230052, Anhui, China

* Correspondence: zjlin@ustc.edu.cn; Tel.: +86-551-63606345; Fax: +86-551-63606348

Received: 19 December 2019; Accepted: 25 March 2020; Published: 27 March 2020

**Abstract:** Dehydrogenation of $H_3COH$ and $H_2O$ are key steps of methanol steam reforming on transition metal surfaces. Oxhydryl dehydrogenation reactions of $H_xCOH$ ($x$ = 0–3) and OH on Ni (111) were investigated by DFT calculations with the OptB88-vdW functional. The transition states were searched by the climbing image nudged elastic band method and the dimer method. The activation energies for the dehydrogenation of individual $H_xCOH^*$ are 68 to 91 kJ/mol, and reduced to 12–17 kJ/mol by neighboring OH*. Bader charge analysis showed the catalysis role of OH* can be attributed to the effect of hydrogen bond (H-bond) in maintaining the charge of oxhydryl H in the reaction path. The mechanism of H-bond catalysis was further demonstrated by the study of OH* and N* assisted dehydrogenation of OH*. Due to the universality of H-bond, the H-bond catalysis shown here, is of broad implication for studies of reaction kinetics.

**Keywords:** Reaction mechanism; first-principle calculation; Bader charge analysis; activation energy; transition state structure

## 1. Introduction

Methanol steam reforming (MSR), $H_3COH + H_2O \rightarrow CO_2 + 3H_2$, is frequently used for the generation of $H_2$ at temperatures of 200–300 °C and atmospheric pressure [1–4]. MSR can be catalysed by a number of metals and metal oxides. Cu is the most common commercial catalyst but suffers from pyrophoricity and catalyst sintering, limiting its long-term applications [5,6]. Noble metals, such as Pd and Pt, that possess long term stability and no pyrophoric behavior [3,7] suffer from high price, limiting their large-scale industrial applications. On balance, Ni is a low price and highly effective MSR catalyst. The high activity of Ni for catalyzing MSR at T ≥ 300°C has been demonstrated in recent experiments [5,6], but a convincing theoretical explanation of the reaction mechanism is lacking.

There have been numerous theoretical studies on the reaction mechanism of MSR over a number of catalysts. Lin et al. conducted density functional theory (DFT) calculations and built a kinetic model of MSR on Cu(111) [8]. They found kinetic relevant steps with high activation energies include $H_3COH^*$ + * → $H_3CO^*$ + H* and, where * denotes an active surface site and R* means a surface adsorbed species. R. Zuo et al. discussed the mechanisms of methanol decomposition, methanol oxidation and steam reforming of methanol on Cu(111) [9]. Wang et al. explained the differentiation of intrinsic reactivity of MSR on Cu, CuZn and Cu/ZnO [10]. In addition, they proposed a microkinetic model for more in-depth mechanics research of MSR on Cu [11]. Smith et al. conducted DFT studies on the initial steps of MSR on PdZn and ZnO surfaces, and found defect sites lower the barrier significantly [12]. Also based on DFT calculations, Krajčí et al. demonstrated the CO/$CO_2$ selectivity of MSR on many alloys, e.g., PdZn, PtZn and NiZn [13]. Chen et al. [14], Lausche et al. [15], and Kramer et al. [16] studied the selectivity of the dehydrogenation of methanol on Cu(110), Ni(100) and Ni(111), respectively. MSRs on Pt [16], Pd [16], Pt$_3$Ni alloy [17], Pt-Skinned PtNi Bimetallic Clusters [18] and Co [19] have also been investigated.

Summarizing the existing results of MSR studies, there are two kinds of MSR kinetics. The first was deduced by considering the dehydrogenation of isolate adsorbed H$_x$COH ($x$ = 3, 2, 1, 0). As the activation energy for the bond breaking of H$_3$CO-H* is high, H$_3$COH* + * → H$_3$CO* + H* was found to be a likely rate determining step (RDS) [8,10,11,17]. For example, Wang et al. [11] showed in their DFT study that the activation energy for CH$_3$OH* + * → CH$_3$O* + H* was 103 kJ/mol and MSR on Cu(111) was mostly limited by the dehydrogenation of CH$_3$OH*. The second considered the interaction of H$_3$COH* and OH* where the co-adsorbed OH* significantly reduces the activation energy of H$_3$COH* + OH* → H$_3$CO* + H$_2$O* and the C–H scission steps were found to be rate limiting [9,13,19]. The resulting kinetics with a lower activation energy agrees better with the experiments [20,21], and is a clear improvement of the first one. Unfortunately, the improved understanding has so far been mainly limited to a computational detail based deduction. In-depth understanding based on general physical concept and/or generable mechanism is highly desirable.

This work focuses on the role of hydrogen bond (H-bond) on reducing the activation energies of oxhydryl dehydrogenation that are important for determining the kinetics of MSR on Ni(111), the dominant catalyzing surface of micron-sized commercial Ni catalysts [22]. DFT calculations were performed to examine the oxhydryl dehydrogenation of H$_x$CO–H* ($x$ = 3, 2, 1, 0), with and without the assistance of the co-adsorbed OH*. Combined analysis of Bader charges and transition state structures showed that H-bond is the root cause for the observed high MSR activity. The catalytic role of H-bond was further supported by investigating the dehydrogenation of O–H* assisted by co-adsorbed N*.

## 2. Computational Methods

DFT calculations were performed using Vienna Ab-initio Simulation Package (VASP) [23–26], a plane wave computational software. The projector augmented wave (PAW) method [27,28] was used to describe the electron-ion interaction between core ion and valence electrons. The Kohn-Sham equations were solved with a 380eV cutoff energy for the wavefunctions of valence electrons. The exchange correlation interaction was described by the functional of OptB88-vdW [29,30]. OptB88-vdW was chosen as it best describes the van der Waals (vdW) interaction on metal surface [31]. The computations were performed on a three-layer slab of 3 × 3 unit cell surface model of Ni(111), with a vacuum region of 10 Å thickness. The surface layer of the slab was allowed to relax, while the bottom two layers were fixed. Spin polarization and dipole correction were considered by setting SPIN = 2 and LDIPOL = .True in all calculations. The Brillouin zone was sampled by a 5 × 5 × 1 k-point Monkhorst-Pack grid. All stable structures were optimized with an energy-based conjugate gradient algorithm [32]. The convergence criteria for electronic and ionic energies were $10^{-6}$ eV/atom and $10^{-5}$ eV/atom, respectively. The cutoff energy and the k-point grid were tested to be appropriate, e.g., the differences in the obtained adsorption energy of H$_3$CO are less than 0.4 kJ/mol and 0.7 kJ/mol when compared to a cutoff energy of up to 460 eV and a k-point grid of up to 8 × 8 × 1, respectively.

Saddle points were determined by combining the climbing image nudged elastic band (CL-NEB) method [33] and the dimer method [34]. First, the less computing intensive CL-NEB was used to find the minimum energy path and the transition state. In the CL-NEB calculations, 7 images were inserted between reactants and products, and the electronic energies and the forces were converged to $10^{-4}$ eV/atom and 0.03 eV/Å, respectively. Second, the transition states obtained by the CL-NEB searches were used as the inputs for the high-precision dimer method [34] to find the accurate transition states efficiently. In the dimer method calculations, the electronic energy and force were converged respectively to $10^{-7}$ eV/atom and 0.01 eV/Å.

The Bader charge was calculated by the method of partitioning charge density grids into Bader volumes, as proposed by Henkelman's group [35,36].

## 3. Results and Discussion

### 3.1. Oxhydryl Dehydrogenation of HxCOH*

Oxhydryl dehydrogenation of isolated $H_xCOH^*$ ($x$ = 3, 2, 1, 0), $H_xCOH^* + ^* \rightarrow H_xCO^* + H^*$, and that assisted by co-adsorbed OH, $H_xCOH^* + OH^* \rightarrow H_xCO^* + H_2O^*$, were considered. The activation barriers of the two types of dehydrogenation reactions were compared in Figure 1.

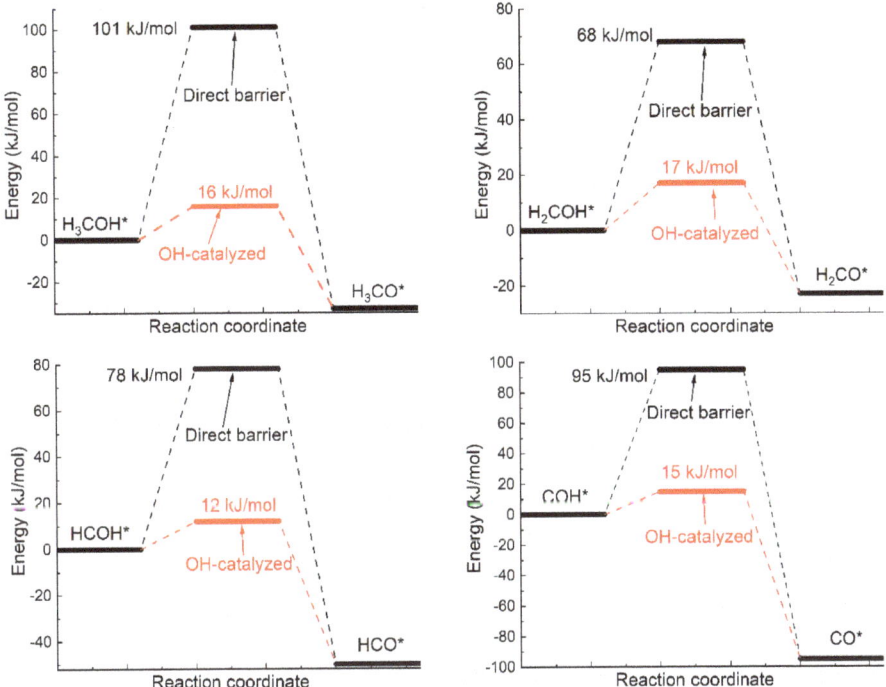

**Figure 1.** Activation energies of dual path dehydrogenation of oxhydryl in $H_xCO$–$H^*$ ($x$ = 3, 2, 1, 0).

Notice that the direct barrier of $H_3COH^*$ dehydrogenation shown Figure 1 is about 91 kJ/mol. In comparison, the barrier was found to range from 39 to 75 kJ/mol in a few early studies [16,37,38]. The difference, due to various factors such as the use of different functionals, surface slab model and transition state search methods, is quite substantial, but is also seen in similar cases. For example, the direct barrier of $H_3COH^*$ dehydrogenation on Cu(111) was found to vary from 62 to 138 kJ/mol by different DFT studies [8–11,39]. That is, the difference known for Cu(111) is comparable to that known for Ni(111). Although it is premature to draw any conclusion, the result here may be preferable due to the demonstrated quality of OptB88-vdW for similar systems [31] and the widely accepted slab model and transition state search method.

As shown in Figure 1, the activation energies ($E_a$'s) of oxhydryl dehydrogenation of $H_xCOH^*$ are reduced from 68–91 kJ/mol for isolated $H_xCOH^*$ to 12–17 kJ/mol for $H_xCOH^*$ with co-adsorbed OH ($x$ = 3, 2, 1, 0). The reduction of $E_a$ for the oxhydryl dehydrogenation of $H_3COH^*$ due to the presence of neighboring $OH^*$ is known in literatures. For example, $E_a$ is reduced from 62 kJ/mol to 32 kJ/mol for Cu(111) [9] and from 80 kJ/mol to 22 kJ/mol for Co(111) [19]. The results here concerning the oxhydryl dehydrogenation of $H_xCOH^*$ for $x$ = 2, 1, 0 indicate that the effect is quite general. The low activation energies mean that all oxhydryl dehydrogenation processes of $H_xCOH^*$ in MSR should be sufficiently fast. Besides, the energy cost for a close proximity of $H_xCOH^*$ and $OH^*$ as compared to isolated adsorbates is low, at 7.7, 14, 11, 5.1 kJ/mol for $x$ = 3, 2, 1, 0, respectively. Therefore, there is no need to consider the oxhydryl dehydrogenation processes of $H_xCOH^*$ when examining the possible RDS in MSR. This result can be used to simplify the elementary reaction step study in many relevant problems. A low $E_a$ for $H_3COH^*$ dehydrogenation is also necessary for the understanding of the high MSR activities of Ni catalysts observed experimentally [5,6].

Notice that, while OH* facilitates the O–H scission process, OH* provides no help for C-H scission. The activation energies for $CH_3O^* + OH^* \rightarrow CH_2O^* + H_2O^*$ and $CH_3OH^* + OH^* \rightarrow CH_2OH^* + H_2O^*$ are 166 and 149 kJ/mol, respectively. Both the activation energies are higher than the corresponding activation energies of 86.8 and 91.5 kJ/mol for $CH_3O^* \rightarrow CH_2O^* + H^*$ and $CH_3OH^* \rightarrow CH_2OH^* + H^*$, respectively. Similar result has also been observed for the C-H scission on PdZn(111) [40]. As reactions prefer the least resistant paths and the fractional coverage of OH* is in the order of 1% and very low coverages for $CH_3OH^*$ and $CH_xO^*$ on Ni(111) [41], the C-H scission is not expected to be adversely impacted by OH*. Due to the high $E_a$ involved in $CH_3O^* \rightarrow CH_2O^* + H^*$ or $CH_3OH^* \rightarrow CH_2OH^* + H^*$. However, the C-H scission step is expected to be rate limiting for MSR on Ni(111).

To reveal the common feature of oxhydryl dehydrogenation in different $H_xCOH^*$, Figure 2 shows the structures and Bader charges of $H_xCOH^*$, with and without OH* co-adsorption, at their initial local minimum geometries and reaction transition states. As seen in Figure 2, the Bader charges of oxhydryl H for isolated $H_xCOH^*$ at their local minimum and transition state structures are on average +0.63 e and +0.14 e, respectively. Clearly, the Bader charges of oxhydryl H of isolated $H_xCOH^*$ at their local minimum and transition state structures are quite different. That is, a significant charge density redistribution is required in the reaction path going from a local minimum energy structure to a transition state configuration. A large electronic energy changes due to the orbital reorganization, or correspondingly a high energy barrier, is expected in the O–H scission process. In comparison, the Bader charges for oxhydryl H of $H_xCOH^*$ with co-adsorbed OH* are on average +0.59 e and +0.65 e at the minimum energy and transition state structures, respectively. There are little charge redistributions required in the bond scission reaction paths. Combined with the fact that the O–H bond of $H_xCOH^*$ is floppy, a very low activation energy is encountered for each of the reactions.

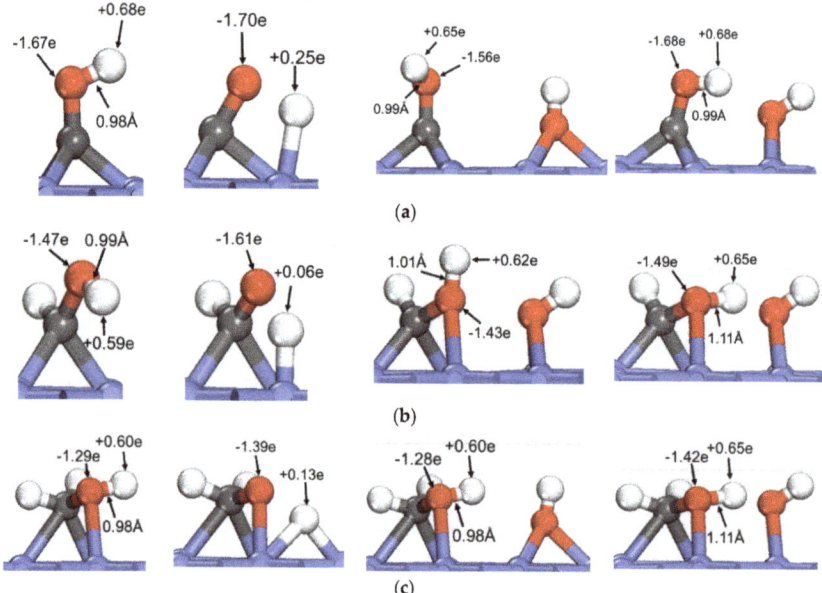

Figure 2. Cont.

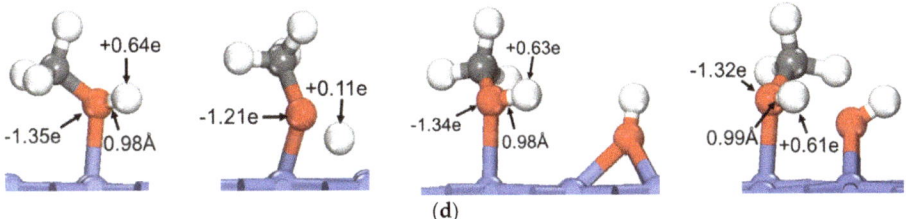

(d)

**Figure 2.** Structures and Bader charges in the oxhydryl dehydrogenation of $H_xCOH$ with and without co-adsorbed OH: (**a**) $x = 0$, (**b**) $x = 1$, (**c**) $x = 2$, (**d**) $x = 3$. From left to right: local minimum and transition state structures of $H_xCOH$ without and with co-adsorbed OH.

Based on the above analysis, it is clear that OH* plays a catalyzing role in the oxhydryl dehydrogenation of $H_xCOH^*$ on Ni(111). The catalysis effect is realized by minimizing the charge redistribution requirement in the reaction path. The charge of oxhydryl H of $H_xCOH^*$ in the reaction process is maintained by interacting with OH*. The H···OH distance at the transition structure is 1.86, 1.61, 1.67 and 2.05 Å for $x = 0$, 1, 2, and 3, respectively. The distances are characteristics of H-bonds. Therefore, the reduced reaction barrier for the oxhydryl dehydrogenation of $H_xCOH^*$ can be attributed to the catalyzing effect of H-bond interaction.

### 3.2. Dehydrogenation of OH* Assisted by H-Bond of O–H···OH and O–H···N

The catalyzing effect of H-bond on oxhydryl dehydrogenation can also be seen in the dehydrogenation of OH* adsorbed on Ni(111) [42]. Figure 3 compares the activation energies, initial and transition state structures and charge distributions of OH* dehydrogenations with and without co adsorbed neighboring OH*. As seen in Figure 3, the activation energy of dehydrogenation is 105 kJ/mol for isolated OH*, but is reduced by 37 kJ/mol to 68 kJ/mol for OH* with co-adsorbed OH* due to the O–H···OH interaction. The reduction of activation energy is also observed for other transition metals. The activation energy for the corresponding reaction is reduced from 1.11 eV to 0.3 eV on Co(111) [19] and from 1.88 eV to 0.82 eV on Cu(111) [9]. Moreover, the higher energy of O* + $H_2O^*$ in comparison with that of OH* + OH* is in qualitative agreement with both the theoretical and experimental results of Che et al [43].

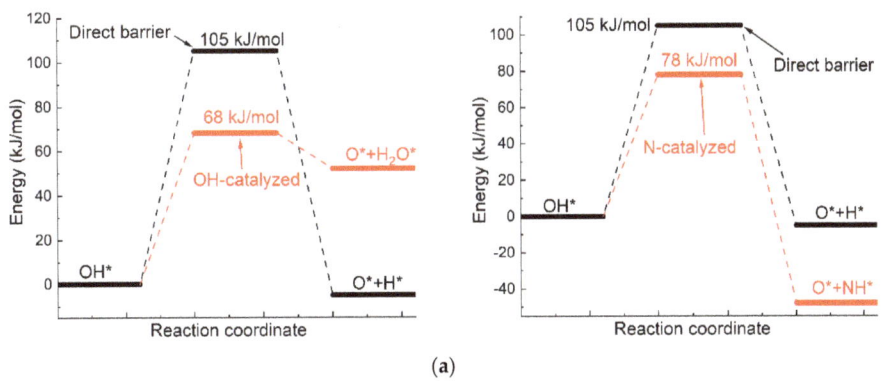

(a)

**Figure 3.** Cont.

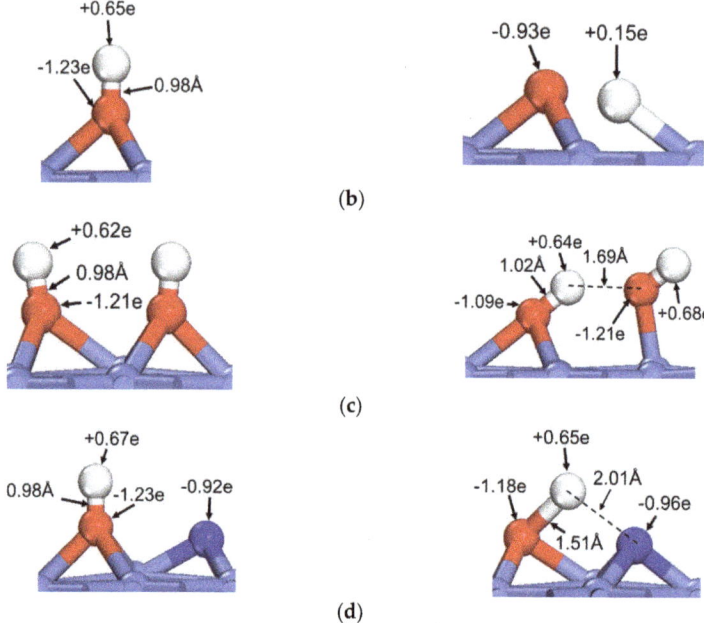

**Figure 3.** The dehydrogenation of O–H on Ni(111) surface: (**a**) The activation energies of dual path dehydrogenation of OH*, (**b**) The initial and transition state structures and charge distributions of individual OH* dehydrogenation, (**c**) The initial and transition state structures and charge distributions of OH-catalyzed OH* dehydrogenation, (**d**) The initial and transition state structures and charge distributions of N-catalyzed OH* dehydrogenation.

Like the case shown in Figure 2, the Bader charges of H for the energy minimum and transition state structures in the dehydrogenation of individual OH* are quite different, at +0.65e and +0.15e, respectively. The corresponding Bader charges are respectively +0.62e and +0.64e in the dehydrogenation of OH* with the presence of the O–H···OH interaction. Once again, the charge of oxhydryl H remains almost constant in the reaction process due to the influence of the O–H···OH H-bond.

H-bonds are ubiquitous in nature and normally exist between electronegative atoms and H atoms covalently bound to similar electronegative atoms. In addition to the OH···O H-bond discussed above, OH···N is another type of commonly seen H-bond. Even though OH···N is not involved in MSR, it may exist in other reaction processes. To test the conceptual generality of H-bond catalysis, the effect of OH···N H-bond on OH* dehydrogenation is examined here.

The activation energy, initial and transition state structures and charge distributions of OH* dehydrogenations with neighboring N* are also shown in Figure 3. As shown in Figure 3, the activation energy of OH* dehydrogenation is reduced by 27 kJ/mol due to the presence of OH···N interaction. Like the cases shown for OH···O interactions, the charge of oxhydryl H changes little in going from the initial local minimum structure to the transition state, even though the direction and distance of the O–H bond are substantially changed in the process.

Combining Figures 2 and 3, a general conclusion may be drawn: the charge of oxhydryl H is kept unchanged in the initial and transition states of dehydrogenation by the presence of H-bond. As a result, the activation energy for the dehydrogenation reaction is reduced in comparison to that in the absence of the H-bond.

It is worth noting that the activation energy is reduced to a very low value of around 15 kJ/mol, for the dehydrogenation of H$_x$COH*, but to 68–78 kJ/mol for the OH* dehydrogenation. The notable difference is attributable to the rigidity of the O-H bond in the species: The O–H bond directionality is quite weak in H$_x$COH*, while relatively strong in OH*. Another point to note is that the activation energy for the H$\cdots$N assisted reaction is 10 kJ/mol higher than that of H$\cdots$O assisted process, even though the H$\cdots$N interaction is known to be stronger than the H$\cdots$O interaction on average. There is no surprise, though, as a specific case does not correspond to the average. As shown in Figure 3, the charge of N* for the H-bond is only −0.92 to −0.96 e, while the charge of O for the H-bond is −1.21 e. Moreover, the H-bond distance of OH$\cdots$N is larger than that of OH$\cdots$O. Consequently, the OH$\cdots$N H-bond here is weaker and less effective in reducing the activation energy than the OH$\cdots$O H-bond. Nevertheless, it is interesting to see the relatively weak OH$\cdots$N H-bond is very effective in maintaining the charge of H in the reaction path that the O-H bond is stretched from the initial 0.98 Å to 1.51 Å at the transition state. Overall, a near constant charge of H during the O–H scission process as maintained by the H-bond interaction is a common feature for the H-bond catalyzed reactions. The effectiveness of an H-bond on lowering the activation energy is, however, dependent on numerous factors, such as the H-bond strength, the O–H bond strength, and the surface material, and thus requires further studies.

## 4. Conclusions

Oxhydryl dehydrogenations of H$_x$COH* ($x$ = 3, 2, 1, 0) on Ni(111) were investigated by DFT calculations and Bader charge analysis. The activation energies are 68 to 91 kJ/mol for isolated H$_x$COH* and much reduced to 12–17 kJ/mol if assisted by neighboring OH*. The catalyzing effect of OH* is attributed to the OH$\cdots$O H-bond that maintains the charge of oxhydryl H in the O–H bond breaking process. The catalytic mechanism of H-bond is further supported by the results of OH* and N* assisted dehydrogenation of OH*. Due to the universality of H-bond, the catalytic mechanics revealed here are of broad implication to the study of reaction kinetics of many systems.

**Author Contributions:** Funding acquisition, Z.L.; Investigation, C.K.; Methodology, C.K.; Resources, Z.L.; Software, Z.L.; Supervision, Z.L.; Writing, C.K. and Z.L. All authors have read and agreed to the published version of the manuscript.

**Funding:** This research was funded by National Natural Science Foundation of China (11774324 & 11574284).

**Acknowledgments:** The computing time of the Super-computing Center of the University of Science and Technology of China are gratefully acknowledged.

**Conflicts of Interest:** The authors declare no conflict of interest.

## References

1. Herdem, M.S.; Sinaki, M.Y.; Farhad, S.; Hamdullahpur, F. An overview of the methanol reforming process: Comparison of fuels, catalysts, reformers, and systems. *Int. J. Energy Res.* **2019**, *43*, 5076–5105. [CrossRef]
2. Khzouz, M.; Gkanas, E.I.; Du, S.; Wood, J. Catalytic performance of Ni-Cu/Al2O3 for effective syngas production by methanol steam reforming. *Fuel* **2018**, *232*, 672–683. [CrossRef]
3. Papavasiliou, J.; Paxinou, A.; Słowik, G.; Neophytides, S.; Avgouropoulos, G. Steam reforming of methanol over nanostructured Pt/TiO2 and Pt/CeO2 catalysts for fuel cell applications. *Catalysts* **2018**, *8*, 544. [CrossRef]
4. Trimm, D.L.; Önsan, Z.I. Onboard fuel conversion for hydrogen-fuel-cell-driven vehicles. *Catal. Rev.* **2001**, *43*, 31–84. [CrossRef]
5. Liu, Z.; Yao, S.; Johnston-Peck, A.; Xu, W.; Rodriguez, J.A.; Senanayake, S.D. Methanol steam reforming over Ni-CeO2 model and powder catalysts: Pathways to high stability and selectivity for H2/CO2 production. *Catal. Today* **2018**, *311*, 74–80. [CrossRef]
6. Lu, J.; Li, X.; He, S.; Han, C.; Wan, G.; Lei, Y.; Chen, R.; Liu, P.; Chen, K.; Zhang, L.; et al. Hydrogen production via methanol steam reforming over Ni-based catalysts: Influences of Lanthanum (La) addition and supports. *Int. J. Hydrogen Energy* **2017**, *42*, 3647–3657. [CrossRef]

7. Kaftan, A.; Kusche, M.; Laurin, M.; Wasserscheid, P.; Libuda, J. KOH-promoted Pt/Al2O3 catalysts for water gas shift and methanol steam reforming: An operando DRIFTS-MS study. *Appl. Catal. B* **2017**, *201*, 169–181. [CrossRef]
8. Lin, S.; Xie, D.; Guo, H. Methyl formate pathway in methanol steam reforming on copper: Density functional calculations. *ACS Catalysis* **2011**, *1*, 1263–1271. [CrossRef]
9. Zuo, Z.-J.; Wang, L.; Han, P.-D.; Huang, W. Insights into the reaction mechanisms of methanol decomposition, methanol oxidation and steam reforming of methanol on Cu(111): A density functional theory study. *Int. J. Hydrogen Energy* **2014**, *39*, 1664–1679. [CrossRef]
10. Wang, S.-S.; Su, H.-Y.; Gu, X.-K.; Li, W.-X. Differentiating intrinsic reactivity of Copper, Copper–Zinc Alloy, and Copper/Zinc Oxide interface for methanol steam reforming by first-principles theory. *J. Phys. Chem. C* **2017**, *121*, 21553–21559. [CrossRef]
11. Wang, S.-S.; Gu, X.-K.; Su, H.-Y.; Li, W.-X. First-principles and microkinetic simulation studies of the structure sensitivity of Cu catalyst for methanol steam reforming. *J. Phys. Chem. C* **2018**, *122*, 10811–10819. [CrossRef]
12. Smith, G.K.; Lin, S.; Lai, W.; Datye, A.; Xie, D.; Guo, H. Initial steps in methanol steam reforming on PdZn and ZnO surfaces: Density functional theory studies. *Surf. Sci.* **2011**, *605*, 750–759. [CrossRef]
13. Krajčí, M.; Tsai, A.P.; Hafner, J. Understanding the selectivity of methanol steam reforming on the (1 1 1) surfaces of NiZn, PdZn and PtZn: Insights from DFT. *J. Catal.* **2015**, *330*, 6–18. [CrossRef]
14. Chen, W.; Cubuk, E.D.; Montemore, M.M.; Reece, C.; Madix, R.J.; Friend, C.M.; Kaxiras, E. A comparative ab initio study of anhydrous dehydrogenation of linear-chain alcohols on Cu(110). *J. Phys. Chem. C* **2018**, *122*, 7806–7815. [CrossRef]
15. Lausche, A.C.; Abild-Pedersen, F.; Madix, R.J.; Nørskov, J.K.; Studt, F. Analysis of sulfur-induced selectivity changes for anhydrous methanol dehydrogenation on Ni(100) surfaces. *Surf. Sci.* **2013**, *613*, 58–62. [CrossRef]
16. Kramer, Z.C.; Gu, X.-K.; Zhou, D.D.Y.; Li, W.-X.; Skodje, R.T. Following molecules through reactive networks: Surface catalyzed decomposition of methanol on Pd(111), Pt(111), and Ni(111). *J. Phys. Chem. C* **2014**, *118*, 12364–12383. [CrossRef]
17. Du, P.; Wu, P.; Cai, C. Mechanism of methanol decomposition on the Pt3Ni(111) surface: DFT study. *J. Phys. Chem. C* **2017**, *121*, 9348–9360. [CrossRef]
18. Liao, T.-W.; Yadav, A.; Ferrari, P.; Niu, Y.; Wei, X.-K.; Vernieres, J.; Hu, K.-J.; Heggen, M.; Dunin-Borkowski, R.E.; Palmer, R.E.; et al. Composition-tuned Pt-skinned PtNi bimetallic clusters as highly efficient methanol dehydrogenation catalysts. *Chem. Mater.* **2019**, *31*, 10040–10048. [CrossRef]
19. Luo, W.; Asthagiri, A. Density functional theory study of methanol steam reforming on Co(0001) and Co(111) surfaces. *J. Phys. Chem. C* **2014**, *118*, 15274–15285. [CrossRef]
20. Kim, D.K.; Iglesia, E. Isotopic and kinetic assessment of the mechanism of CH3OH-H2O catalysis on supported Copper clusters. *J. Phys. Chem. C* **2008**, *112*, 17235–17243. [CrossRef]
21. Lee, J.K.; Ko, J.B.; Kim, D.H. Methanol steam reforming over Cu/ZnO/Al2O3 catalyst: Kinetics and effectiveness factor. *Appl. Catal., A* **2004**, *278*, 25–35. [CrossRef]
22. Blaylock, D.W.; Zhu, Y.-A.; Green, W.H. Computational investigation of the thermochemistry and kinetics of steam methane reforming over a multi-faceted nickel catalyst. *Top. Catal.* **2011**, 828–844. [CrossRef]
23. Kresse, G.; Furthmuller, J. Efficient iterative schemes for ab initio total-energy calculations using a plane-wave basis set. *Phys. Rev. B* **1996**, *54*, 11169–11186. [CrossRef] [PubMed]
24. Kresse, G.; Furthmuller, J. Efficiency of ab-initio total energy calculations for metals and semiconductors using a plane-wave basis set. *Comput. Mater. Sci.* **1996**, *6*, 15–50. [CrossRef]
25. Kresse, G.; Hafner, J. Ab initio molecular dynamics for liquid metals. *Phys. Rev. B* **1993**, *47*, 558–561. [CrossRef]
26. Kresse, G.; Hafner, J. Ab initio molecular-dynamics simulation of the liquid-metal-amorphous-semiconductor transition in germanium. *Phys. Rev. B* **1994**, *49*, 14251–14269. [CrossRef]
27. Kresse, G.; Joubert, D. From ultrasoft pseudopotentials to the projector augmented-wave method. *Phys. Rev. B* **1999**, *59*, 1758–1775. [CrossRef]
28. Blochl, P.E. Projector augmented-wave method. *Phys. Rev. B* **1994**, *50*, 17953–17979. [CrossRef]
29. Klimes, J.; Bowler, D.R.; Michaelides, A. Chemical accuracy for the van der Waals density functional. *J. Phys. Condens. Matter* **2010**, *22*, 022201. [CrossRef]
30. Klimes, J.; Bowler, D.R.; Michaelides, A. Van der Waals density functionals applied to solids. *Phys. Rev. B* **2011**, *83*. [CrossRef]

31. Carrasco, J.; Klimes, J.; Michaelides, A. The role of van der Waals forces in water adsorption on metals. *J. Chem. Phys.* **2013**, *138*, 024708. [CrossRef] [PubMed]
32. Press, W.H.; Flannery, B.P.; Teukolsky, S.A.; Vetterling, W.T. *Numerical Recipes: The Art of Scientific Computing*; Cambridge University Press: Cambridge, UK, 1986.
33. Henkelman, G.; Jonsson, H. Improved tangent estimate in the nudged elastic band method for finding minimum energy paths and saddle points. *J. Chem. Phys.* **2000**, *113*, 9978–9985. [CrossRef]
34. Henkelman, G.; Jonsson, H. A dimer method for finding saddle points on high dimensional potential surfaces using only first derivatives. *J. Chem. Phys.* **1999**, *111*, 7010–7022. [CrossRef]
35. Sanville, E.; Kenny, S.D.; Smith, R.; Henkelman, G. Improved grid-based algorithm for Bader charge allocation. *J. Comput. Chem.* **2007**, *28*, 899–908. [CrossRef]
36. Tang, W.; Sanville, E.; Henkelman, G. A grid-based Bader analysis algorithm without lattice bias. *J. Phys. Condens. Matter* **2009**, *21*, 084204. [CrossRef]
37. Wang, G.C.; Zhou, Y.H.; Morikawa, Y.; Nakamura, J.; Cai, Z.S.; Zhao, X.Z. Kinetic mechanism of methanol decomposition on Ni(111) surface: A theoretical study. *J. Phys. Chem. B* **2005**, *109*, 12431–12442. [CrossRef]
38. Zhou, Y.-H.; Lv, P.-H.; Wang, G.-C. DFT studies of methanol decomposition on Ni(100) surface: Compared with Ni(111) surface. *J. Mol. Catal. A Chem.* **2006**, *258*, 203–215. [CrossRef]
39. Greeley, J.; Mavrikakis, M. Methanol decomposition on Cu(111): A DFT study. *J. Catal.* **2002**, *208*, 291–300. [CrossRef]
40. Huang, Y.; He, X.; Chen, Z.X. Density functional study of methanol decomposition on clean and O or OH adsorbed PdZn(111). *J. Chem. Phys.* **2013**, *138*, 184701. [CrossRef]
41. Ke, C.; Lin, Z. Density functional theory based micro- and macro-kinetic studies of Ni-catalyzed methanol steam reforming. *Catalysts* **2020**, *10*, 349. [CrossRef]
42. Ke, C.; Lin, Z. Elementary reaction pathway study and a deduced macrokinetic model for the unified understanding of Ni-catalyzed steam methane reforming. *React. Chem. Eng.* **2020**. [CrossRef]
43. Che, F.L.; Gray, J.T.; Ha, S.; McEwen, J.S. Catalytic water dehydrogenation and formation on nickel: Dual path mechanism in high electric fields. *J. Catal.* **2015**, *332*, 187–200. [CrossRef]

**Sample Availability:** Samples of the compounds are available from the authors.

© 2020 by the authors. Licensee MDPI, Basel, Switzerland. This article is an open access article distributed under the terms and conditions of the Creative Commons Attribution (CC BY) license (http://creativecommons.org/licenses/by/4.0/).

Article

# Phosphine Oxides as Spectroscopic Halogen Bond Descriptors: IR and NMR Correlations with Interatomic Distances and Complexation Energy

Alexei S. Ostras', Daniil M. Ivanov, Alexander S. Novikov and Peter M. Tolstoy *

Institute of Chemistry, St. Petersburg State University, 198504 St. Petersburg, Russia;
st052055@student.spbu.ru (A.S.O.); dan15101992@gmail.com (D.M.I.); a.s.novikov@spbu.ru (A.S.N.)
* Correspondence: peter.tolstoy@spbu.ru; Tel.: +7-921-430-8191

Academic Editor: Ilya G. Shenderovich
Received: 27 February 2020; Accepted: 16 March 2020; Published: 19 March 2020

**Abstract:** An extensive series of 128 halogen-bonded complexes formed by trimethylphosphine oxide and various F-, Cl-, Br-, I- and At-containing molecules, ranging in energy from 0 to 124 kJ/mol, is studied by DFT calculations in vacuum. The results reveal correlations between R–X···O=PMe$_3$ halogen bond energy $\Delta E$, X···O distance $r$, halogen's σ-hole size, QTAIM parameters at halogen bond critical point and changes of spectroscopic parameters of phosphine oxide upon complexation, such as $^{31}$P NMR chemical shift, $\Delta\delta$P, and P=O stretching frequency, $\Delta\nu$. Some of the correlations are halogen-specific, i.e., different for F, Cl, Br, I and At, such as $\Delta E(r)$, while others are general, i.e., fulfilled for the whole set of complexes at once, such as $\Delta E(\Delta\delta P)$. The proposed correlations could be used to estimate the halogen bond properties in disordered media (liquids, solutions, polymers, glasses) from the corresponding NMR and IR spectra.

**Keywords:** halogen bond; phosphine oxide; $^{31}$P NMR spectroscopy; IR spectroscopy; non-covalent interactions; spectral correlations

---

## 1. Introduction

Halogen bonding is one of the most abundant non-covalent interactions in chemistry [1,2]. Due to the anisotropic distribution of electron density around the covalently bound halogen atom, it has two distinct regions: (a) the region of increased electron density (nucleophilic site), located perpendicular to the covalent bond and corresponding to negative values of electrostatic potential (ESP) and (b) the region of decreased electron density (electrophilic site), also called σ-hole [3], located along the covalent bond. It is the existence of the electron-depleted σ-hole that determines the ability of the halogen atom to participate in attractive interactions with electron-donating atoms or groups [4–6], forming the so-called halogen bond R–X···Y (X—halogen atom). In halogen-bonded complexes, the X ··· Y distances are usually shorter than the sum of van der Waals radii of X and Y atoms, while the RXY angles tend to be close to 180°, because the σ-hole region is located on the continuation of the R–X axis. The range of halogen bond interaction energies is similar to that for hydrogen bonds, spanning from a fraction of kJ/mol up to 150 kJ/mol [7,8].

The formation of halogen bonds can be detected in solids [9–13], in liquids and solutions [14–18] and in gas phase [19–21]. Halogen bonding is being actively studied, and it has been demonstrated that it plays a significant role in biochemistry [22,23], in crystal design and design of functional materials (liquid crystals, molecular receptors, conductors, luminescence emitters, non-linear optical materials, etc.) [24–32], in organocatalysis [33–38] and in design of pharmaceuticals (here, halogen bonds are considered primarily as a type of hydrophobic functional groups, increasing the lipophilic properties of molecules and allowing them to pass through cell membranes) [39,40]. Besides, the halogen bond has been a subject of numerous theoretical works [41–47].

The main characteristics of halogen bond include (a) the geometric parameters (interatomic distances and angles, which could be estimated from diffraction studies) and (b) the complexation energy (experimentally available using, for example, calorimetric methods, though in condensed phases and for intramolecular interactions the definition of complexation energy might be ambiguous). In practice, it is often needed to characterize halogen bonds in disordered states of the matter, where in many cases it is done by indirect approaches, such as analysis of spectroscopic data based on previously established correlations. As a spectroscopic descriptor in such correlational methods, it is often convenient to take the change of a given parameter upon complexation, i.e., the difference between that parameter for the complex and the free molecule (which could be a probe molecule).

The main goal of this work is to propose a new spectroscopic method for the quantitative characterization of geometry and energy of halogen bonds. For this purpose we have selected trimethylphosphine oxide, $Me_3PO$, as a model probe molecule. As spectroscopic descriptors of the halogen-bonded complexes formed by $Me_3PO$, we have chosen (a) the change of the frequency of P=O stretching vibration upon complexation, $\Delta\nu$, and (b) the change of the $^{31}P$ NMR chemical shift upon complexation, $\Delta\delta P$. The choice of $Me_3PO$ as a probe molecule is stipulated by the fact that, due to the high polarization of the P=O bond, the terminal oxygen atom is an effective electron donor in non-covalent interactions, i.e., it is an effective halogen bond acceptor. For the same reason, it is expected that P=O coordination leads to substantial displacement of vibrational bands in IR spectra. Besides, the relatively rigid skeleton of $Me_3PO$ minimizes the influence on $\Delta\delta P$ of factors such as conformation of substituents (which becomes larges already for more flexible triethylphosphine oxide, let alone triphenylphosphine oxide).

Phosphine oxides in general are perspective probes for the diagnostics of non-covalent interactions. Intermolecular hydrogen-bonded complexes of phosphine oxides and their related compounds are discussed in several publications [48–51]. Moreover, there are several publications where the participation of phosphine oxides in halogen bonds is considered. For example, in [52,53], the crystal adducts of various phosphine oxides ($Ph_2(Me)P=O$, $Ph_3PO$) with strong halogen donors such as pentafluoroiodoebenzene $C_6F_5I$ and 1,4-diaryl-5-iodotriazole are described. Finally, for a series of crystals containing molecules with P=O groups, the presence of short X···O contacts (X = Cl, Br, I), which can be interpreted as halogen bonds, has been established with the help of X-ray analysis and quantum-chemical calculations [54–61]. Many other examples of XBs with phosphine oxides and related compounds can be found in CCDC data, which allowed us to perform an extensive database search and analyze the distributions of geometric parameters, as described in Section 4. The high electron-donating ability of phosphine oxides previously has been employed in the Gutmann–Beckett method to characterize the acceptor properties of solvents [62] and other compounds exhibiting Lewis acidity [63,64], expressed in so-called acceptor numbers, AN [65]. The main experimental parameter in this method is the change of the $^{31}P$ NMR chemical shift of triethylphosphine oxide upon complexation, $\Delta\delta P$, normalized to 100 for the complex with $SbCl_5$.

In this work we have used the following criteria for the selection of halogen donor molecules: (i) relatively simple structure; (ii) presence of an electron-accepting group, which increases the positive ESP value in the σ-hole region and, consequently, enhances the electron-accepting ability of the halogen; (iii) absence of acidic hydrogen atoms that could compete with halogen bond by forming hydrogen bonds. Based on these criteria, we have selected an extensive set of 128 halogen-containing neutral molecules belonging to different classes of inorganic and organic chemical compounds R–X, where for simplicity RX stands also for $R_3X$ and $R_5X$ in case of halogen(III) and halogen(V) compounds, respectively:

(1) halogens ($F_2$, $Cl_2$, $Br_2$, $I_2$, $At_2$);
(2) interhalides (ClF, $ClF_3$, $ClF_5$, BrF, $BrF_3$, $BrF_5$, BrCl, IF, $IF_3$, $IF_5$, ICl, $ICl_3$, IBr, AtCl, AtBr, AtI);

(3) oxohalides (OF$_2$, ClO$_3$OF, Cl$_2$O, ClO$_2$, ClO$_2$F, ClO$_3$OCl, Br$_2$O, BrO$_2$, BrO$_2$F, ClO$_3$OBr, IO$_2$F, ClO$_3$OI);
(4) pseudohalides (FCN, FN$_3$, FCNO, ClCN, ClN$_3$, ClNCO, ClSCN, BrCN, BrN$_3$, BrNCO, BrSCN, ICN, IN$_3$, INCO, ISCN);
(5) halogenated methanes and their derivatives (CF$_3$OF, CF$_3$SO$_2$OF, CF$_3$Cl, CCl$_2$F$_2$, CCl$_3$F, CCl$_4$, CF$_3$OCl, CF$_3$SO$_2$OCl, CF$_3$Br, CBr$_2$F$_2$, CBr$_3$F, CBrCl$_3$, CBrClF$_2$, CBr$_4$, CF$_3$OBr, CF$_3$SO$_2$OBr, CF$_3$I, CI$_2$F$_2$, CI$_3$F, CIClF$_2$, CI$_4$);
(6) halogenated ethylene, halogenated acetylene and their derivatives (C$_2$F$_4$, C$_2$Cl$_4$, C$_2$F$_3$Cl, C$_2$Br$_4$, C$_2$F$_3$Br, C$_2$I$_4$, C$_2$F$_3$I, C$_2$(CN)$_3$Cl, C$_2$F$_2$, C$_2$Cl$_2$);
(7) phosgene and its derivatives (COF$_2$, COClF, COCl$_2$, COBrCl, COBr$_2$, COBrF, COIF);
(8) thionyl- and sulfurylhalides (SOF$_2$, SO$_2$ClF, SOCl$_2$, SOBr$_2$, SO$_2$Cl$_2$, SO$_2$BrF);
(9) sulfur halides and sulfur hypohalites (SF$_6$, SF$_5$OF, SF$_5$Cl, SF$_5$OCl, S$_2$Cl$_2$, SCl$_2$, SF$_5$Br, S$_2$Br$_2$, SBr$_2$);
(10) halogenated nitrogen-containing inorganic compounds (NF$_3$, NOF, NO$_2$F, NO$_2$OF, NCl$_3$, NF$_2$Cl, NOCl, NO$_2$Cl, NO$_2$OCl, NBr$_3$, NF$_2$Br, NOBr, NO$_2$Br, NO$_2$OBr, NI$_3$);
(11) assorted organic compounds (tetrafluoro-1,4-benzoquinone, tetrachloro-1,4-benzoquinone, tetrabromo-1,4-benzoquinone, tetraiodo-1,4-benzoquinone, C$_6$H$_5$(C≡C)Cl, C$_6$H$_5$(C≡C)Br, C$_6$H$_5$(C≡C)I, CCl(CN)$_3$, CBr(CN)$_3$, N-chlorosuccinimide, N-bromosuccinimide, N-iodosuccinimide).

For each 1:1 complex formed by Me$_3$PO molecule and one of the abovementioned halogen donors, we have quantum-chemically calculated (M06-2X/def2-TZVPPD level of theory [66]) the equilibrium geometry, the harmonic vibrational frequencies, the nuclear chemical shielding constants, the complexation energies (corrected for BSSE, the basis set superposition error) and the value of ESP in the σ-hole region for the free halogen donor molecule. Besides, we have calculated and analyzed the electron density properties at the (3; −1) halogen bond critical point (BCP) using the Bader's quantum theory of atoms in molecules (QTAIM) methodology [67]. The parameters which are determined and analyzed in this work are shown in Scheme 1.

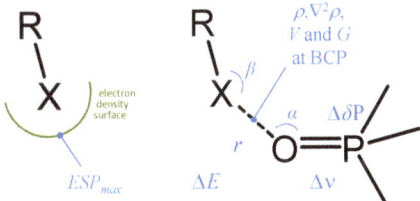

Scheme 1. Schematic structure of 1:1 halogen-bonded complexes formed by Me$_3$PO with RX (X—halogen) and the list of calculated parameters studied in this work: $r$—X···O distance; $α$ and $β$—X···O=P and R–X···O angles, respectively; $ΔE$—complexation energy (BSSE corrected); $ρ$, $\nabla^2ρ$, $V$ and $G$ are electron density, Laplacian of electron density, local electron potential and kinetic energies densities at halogen bond critical point (3; −1), respectively; $ρ$ and $\nabla^2ρ$ are electron density and Laplacian of electron density; $ΔδP$—change of $^{31}$P NMR chemical shift upon complexation; $Δν$ –change of harmonic P=O stretching wavenumber upon complexation; $ESP_{max}$—the extremal value of electrostatic potential in the region of σ-hole on the surface of equal electron density taken at 0.001 electron/Bohr$^3$ level.

## 2. Results and Discussion

The calculated geometries of all complexes are shown in Figure S1, and all the parameters considered in this work (see Scheme 1) are collected in Tables S1 and S2. In this section we will present only the plots and fitting equations that are necessary for the discussion.

## 2.1. Angular Distribution

Based on the calculated optimized geometries of halogen-bonded R–X···O=P complexes (X = F, Cl, Br, I, At), the distributions of angles $\alpha$ (angle X···O=P) and $\beta$ (angle R–X···O) were constructed. In case of multivalent halogen bond donors, such as some of interhalides, the angle $\beta$ was chosen as the largest one among all possible R–X···O values. In Figure 1, the resulting distributions are shown as histograms in which the bar height is proportional to the number of complexes having the corresponding angle within the given 5° range.

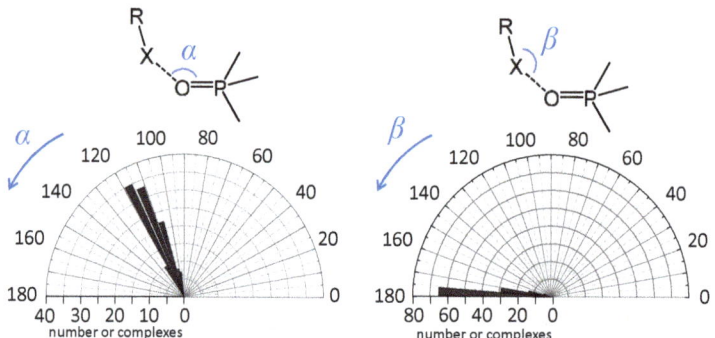

**Figure 1.** Distribution of angles $\alpha$ (X···O=P) and $\beta$ (R–X···O) for 128 complexes studied in this work (X = F, Cl, Br, I, At). Histogram bars are taken every 10° for $\alpha$ and every 5° for $\beta$.

The values of angle $\alpha$ lie primarily in the range 100–120°. This indicates that the halogen bonds are expectedly formed along the direction of oxygen lone pairs. For the $sp^2$-hybridized oxygen, the angle between the P=O bond and the oxygen lone pair should be 120°, though the high polarization of the PO bond increases the contribution of the P$^+$–O$^-$ resonance structure, corresponding to the $sp^3$ hybridization, which makes $\alpha$ closer to the tetrahedral angle 109.5°. Thus, the distribution of angles $\alpha$ might be considered as an indication that the phosphine oxide structure is intermediate between the neutral one, P=O, and the zwitterionic one, P$^+$–O$^-$. Further deviations of angles $\alpha$ are most probably due to the presence of secondary interactions or steric hindrance.

The angle $\beta$ describes how linear the halogen bond is. The values of $\beta$ are influenced by the shape and size of the σ-hole on the halogen atom. As σ-holes are located along the continuation of the R–X bond, the values of $\beta$ lie close to 180°. Overall, the stronger and shorter the halogen bond is, the more linear it is (see Figure S2). It stands to reason that the complexation energy increases with the increase of the σ-hole, which in turn depends on the polarizability of the halogen atom; thus, it could be expected that, in general, the broader range of angles $\beta$ would correspond to the weakest R–F···O=P complexes, and the narrow range of $\beta$ values close to 180° would be observed for stronger R–I···O=P and R–At···O=P complexes. Indeed, this is the case within the series of complexes considered in this work (see Figure S2).

The abovementioned angular distributions could be compared with the results of the statistical analysis of CCDC 2020 data for phosphine oxides and related PO-containing compounds as O-nucleophiles for halogen bond. Note that only three halogens (Cl, Br, I) were considered, because (i) no results for At were found due to all astatine isotopes being radioactive; and (ii) although we found two structures (XAQYOF and XUFQOF) for fluorine, this information is definitely not enough to collect statistics. The values of angle $\alpha$ for this statistical data lie primarily in the range 120–140°, with a median value of 129.4°, as shown in Figure 2. This indicates that the halogen bonds are still preferably formed along the direction of oxygen lone pairs, but, in case of bulk substituents around P=O functionality (larger than CH$_3$ groups in the model trimethylphosphine oxide), the halogen donors have to occupy positions with larger values of $\alpha$, which is less optimal for the halogen bond stabilization; even almost linear complexes are possible (the largest $\alpha$ value is 173.2° in OBUCAQ).

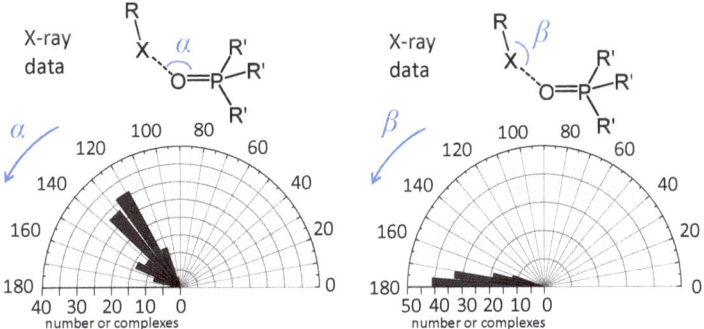

**Figure 2.** Distribution of angles $\alpha$ (X···O=P) and $\beta$ (R–X···O) for CCDC 2020 structures containing R–X···O=P fragment (X = Cl, Br, I). The see complete list of entries in Table S3. Histogram bars are taken every 10° for $\alpha$ and every 5° for $\beta$.

The R–X···O fragments are less strictly linear for the CCDC data than for theoretically investigated R–X···OPMe$_3$ models (see Figure 2). However, the median value of angle $\beta$ is 170.1°, which is still rather close to 180°. As seen from the calculated data, the stronger and shorter the halogen bond is, the more linear it is (see Figure 3). Note that for all halogens under consideration (Cl, Br, I) combined and for each halogen individually it is possible to draw the one cut-off line in Figure 3 above which almost all data points will be located.

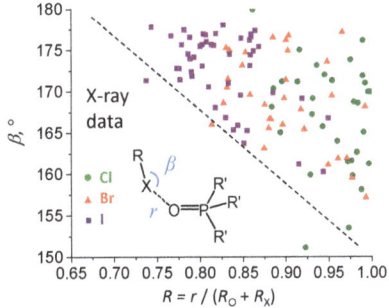

**Figure 3.** Correlation between angles $\beta$ (R–X···O) and normalized distance parameter $R = r/(R_O + R_X)$, where $r$ is X···O distance and $R_O$ and $R_X$ are van der Waals radii of oxygen and halogen atoms, respectively, for CCDC 2020 structures containing R–X···O=P fragment (X = Cl, Br, I). For the complete list of entries see Table S3.

## 2.2. Complexation Energy Dependence On Intermolecular Distance

There seems to be no universal dependence which can be used to estimate the halogen bond energy using the interatomic distances and bond angles. The previously reported correlations were tested on relatively homologous sets of intermolecular complexes [47,68]. Here, using our set of 128 complexes with participation of various halogens, we attempt to construct energy–structure correlations which on one hand would be general enough (i.e., would have similar form for all halogen donor types) and on the other hand would be simple enough, allowing for a rapid energy estimation.

Figure 4 shows the correlations between calculated complexation energies $\Delta E$ and X···O interatomic distances for all complexes studied in this work. The majority of data points in the plot are located in the range of X···O distances from 2.4 to 3.1 Å, while the overall span of complexation energies is from 3.6 up to 124 kJ/mol.

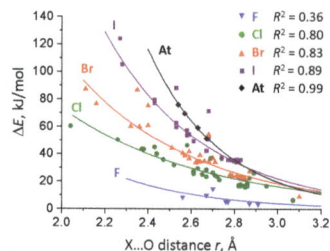

**Figure 4.** Correlation between calculated complexation energy $\Delta E$ and interatomic distance $r$ (X···O, where X = F, Cl, Br, I, At) for R–X···O=PMe$_3$ halogen-bonded complexes studied in this work. The solid curves correspond to Equation (1).

Quite expectedly, for each given halogen (F, Cl, Br, I or At) the decrease of the X···O distance $r$ is accompanied by the increase of complexation energy $\Delta E$. The shape of this correlation could be approximated by exponentially decaying functions such as:

$$\Delta E(\text{kJ/mol}) = A \cdot e^{-b \cdot r}, \qquad (1)$$

where $A$ (in kJ/mol) and $b$ (in Å$^{-1}$) are fitting coefficients, individual for each halogen. The corresponding fitted curves are added to Figure 4 as solid lines, while the numerical values of parameters $A$ and $b$ are collected in Table 1. One can see that the fitted parameters found for complexes with fluorine-containing molecules are falling out of the overall trend, which is probably associated not only with the small number of data points, but also with the weak ability of fluorine to act as an electron acceptor (due to the small—sometimes almost absent—σ-hole).

**Table 1.** The numerical values of fitted parameters $A$, $b$, $B$, $D$, $K$ and $M$ used in Equations (1)–(6). The resulting curves are shown in Figures 4–7.

| Donor Atom | Equations (1) and (3) | Equation (1) | Equation (3) | Equation (4) | Equation (5) | Equation (7) |
|---|---|---|---|---|---|---|
| | $A$, kJ/mol | $b$, Å$^{-1}$ | $B$ | $D$ | $K$, kJ/mol/cm$^{-1}$ | $M$, kJ/mol/ppm |
| F | 5490 | 2.4 | 7.2 | 0.20 | 1.30 | 2.9 |
| Cl | 1470 | 1.5 | 4.9 | 0.25 | 0.75 | 2.45 |
| Br | 4100 | 1.8 | 6.1 | 0.30 | 0.85 | 2.65 |
| I | 16,300 | 2.2 | 7.7 | 0.34 | 0.90 | 3.05 |
| At | 138,000 | 2.95 | 10.6 | 0.35 | 1.00 | 3.30 |

It should be noted that the quality of the correlation increases in the order from chlorine- to astatine-containing halogen donors (see the corresponding $R^2$ values added to Figure 2), because for heavier atoms with larger σ-hole (Br, I, At) the halogen bond gets generally stronger and it becomes the dominant factor determining the interatomic distance X···O.

Partially, the differences between fitted parameters in Equation (1) for different halogens are explained by the fact that halogens simply increase in size going from F to At. Thus it is interesting to replot Figure 4, using for abscissa not the absolute values or $r$ but the "reduced" values, normalized to the sum of van der Waals radii of oxygen and halogen:

$$R = \frac{r}{R_O + R_X}, \qquad (2)$$

where $R_O$ and $R_X$ are van der Waals radii of oxygen and halogen, respectively. For this purpose we have used Bondi's radii [69] (in Å: $R_O$ = 1.52, $R_F$ = 1.47, $R_{Cl}$ = 1.75, $R_{Br}$ = 1.85, $R_I$ = 1.98, $R_{At}$ = 2.02, the latter value taken from [70]). The result is shown in Figure 5. The $R$ values could be considered as a measure of an overlap of atomic electron shells upon complexation: the closer $R$ values are to unity, the weaker the halogen bond is. For example, as fluorine-containing molecules are weaker halogen donors in the series, the $R$ values for their complexes are the largest; in contrast, for astatine-containing molecules, the $R$ values are significantly smaller.

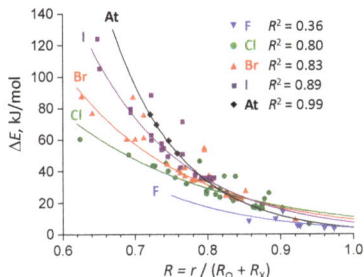

**Figure 5.** Correlation between calculated complexation energy $\Delta E$ and normalized distance parameter $R = r/(R_O + R_X)$, where $r$ is X⋯O distance and $R_O$ and $R_X$ are van der Waals radii of oxygen and halogen atoms, respectively, for R–X⋯O=PMe$_3$ halogen-bonded complexes studied in this work (X = F, Cl, Br, I, At). The solid curves correspond to Equation (3).

Due to the fact that $R$ is a simple normalization of initial $r$ values, the data sets in Figure 5 could be fitted by the same type of functions, as in Equation (1), with the same quality of the fit:

$$\Delta E(\text{kJ/mol}) = A \cdot e^{-B \cdot R}, \tag{3}$$

where the values of unitless parameters $B$ are as listed in Table 1. The spread of fitting curves in Figure 5 is significantly smaller than that in Figure 4, though there is still some difference in behavior of different halogens. Due to the increase of polarizability the decrease of electronegativity and, therefore, increase of maximal σ-hole potential of halogens from F to At, the shortening of the X⋯O bond leads to a more effective increase of the interaction energy for At and I, as compared to other halogens. This effect is still noticeable even if $R$ values (Figure 5) are taken instead of $r$ values (Figure 4) for the correlation.

The proposed correlational functions could be used to estimate the halogen bond energy in crystals in case the interatomic distance is known. Considering the scattering of data points, the precision of such estimation would be roughly ±20% over the whole range of energies. The applicability of the correlations is probably limited to halogen-bonded complexes of X⋯O type; for X⋯N, X⋯S, X⋯P and X⋯X interactions the fitted parameters might differ. Nevertheless, it could be expected that qualitatively the type fitting function would remain unchanged.

## 2.3. Complexation Energy Dependence On σ-hole Electrostatic Characteristics

The halogen's σ-hole could be characterized by $ESP_{max}$, which is the maximal value of electrostatic potential measured on the surface of equal electron density. Usually for this purpose the electron density level of 0.001 electron/Bohr$^3$ is selected. It is convenient to represent $ESP_{max}$ values in the units of energy, kJ/mol, as the energy of interaction with a virtual unit probe charge. The $ESP_{max}$ values for the halogen donor molecules considered in this work are listed in Table S1 and plotted versus $\Delta E$ in Figure 6. It is possible to fit the data sets with linear equations:

$$\Delta E(\text{kJ/mol}) = D \cdot ESP_{max}, \tag{4}$$

where the unitless parameters $D$ are again individual for a given halogen. The numerical values of coefficients $D$ are collected in Table 1. The correlation lines (see Figure 6) pass through the origin, which means that the halogen bond does not form in the absence of σ-hole ($\Delta E = 0$). Some of the deviations of the data points from linear correlations are possibly due to the presence of secondary non-covalent interactions, other than halogen bonding, i.e., not associated with the σ-hole on the halogen atom, such as various electrostatic and dispersion interactions.

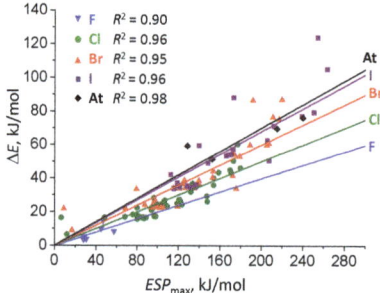

**Figure 6.** Correlation between calculated complexation energy $\Delta E$ and the extremal value of electrostatic potential $ESP_{max}$, measured in the region of σ-hole of X atom on the surface of equal electron density taken at 0.001 electron/Bohr$^3$ level for R–X···O=PMe$_3$ halogen-bonded complexes studied in this work (X = F, Cl, Br, I, At). The solid lines correspond to Equation (4).

### 2.4. Correlation Between Complexation Energy And P=O Stretching Frequency

Vibrational spectroscopy is one the most widely used methods for identification and characterization of non-covalent interactions. The formation of an intermolecular complex is usually accompanied by the decrease of the stretching vibrational frequency of the electron-donating group. This frequency shift could be used to construct correlations, allowing one to indirectly estimate other parameters of the complex, such as its energy or geometry. In Figure 7 we show the $\Delta E$ dependence on the calculated P=O stretching vibration frequency shift upon complexation, $\Delta \nu$. The overall trend is apparent: the stronger is the complex, the larger is the frequency shift. This could be rationalized as follows: the formation of halogen-bonded complex leads to the electron density transfer on the σ* molecular orbital of the electron-accepting molecule. This in turn leads to the weakening and lengthening of the P=O bond, which results in the lowering of the harmonic force constant for the P=O stretching vibration.

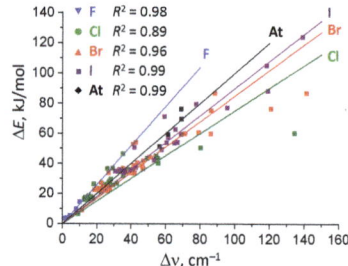

**Figure 7.** Correlation between calculated complexation energy $\Delta E$ and absolute value of the change of the P=O stretching vibration frequency upon complexation, $\Delta \nu$, for R–X···O=PMe$_3$ halogen-bonded complexes studied in this work (X = F, Cl, Br, I, At). The solid lines correspond to Equation (5).

The data sets shown in Figure 7 could be approximated by linear functions, passing through the origin:

$$\Delta E (\text{kJ/mol}) = K \cdot \Delta v, \tag{5}$$

where $K$ is the proportionality coefficient expressed in kJ/mol/cm$^{-1}$ units. The numerical values of coefficients $K$, individual for each halogen, are collected in Table 1. As was the case for other correlations presented above, the data set for fluorine-containing molecules is subject to the largest approximation errors and is likely to perform poorly beyond the complexation energy of ca. 15 kJ/mol. In order to further simplify Equation (5)—admittedly with some loss of its accuracy—it is possible to propose one universal "average" equation, suited for rough energy estimations:

$$\Delta E (\text{kJ/mol}) = 0.85 \cdot \Delta v, \tag{6}$$

where $\Delta v$ is measured in cm$^{-1}$. Here, as the average coefficient, we take the coefficient obtained by a linear fitting of the entire data set shown in Figure 7.

2.5. Correlation Between Complexation Energy And $^{31}$P NMR Chemical Shift

The $^{31}$P NMR chemical shift is very sensitive to the chemical structure of the phosphorous-containing compounds and to their environment. The "rule of thumb" is that the $^{31}$P NMR signal shifts to the high field due to the presence of electron-donating substituents and to the low field due to electron-accepting ones. This is consistent with what we observe in calculations: formation of halogen bond increases the $^{31}$P NMR chemical shift of Me$_3$PO. This is illustrated in Figure 8, where a correlation between $\Delta E$ and $\Delta \delta P$ is shown: the stronger the halogen bond is, the larger the signal shift to the low field is. The data sets plotted in Figure 6 can be fitted by linear functions:

$$\Delta E (\text{kJ/mol}) = M \cdot \Delta \delta P, \tag{7}$$

where coefficients $M$ (in kJ/mol/ppm) slightly differ for F, Cl, Br, I and At. The numerical values are collected in Table 1. In a similar way as it was done for $\Delta v$, for $\Delta \delta P$ it is possible to construct one universal correlation for rough estimations of complexation energies based of $^{31}$P NMR spectra:

$$\Delta E (\text{kJ/mol}) = 2.7 \cdot \Delta \delta P, \tag{8}$$

where $\Delta \delta P$ is measured in ppm. Again, the average coefficient refers to the fitting of the entire data set shown in Figure 8. It should be noted that despite the generally high sensitivity of $\delta P$ to the molecular structure and to non-covalent interactions [71], in the literature there are only few attempts to use $\delta P$ for the solution of the reverse spectroscopic problem for non-covalent complexes, i.e., for the finding of the complex's energy and structure based on the phosphorous chemical shift value [72–75]. Partially the reason for this might be in the high sensitivity itself, because contributions to $\delta P$ from various weak secondary non-covalent interactions might "smudge" the effect of the halogen bonding, thus strongly reducing the diagnostic value of the spectroscopic marker. In case of phosphine oxides R$_3$PO, we hope that such probe molecules are reasonably rigid and less prone to secondary interactions than, for example, phosphinic acids; if so, the correlation proposed in this work would be of practical value.

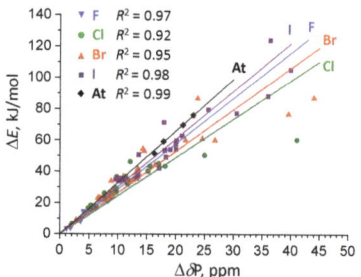

**Figure 8.** Correlation between calculated complexation energy $\Delta E$ and change of the $^{31}$P NMR chemical shift upon complexation, $\Delta\delta P$, for R–X···O=PMe$_3$ halogen-bonded complexes studied in this work (X = F, Cl, Br, I, At). The solid lines correspond to Equation (7).

In Figure S7 we test the possibility of using $\Delta\delta P$ to estimate halogen atom's electrophilicity, measured as $ESP_{max}$ value. There is significant scattering of data points which has prevented us from proposing a correlation, though it could be speculated that the overall dependence is independent of the type of halogen. A much better universal correlation could be proposed between the reduced interatomic distance $R$ (see Equation (2) for the definition) and $\Delta\delta P$ (see Figure 9).

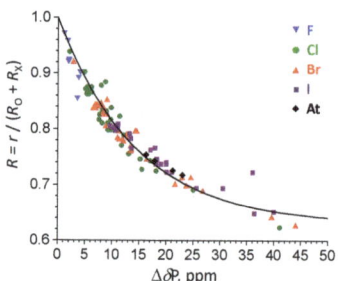

**Figure 9.** Correlation between the normalized distance parameter $R = r/(R_O + R_X)$, where $r$ is X···O distance and $R_O$ and $R_X$ are van der Waals radii of oxygen and halogen atoms, respectively, and the $^{31}$P NMR chemical shift upon complexation, $\Delta\delta P$, for R–X···O=PMe$_3$ halogen-bonded complexes studied in this work (X = F, Cl, Br, I, At). The solid line corresponds to Equation (9). $R^2$ (coefficient of determination) of the fitted curve equals to 0.933.

The data points in Figure 9 could be fit reasonably well with a single exponential function:

$$R = 0.63 + 0.37 \cdot e^{-0.07 \cdot \Delta\delta P}, \qquad (9)$$

where $\Delta\delta P$ is measured in ppm.

From the dependencies shown in Figures 7 and 10, one could expect that the effects of complexation on IR and NMR spectra are correlated. This is indeed the case, as shown in Figure 10; for all types of halogens, the data points could be approximated with a single linear dependence:

$$\Delta\delta P = 0.3 \cdot \Delta\nu, \qquad (10)$$

where $\Delta\delta P$ is in ppm and $\Delta\nu$ is in cm$^{-1}$. The high degree of correlation between two independent spectral characteristics could serve as an indication of the robustness of the selected probe molecule (Me$_3$PO in our case), as well as an indication that both in IR and NMR spectra the main effects (the change of the P=O stretching frequency and the change of the $^{31}$P NMR chemical shift) are caused by the donation of the electron density from one of the oxygen lone pairs to the halogen donor molecule.

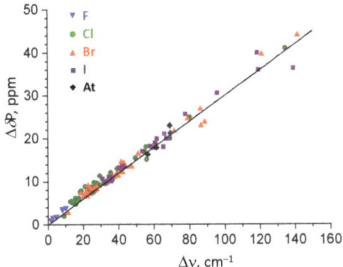

**Figure 10.** Correlation between the change of the P=O stretching vibration frequency upon complexation, $\Delta\nu$, and the change of the $^{31}$P NMR chemical shift upon complexation, $\Delta\delta$P, for R–X···O=PMe$_3$ halogen-bonded complexes studied in this work (X = F, Cl, Br, I, At). The solid line corresponds to Equation (10). $R^2$ (coefficient of determination) of the fitted line equals to 0.993.

### 2.6. QTAIM Analysis of the Electronic Structure of Complexes

Figures S3–S6 show the dependencies of complexation energy $\Delta E$ on QTAIM parameters at BCP, namely $G$, $V$, $\rho$ and $\nabla^2\rho$ (see Section 2 for more information). In this section, we summarize only the main observations. In all cases, the scattering of data points in plots is substantial, and no universal trends applicable for all halogens could be noticed. Nevertheless, it is possible to construct linear approximations such as:

$$\begin{aligned}
\Delta E(\text{kJ/mol}) &= C_G \cdot G, \\
\Delta E(\text{kJ/mol}) &= C_V \cdot V, \\
\Delta E(\text{kJ/mol}) &= C_{rho} \cdot \rho, \\
\Delta E(\text{kJ/mol}) &= C_{Lap} \cdot \nabla^2\rho,
\end{aligned} \quad (11)$$

where independent variables are expressed in the following units: $G$ in kJ/mol/Å$^3$, $V$ in kJ/mol/Å$^3$ (both $V$ and $G$ were taken as absolute values), $\rho$ in a.u./Å$^3$ and $\nabla^2\rho$ in a.u./Å$^5$. The resulting fitting coefficients are collected in Table 2. In all cases, the quality of the fit ($R^2$) was 0.8–0.9 for F-containing halogen bonded complexes and 0.91–0.99 for Cl$^-$, Br$^-$, I$^-$ or At-containing ones. The proportionality coefficients in Equation (9) are not universal: they vary for different halogens; usage of one average coefficient would lead to significant errors in energy estimations. This observation is similar to the one made in [76], where similar coefficients were proposed for Cl-, Br- and I-containing halogen bonds (see the values listed in Table 2).

**Table 2.** The numerical values of fitted parameters $C_G$, $C_V$, $C_{rho}$ and $C_{Lap}$ used in Equation (9) to fit data points shown in Figures S3–S6. For comparison, several correlation coefficients previously reported in [77] are added to the last two columns.

| Halogen Donor | This Work | | | | Ref. [77] | |
|---|---|---|---|---|---|---|
| | $C_G$, Å$^3$ | $C_V$, Å$^3$ | $C_{rho}$, Å$^3$·kJ/mol/a.u. | $C_{Lap}$, Å$^5$·kJ/mol/a.u. | $C_G$, Å$^3$ | $C_V$, Å$^3$ |
| F | 0.18 | 0.14 | 600 | 150 | | |
| Cl | 0.47 | 0.43 | 1550 | 320 | 0.47 | 0.49 |
| Br | 0.57 | 0.60 | 1300 | 400 | 0.57 | 0.58 |
| I | 0.74 | 0.70 | 1600 | 510 | 0.67 | 0.68 |
| At | 0.78 | 0.75 | 1900 | 530 | | |

## 3. Materials and Methods

### 3.1. Computational Details

The full geometry optimization of all model structures in vacuum, complexation energies (including the relaxation energy of optimized monomers and corrected for the basis set superposition error by counterpoise method [77] as a single point calculation after the geometry optimization), ESP values and spectroscopic characteristics were calculated at the DFT level of theory using Gaussian 16 software (Gaussian Inc. Wallingford, CT, USA) [78]. The visualization was done using GaussView 6.0 (Gaussian Inc. Wallingford, CT, USA) [79] and ChemCraft [80] software.

For all calculations, we used the hybrid functional M06-2x, which was previously shown to perform well for the investigation of non-covalent interaction of small molecules [81–85]. Due to the correction for dispersion interactions, this functional is well suited for the estimation of geometry and energy of halogen bonds [86]. The basis set def2-TZVPPD was selected because it includes (i) polarization functions allowing for a better description of asymmetric electron distributions in halogens; (ii) diffuse functions which describe well the relatively long-distance non-covalent bonds and (iii) parametrization of pseudopotentials necessary to describe the relativistic effects for heavy halogens, such as I and At [87].

The calculations of harmonic vibrational frequencies were used to define the shift of the P=O stretching band upon complexation as $\Delta\nu = \nu_0 - \nu$, where $\nu_0$ are $\nu$ are vibrational frequencies for the free Me$_3$PO and Me$_3$PO in complex, respectively. For the free Me$_3$PO, $\nu_0 = 1268$ см$^{-1}$. Taking into account that for all studied complexes $\nu < \nu_0$, the definition used for $\Delta\nu$ makes its values positive, which is done for convenience. The calculations of chemical shielding were performed using GIAO method. The change of $^{31}$P NMR chemical shift upon complexation was defined as $\Delta\delta P = \sigma_0(^{31}P) - \sigma(^{31}P)$, where $\sigma_0(^{31}P)$ and $\sigma(^{31}P)$ are shielding constants of $^{31}$P nucleus in the free Me$_3$PO and Me$_3$PO in complex, respectively. For the free Me$_3$PO, $\sigma_0(^{31}P) = 268$ ppm.

The topological analysis of electron density at halogen bond critical points (BCPs) was carried out within the framework of QTAIM methodology using MultiWFN software [88]. The following BCP parameters were considered: local electron kinetic ($G$) and potential ($V$) energy densities, electron density $\rho$ as well as its Laplacian $\nabla^2\rho$.

All of the abovementioned calculated halogen bond characteristics were correlated with complexation energy and in some cases between each other; the proposed linear and non-linear correlation functions were fitted by least squares method using Origin software [89]. The complexes in which the dominant interaction was not the R–X···O=P halogen bond (e.g., pnictogen bond, chalcogen bond or π-hole interaction) were not included into the regression analysis. Such complexes are marked with color in Figure S1 and Tables S1 and S2. Throughout the paper, in all the plots, we show only those data points that were actually included in the regression analysis; in other words the data for complexes for which the dominant interaction was not the halogen bond can be found only in the Supplementary Materials.

### 3.2. CCDC Data Search

The search of the relevant R–X···O=P interactions was performed using the CCDC 2020 offline data (program ConQuest 2.0.4). Search criteria: NM~X···O~P(~NM)$_3$ fragment, where (i) symbol ~ stands for any bond; (ii) X = Cl, Br, I; (iii) NM is any nonmetal; (iv) d(X···O) ≡ $r$ is less than the Bondi's vdW radii sums; (v) ∠(NM~X···O) ≡ $\beta$, 150° ≤ $\beta$ ≤ 180°; (vi) number of bonded atoms for X is 1; (vii) number of bonded atoms for P is 4; (viii) structures are non-disordered; (ix) final R1 index [I ≥ 2σ (I)] is less or equal 10%.

## 4. Conclusions

In this work, we have considered a large set of 128 halogen-bonded complexes formed by trimethylphosphine oxide and halogen donors belonging to various classes of chemical compounds.

The energies of these complexes span from 3.6 to 124 kJ/mol, while the halogen bond distances $R$, measured as a percentage of the sum of van der Waals radii of participating atoms, ranged from 100% down to 62%. The obtained distributions of interatomic distances and angles are rather similar to those obtained from the comprehensive search in the CCDC 2020 database of various RX···PO short contacts (compare Figures 1 and 2). On the one hand, the Me$_3$P=O molecule could be considered as a probe used to characterize the halogen-donating ability of isolated F-, Cl-, Br-, I- and At-containing species (the size of the σ-hole on halogen atom). On the other hand, the spectroscopic parameters of phosphine oxide involved in a R–X···O=P complex were used to determine the energy and geometry of the halogen bond. We showed that the change of $^{31}$P NMR chemical shift and P=O stretching frequency upon complexation have practically the same diagnostic value: they are well correlated with each other (Figure 10), linearly correlated to the halogen bond energy $\Delta E$ (Figures 7 and 8) and exponentially related to halogen bond geometry (Figure 10). The overall spans of spectroscopic parameters are substantial: ca. 45 ppm for the $^{31}$P NMR chemical shift and ca. 50 cm$^{-1}$ for the P=O frequency. Interestingly, the spectroscopic correlation with $R$ values is general, i.e., it is fulfilled for the whole set of complexes at once, while in many other cases correlations remain halogen-specific, i.e., different for F-, Cl, Br-, I- or At-containing halogen donors. We believe that the interdependences between halogen bond descriptors and spectroscopic markers—Equations (5)–(9)—would be useful in case direct crystallographic or calorimetric data are not available, as in the case of halogen-bonded complexes in liquids, in solutions and in other kinds of disordered media.

**Supplementary Materials:** The following are available online at http://www.mdpi.com/1420-3049/25/6/1406/s1, Figure S1: optimized structures of Me$_3$P=O···XR complexes, Table S1: calculated halogen bond geometries, complexation energies and spectroscopic parameters, Table S2: QTAIM parameters ($\rho$, $\nabla^2\rho$, $V$ and $G$) at halogen bond BCP, Figure S2: the correlation between angles $\beta$ (angle O···X–R) and the complexation energy, Table S3: geometric parameters of the R–X···O=P halogen bonds found in CCDC 2020 database for X = Cl, Br, I, Figure S3: correlation between $\Delta E$ and local kinetic energy density $G$ at BCP, Figure S4: correlation between $\Delta E$ and local potential energy density $V$ at BCP, Figure S5: correlation between $\Delta E$ and electron density $\rho$ at BCP, Figure S6: correlation between energy $\Delta E$ and Laplacian of electron density $\nabla^2\rho$ at BCP, Figure S7: correlation between the extremal value of electrostatic potential $ESP_{max}$ and change of the $^{31}$P NMR chemical shift upon complexation, $\Delta\delta P$, for Me$_3$P=O···XR complexes.

**Author Contributions:** Conceptualization, P.M.T.; Formal analysis, D.M.I.; Funding acquisition, P.M.T.; Investigation, A.S.O. and D.M.I.; Methodology, A.S.N.; Supervision, P.M.T.; Validation, A.S.O.; Writing—original draft, A.S.O. and P.M.T. All authors have read and agreed to the published version of the manuscript.

**Funding:** This research was funded by RSF garnt 18-13-00050.

**Acknowledgments:** Quantum-chemical calculations were performed at the Computing Center of St. Petersburg State University Research Park.

**Conflicts of Interest:** The authors declare no conflict of interest.

## References

1. Cavallo, G.; Metrangolo, P.; Milani, R.; Pilati, T.; Priimagi, A.; Resnati, G.; Terraneo, G. The halogen dond. *Chem. Rev.* **2016**, *116*, 2478–2601. [CrossRef]
2. Desiraju, G.R.; Ho, P.S.; Kloo, L.; Legon, A.C.; Marquardt, R.; Metrangolo, P.; Politzer, P.; Resnati, G.; Rissanen, K. Definition of the halogen bond (IUPAC Recommendations 2013). *Pure Appl. Chem.* **2013**, *85*, 1711–1713. [CrossRef]
3. Metrangolo, P.; Murray, J.S.; Pilati, T.; Politzer, P.; Resnati, G.; Terraneo, G. The fluorine atom as a halogen bond donor, viz. a positive site. *CrystEngComm.* **2011**, *13*, 6593–6896. [CrossRef]
4. Metrangolo, P.; Resnati, G. *Halogen Bonding. Fundamentals and Applications*; Springer: Berlin/Heidelberg, Germany, 2008.
5. Murray, J.S.; Lane, P.; Politzer, P. Expansion of the sigma-hole concept. *J. Mol. Model.* **2009**, *15*, 723–729. [CrossRef]
6. Politzer, P.; Murray, J.S.; Clark, T. Halogen bonding and other σ-hole interactions: A perspective. *Phys. Chem. Chem. Phys.* **2013**, *15*, 11178–11189. [CrossRef]

7. Xu, K.; Ho, D.M.; Pascal, R.A. Azaaromatic chlorides: A prescription for crystal structures with extensive nitrogen-chlorine donor-acceptor interactions. *J. Am. Chem. Soc.* **1994**, *116*, 105–110. [CrossRef]
8. Müller, M.; Albrecht, M.; Gossen, V.; Peters, T.; Hoffmann, A.; Raabe, G.; Valkonen, A.; Rissanen, K. Anion–π interactions in salts with polyhalide anions: Trapping of $I_4{}^{2-}$. *Chem. Eur. J.* **2010**, *16*, 12446–12453.
9. Itoh, T.; Nomura, S.; Nakasho, H.; Uno, T.; Kubo, M.; Tohnai, N.; Miyata, M. Halogen bond effect for single-crystal-to-single-crystal transformation: Topochemical polymerization of substituted quinodimethane. *Macromolecules* **2015**, *48*, 5450–5455. [CrossRef]
10. Dong, M.; Miao, K.; Wu, J.; Miao, X.; Li, J.; Pang, P.; Deng, W. Halogen substituent effects on concentration-controlled self-assembly of fluorenone derivatives: Halogen bond versus hydrogen bond. *J. Phys. Chem. C* **2019**, *123*, 4349–4359. [CrossRef]
11. Berski, S.; Ciunik, Z.; Drabent, K.; Latajka, Z.; Panek, J. Dominant role of C−Br⋯N halogen bond in molecular self-organization. Crystallographic and quantum-chemical study of Schiff-base-containing triazoles. *J. Phys. Chem. B* **2004**, *108*, 12327–12332. [CrossRef]
12. Ghiassi, K.B.; Wescott, J.; Chen, S.Y.; Balch, A.L.; Olmstead, M.M. Waves of halogen–halogen bond formation in the cocrystallization of hexabromobenzene and 1,2,4,5-tetrabromobenzene with $C_{70}$. *Cryst. Growth Des.* **2015**, *15*, 2480–2485. [CrossRef]
13. Liu, R.; Wang, H.; Jin, W.J. Soft-cavity-type host–guest structure of cocrystals with good luminescence behavior assembled by halogen bond and other weak interactions. *Cryst. Growth Des.* **2017**, *17*, 3331–3337. [CrossRef]
14. Vanderkooy, A.; Taylor, M.S. Solution-phase self-assembly of complementary halogen bonding polymers. *J. Am. Chem. Soc.* **2015**, *137*, 5080–5086. [CrossRef] [PubMed]
15. Sarwar, M.G.; Dragisic, B.; Salsberg, L.J.; Gouliaras, C.; Taylor, M.S. Thermodynamics of halogen bonding in solution: Substituent, structural, and solvent effects. *J. Am. Chem. Soc.* **2010**, *132*, 1646–1653. [CrossRef]
16. Libri, S.; Jasim, N.A.; Perutz, R.N.; Brammer, L. Metal fluorides form strong hydrogen bonds and halogen bonds: Measuring interaction enthalpies and entropies in solution. *J. Am. Chem. Soc.* **2008**, *130*, 7842–7844. [CrossRef]
17. Benesi, H.A.; Hildebrand, J.H. A spectrophotometric investigation of the interaction of iodine with aromatic hydrocarbons. *J. Am. Chem. Soc.* **1949**, *71*, 2703–2707. [CrossRef]
18. Hassel, O.; Hvoslef, J. The structure of bromine 1,4-dioxanate. *Acta Chem. Scand.* **1954**, *8*, 873. [CrossRef]
19. Pearcy, A.C.; Mason, K.A.; El-Shall, M.S. Ionic hydrogen and halogen bonding in the gas phase association of acetonitrile and acetone with halogenated benzene cations. *J. Phys. Chem. A* **2019**, *123*, 1363–1371. [CrossRef]
20. Gillis, E.A.L.; Demireva, M.; Sarwar, M.G.; Chudzinski, M.G.; Taylor, M.S.; Williams, E.R.; Fridgen, T.D. Structure and energetics of gas phase halogen-bonding in mono-, bi-, and tri-dentate anion receptors as studied by BIRD. *Phys. Chem. Chem. Phys.* **2013**, *15*, 7638–7647. [CrossRef]
21. Wegeberg, C.; Donald, W.A.; McKenzie, C.J. Noncovalent halogen bonding as a mechanism for gas-phase clustering. *J. Am. Soc. Mass Spectr.* **2017**, *28*, 2209–2216. [CrossRef]
22. Danelius, E.; Andersson, H.; Jarvoll, P.; Lood, K.; Gräfenstein, J.; Erdélyi, M. Halogen bonding: A powerful tool for modulation of peptide conformation. *Biochemistry* **2017**, *56*, 3265–3272. [CrossRef] [PubMed]
23. Fanfrlík, J.; Kolář, M.; Kamlar, M.; Hurný, D.; Ruiz, F.X.; Cousido-Siah, A.; Mitschler, A.; Řezáč, J.; Munusamy, E.; Lepšík, M.; et al. Modulation of aldose reductase inhibition by halogen bond tuning. *ACS Chem. Biol.* **2013**, *8*, 2484–2492.
24. Christopherson, J.-C.; Topić, F.; Barrett, J.C.; Friščić, T. Halogen-bonded cocrystals as optical materials: Next-generation control over light–matter interactions. *Cryst. Growth Des.* **2018**, *18*, 1245–1259. [CrossRef]
25. Priimagi, A.; Cavallo, G.; Metrangolo, P.; Resnati, G. The halogen bond in the design of functional supramolecular materials: Recent advances. *Acc. Chem. Res.* **2013**, *46*, 2686–2695. [CrossRef] [PubMed]
26. Chudzinski, M.G.; McClary, C.A.; Taylor, M.S. Anion receptors composed of hydrogen- and halogen-bond donor groups: Modulating selectivity with combinations of distinct noncovalent interactions. *J. Am. Chem. Soc.* **2011**, *133*, 10559–10567. [CrossRef] [PubMed]
27. Zheng, Q.-N.; Liu, X.-H.; Chen, T.; Yan, H.-J.; Cook, T.; Wang, D.; Stang, P.J.; Wan, L.-J. Formation of halogen bond-based 2D supramolecular assemblies by electric manipulation. *J. Am. Chem. Soc.* **2015**, *137*, 6128–6131. [CrossRef]
28. Liao, W.-Q.; Tang, Y.-Y.; Li, P.-F.; You, Y.-M.; Xiong, R.-G. Competitive halogen bond in the molecular ferroelectric with large piezoelectric response. *J. Am. Chem. Soc.* **2018**, *140*, 3975–3980. [CrossRef]

29. Zhang, L.; Liu, X.; Su, J.; Li, J. First-principles study of molecular adsorption on Lead iodide perovskite surface: A case study of halogen bond passivation for solar cell application. *J. Phys. Chem. C* **2016**, *120*, 23536–23541. [CrossRef]
30. Lieffrig, J.; Jeannin, O.; Guizouarn, T.; Auban-Senzier, P.; Fourmigué, M. Competition between the C–H···N hydrogen bond and C–I···N halogen bond in TCNQF$_n$ ($n$ = 0, 2, 4) salts with variable charge transfer. *Cryst. Growth Des.* **2012**, *12*, 4248–4257. [CrossRef]
31. Vanderkooy, A.; Pfefferkorn, P.; Taylor, M.S. Self-assembly of polymer nanostructures through halogen bonding interactions of an iodoperfluoroarene-functionalized polystyrene derivative. *Macromolecules* **2017**, *50*, 3807–3817. [CrossRef]
32. Cariati, E.; Cavallo, G.; Forni, A.; Leem, G.; Metrangolo, P.; Meyer, F.; Pilati, T.; Resnati, G.; Righetto, S.; Terraneo, G.; et al. Self-complementary nonlinear optical-phores targeted to halogen bond-driven self-assembly of electro-optic materials. *Cryst. Growth Des.* **2011**, *11*, 5642–5648. [CrossRef]
33. Takeda, Y.; Hisakuni, D.; Lin, C.-H.; Minakata, S. 2-Halogenoimidazolium salt catalyzed aza-Diels–Alder reaction through halogen-bond formation. *Org. Lett.* **2015**, *17*, 318–321. [CrossRef] [PubMed]
34. Jungbauer, S.H.; Huber, S.M. Cationic multidentate halogen-bond donors in halide abstraction organocatalysis: Catalyst optimization by preorganization. *J. Am. Chem. Soc.* **2015**, *137*, 12110–12120. [CrossRef] [PubMed]
35. Kazi, I.; Guha, S.; Sekar, G. CBr$_4$ as a halogen bond donor catalyst for the selective activation of benzaldehydes to synthesize α,β-unsaturated ketones. *Org. Lett.* **2017**, *19*, 1244–1247. [CrossRef] [PubMed]
36. Kaasik, M.; Metsala, A.; Kaabel, S.; Kriis, K.; Järving, I.; Kanger, T. Halo-1,2,3-triazolium salts as halogen bond donors for the activation of imines in dihydropyridinone synthesis. *J. Org. Chem.* **2019**, *84*, 4294–4303. [CrossRef]
37. Lefèvre, G.; Franc, G.; Adamo, C.; Jutand, A.; Ciofini, I. Influence of the formation of the halogen bond ArX···N on the mechanism of diketonate ligated copper-catalyzed amination of aromatic halides. *Organometallics* **2012**, *31*, 914–920. [CrossRef]
38. Combe, S.H.; Hosseini, A.; Song, L.; Hausmann, H.; Schreiner, P.R. Catalytic halogen bond activation in the benzylic C–H bond iodination with iodohydantoins. *Org. Lett.* **2017**, *19*, 6156–6159. [CrossRef]
39. Xu, Z.; Yang, Z.; Liu, Y.; Lu, Y.; Chen, K.; Zhu, W. Halogen bond: Its role beyond drug–target binding affinity for drug discovery and development. *J. Chem. Inf. Model.* **2014**, *54*, 69–78. [CrossRef]
40. Baldrighi, M.; Cavallo, G.; Chierotti, M.R.; Gobetto, R.; Metrangolo, P.; Pilati, T.; Resnati, G.; Terraneo, G. Halogen bonding and pharmaceutical cocrystals: The case of a widely used preservative. *Mol. Pharm.* **2013**, *10*, 1760–1772. [CrossRef]
41. Wang, W.; Zhang, Y.; Ji, B. On the difference of the properties between the blue-shifting halogen bond and the blue-shifting hydrogen bond. *J. Phys. Chem. A* **2010**, *114*, 7257–7260. [CrossRef]
42. Roper, L.C.; Präsang, C.; Kozhevnikov, V.N.; Whitwood, A.C.; Karadakov, P.B.; Bruce, D.W. Experimental and theoretical study of halogen-bonded complexes of DMAP with di- and triiodofluorobenzenes. A complex with a very short N···I halogen bond. *Cryst. Growth Des.* **2010**, *10*, 3710–3720. [CrossRef]
43. Lu, Y.; Li, H.; Zhu, X.; Zhu, W.; Liu, H. How does halogen bonding behave in solution? A theoretical study using implicit solvation model. *J. Phys. Chem. A* **2011**, *115*, 4467–4475. [CrossRef] [PubMed]
44. Lindblad, S.; Mehmeti, K.; Veiga, A.X.; Nekoueishahraki, B.; Gräfenstein, J.; Erdélyi, M. Halogen bond asymmetry in solution. *J. Am. Chem. Soc.* **2018**, *140*, 13503–13513. [CrossRef] [PubMed]
45. Wang, W.; Wong, N.-B.; Zheng, W.; Tian, A. Theoretical study on the blueshifting halogen bond. *J. Phys. Chem. A* **2004**, *108*, 1799–1805. [CrossRef]
46. Wang, C.; Danovich, D.; Mo, Y.; Shaik, S. On the nature of the halogen bond. *J. Chem. Theory Comput.* **2014**, *10*, 3726–3737. [CrossRef] [PubMed]
47. Wang, L.; Gao, J.; Bi, F.; Song, B.; Liu, C. Toward the development of the potential with angular distortion for halogen bond: A comparison of potential energy surfaces between halogen bond and hydrogen bond. *J. Phys. Chem. A* **2014**, *118*, 9140–9147. [CrossRef]
48. Cuypers, R.; Sudhçlter, E.J.R.; Zuilhof, H. Hydrogen bonding in phosphine oxide/phosphate–phenol complexes. *Chem. Phys. Chem.* **2010**, *11*, 2230–2240. [CrossRef]
49. Kolling, O.W. Triethylphosphine oxide as a probe of weak hydrogen bond donor behavior. *Trans. Kansas Acad. Sci.* **1984**, *87*, 115–118. [CrossRef]

50. Hilliard, C.R.; Kharel, S.; Cluff, K.J.; Bhuvanesh, N.; Gladysz, J.A.; Blümel, J. Structures and unexpected dynamic properties of phosphine oxides adsorbed on silica surfaces. *Chem. Eur. J.* **2014**, *20*, 17292–17295. [CrossRef]
51. Tupikina, E.Y.; Bodensteiner, M.; Tolstoy, P.M.; Denisov, G.S.; Shenderovich, I.G. P=O Moiety as an ambidextrous hydrogen bond acceptor. *J. Phys. Chem. C* **2018**, *122*, 1711–1720. [CrossRef]
52. Oh, S.Y.; Nickels, C.W.; Garcia, F.; Jones, W.; Fris, T. Switching between halogen- and hydrogen-bonding in stoichiometric variations of a cocrystal of a phosphine oxide. *Cryst. Eng. Comm.* **2012**, *14*, 6110–6114. [CrossRef]
53. Maugeri, L.; Lébl, T.; Cordes, D.B.; Slawin, A.M.Z.; Philp, D. Cooperative binding in a phosphine oxide-based halogen bonded dimer drives supramolecular oligomerization. *J. Org. Chem.* **2017**, *82*, 1986–1995. [CrossRef]
54. Shipov, A.E.; Makarov, M.V.; Petrovskii, P.V.; Rybalkina, E.Y.; Nelyubina, Y.V.; Odinets, I.L. 3,5-Bis(arylidene)piperid-4-ones containing 1,3,2-oxazaphosphorinane moieties: Synthesis and antitumor activity. *Heteroat. Chem.* **2013**, *24*, 191–199. [CrossRef]
55. Tarahhomi, A.; Pourayoubi, M.; Golen, J.A.; Zargaran, P.; Elahi, B.; Rheingold, A.L.; Leyva Ramírez, M.A.; Mancilla Percino, T. Hirshfeld surface analysis of new phosphoramidates. *Acta Cryst.* **2013**, *B69*, 260–270. [CrossRef]
56. Kongprakaiwoot, N.; Bultman, M.S.; Luck, R.L.; Urnezius, E. Synthesis and structural characterizations of para-bis(dialkyl/diarylphosphino)phenylenes built around tetrahalogenated benzene cores. *Inorg. Chim. Acta* **2005**, *358*, 3423–3429. [CrossRef]
57. Nayak, S.K.; Terraneo, G.; Forni, A.; Metrangolo, P.; Resnati, G. C–Br···O supramolecular synthon: In situ cryocrystallography of low melting halogen-bonded complexes. *Cryst. Eng. Comm.* **2012**, *14*, 4259–4261. [CrossRef]
58. Nicolas, E.; Cheisson, T.; de Jong, G.B.; Tazelaara, C.J.G.; Slootwega, J.C. A new synthetic route to the electron-deficient ligand tris-(3,4,5-tri bromo pyrazol-1-yl)phosphine oxide. *Acta Cryst.* **2016**, *C72*, 846–849.
59. Matczak-Jon, E.; Slepokura, K.; Kafarskia, P. [(5-bromopyridinium-2-ylamino)(phosphono)methyl]phosphonate. *Acta Cryst.* **2006**, *C62*, 132–135. [CrossRef] [PubMed]
60. Xu, Y.; Champion, L.; Gabidullin, B.; Bryce, D.L. A Kinetic study of mechanochemical halogen bond formation by in-situ $^{31}$P solid-state NMR spectroscopy. *Chem. Comm.* **2017**, *53*, 9930–9933. [CrossRef]
61. Griffiths, D.V.; Harris, J.E.; Miller, R.J. Dimethyl 2-iodobenzoylphosphonate, an unusual example of a crystalline α-ketophosphonate. *Acta Cryst.* **1997**, *C53*, 1462–1464. [CrossRef]
62. Mayer, U.; Gutmann, V.; Gerger, W. The acceptor number – a quantitative empirical parameter for the electrophilic properties of solvents. *Monat. fur Chemie* **1975**, *106*, 1235–1257. [CrossRef]
63. Pahl, J.; Brand, S.; Elsen, H.; Harder, S. Highly Lewis acidic cationic alkaline earth metal complexes. *Chem. Commun.* **2018**, *54*, 8685–8688. [CrossRef]
64. Beckett, M.A.; Brassington, D.S.; Coles, S.J.; Hursthouse, M.B. Lewis acidity of tris(pentafluorophenyl)borane: Crystal and molecular structure of B(C$_6$F$_5$)$_3$·OPEt$_3$. *Inorg. Chem. Commun.* **2000**, *3*, 530–533. [CrossRef]
65. Hamilton, P.A.; Murrells, T.P. Kinetics and mechanism of the reactions of PH$_3$ with O(3P) and N(4S) atoms. *J. Chem. Soc. Faraday Trans.* **1985**, *2*, 1531–1541. [CrossRef]
66. Zhao, Y.; Truhlar, D.G. The M06 suite of density functionals for main group thermochemistry, thermochemical kinetics, noncovalent interactions, excited states, and transition elements: Two new functionals and systematic testing of four M06-class functionals and 12 other functionals. *Theor. Chem. Acc.* **2000**, *120*, 215–241.
67. Bader, R.F.W. A quantum theory of molecular structure and its applications. *Chem. Rev.* **1991**, *91*, 893–928. [CrossRef]
68. Chen, Z.; Wang, G.; Xu, Z.; Wang, J.; Yu, Y.; Cai, T.; Shao, Q.; Shi, J.; Zhu, W. How do distance and solvent affect halogen bonding involving negatively charged donors? *J. Phys. Chem. B* **2016**, *120*, 8784–8793. [CrossRef] [PubMed]
69. Bondi, A. Van der Waals volumes and radii. *J. Phys. Chem.* **1964**, *68*, 441–451. [CrossRef]
70. Mantina, M.; Chamberlin, A.C.; Valero, R.; Cramer, C.J.; Truhlar, D.G. Consistent van der Waals radii for the whole main group. *J. Phys. Chem.* **2009**, *113*, 5806–5812. [CrossRef]
71. Latypov, S.K.; Polyancev, F.M.; Yakhvarov, D.G.; Sinyashin, O.G. Quantum chemical calculations of $^{31}$P NMR chemical shifts: Scopes and limitations. *Phys. Chem. Chem. Phys.* **2015**, *17*, 6976–6987. [CrossRef]

72. Alkorta, I.; Elguero, J. Is it possible to use the $^{31}$P chemical shifts of phosphines to measure hydrogen bond acidities (HBA)? A comparative study with the use of the $^{15}$N chemical shifts of amines for measuring HBA. *J. Phys. Org. Chem.* **2017**, *30*, e3690. [CrossRef]
73. Mulloyarova, V.V.; Giba, I.S.; Kostin, M.A.; Denisov, G.S.; Shenderovich, I.G.; Tolstoy, P.M. Cyclic trimers of phosphinic acids in polar aprotic solvent: Symmetry, chirality and H/D isotope effects on NMR chemical shifts. *Phys. Chem. Chem. Phys.* **2018**, *20*, 4901–4910. [CrossRef] [PubMed]
74. Mulloyarova, V.V.; Giba, I.S.; Denisov, G.S.; Tolstoy, P.M. Conformational mobility and proton transfer in hydrogen-bonded dimers and trimers of phosphinic and phosphoric acids. *J. Phys. Chem A* **2019**, *123*, 6761–6771. [CrossRef] [PubMed]
75. Giba, I.S.; Mulloyarova, V.V.; Denisov, G.S.; Tolstoy, P.M. Influence of hydrogen bonds in 1:1 complexes of phosphinic acids with substituted pyridines on $^1$H and $^{31}$P NMR chemical shifts. *J. Phys. Chem. A* **2019**, *123*, 2252–2260. [CrossRef] [PubMed]
76. Bartashevich, E.V.; Tsirelson, V.G. Interplay between non-covalent interactions in complexes and crystals with halogen bonds. *Russ. Chem. Rev.* **2014**, *83*, 1181–1203. [CrossRef]
77. van Duijneveldt, F.B.; van Duijneveldt-van de Rijdt, J.G.C.M.; van Lenthe, J.H. State of the art in counterpoise theory. *Chem. Rev.* **1994**, *94*, 1873–1885. [CrossRef]
78. Frisch, M.J.; Trucks, G.W.; Schlegel, H.B.; Scuseria, G.E.; Robb, M.A.; Cheeseman, J.R.; Scalmani, G.; Barone, V.; Petersson, G.A.; Nakatsuji, H.; et al. *Gaussian 16, Revision C.01*; Gaussian, Inc.: Wallingford, CT, USA, 2016.
79. Available online: https://gaussian.com/gaussview6/ (accessed on 27 February 2020).
80. Available online: http://www.chemcraftprog.com (accessed on 27 February 2020).
81. Zimin, D.P.; Dar'in, D.V.; Eliseeva, A.A.; Novikov, A.S.; Rassadin, V.A.; Kukushkin, V.Y. Gold-catalyzed functionalization of semicarbazides with terminal alkynes to achieve substituted semicarbazones. *Eur. J. Org. Chem.* **2019**, *2019*, 6094–6100. [CrossRef]
82. Ivanov, D.M.; Kinzhalov, M.A.; Novikov, A.S.; Ananyev, I.V.; Romanova, A.A.; Boyarskiy, V.P.; Haukka, M.; Kukushkin, V.Y. H$_2$C(X)–X···X– (X = Cl, Br) halogen bonding of dihalomethanes. *Cryst. Growth Des.* **2017**, *17*, 1353–1362. [CrossRef]
83. Rozhkov, A.V.; Krykova, M.A.; Ivanov, D.M.; Novikov, A.S.; Sinelshchikova, A.A.; Volostnykh, M.V.; Konovalov, M.A.; Grigoriev, M.S.; Gorbunova, Y.G.; Kukushkin, V.Y. Reverse arene sandwich structures based upon π-Hole···[MII] ($d^8$ M=Pt, Pd) interactions, where positively charged metal centers play the role of a nucleophile. *Angew. Chem. Int. Ed.* **2019**, *58*, 4164–4168. [CrossRef]
84. Burianova, V.K.; Bolotin, D.S.; Mikherdov, A.S.; Novikov, A.S.; Mokolokolo, P.P.; Roodt, A.; Boyarskiy, V.P.; Dar'in, D.; Krasavin, M.; Suslonov, V.V.; et al. Mechanism of generation of closo-decaborato amidrazones. Intramolecular non-covalent B–H···π(Ph) interaction determines stabilization of the configuration around the amidrazone C=N bond. *New J. Chem.* **2018**, *42*, 8693–8703. [CrossRef]
85. Kashina, M.V.; Kinzhalov, M.A.; Smirnov, A.S.; Ivanov, D.M.; Novikov, A.S.; Kukushkin, V.Y. Dihalomethanes as bent bifunctional XB/XB-donating building blocks for construction of metal-involving halogen bonded hexagons. *Chem. Asian J.* **2019**, *14*, 3915–3920. [CrossRef] [PubMed]
86. Kozuch, S.; Martin, J.M.L. Halogen bonds: Benchmarks and theoretical analysis. *J. Chem. Theor. Comput.* **2013**, *9*, 1918–1931. [CrossRef] [PubMed]
87. Kuchle, W.; Dolg, M.; Stoll, H.; Preuss, H. Energy-adjusted pseudopotentials for the actinides. Parameter sets and test calculations for thorium and thorium monoxide. *J. Chem. Phys.* **1994**, *100*, 7535–7542. [CrossRef]
88. Lu, T.; Chen, F. Multiwfn: A multifunctional wavefunction analyzer. *J. Comput. Chem.* **2012**, *33*, 580–592. [CrossRef] [PubMed]
89. Available online: https://www.originlab.com/ (accessed on 27 February 2020).

**Sample Availability:** Calculation output files for all studied complexes are available from the authors.

© 2020 by the authors. Licensee MDPI, Basel, Switzerland. This article is an open access article distributed under the terms and conditions of the Creative Commons Attribution (CC BY) license (http://creativecommons.org/licenses/by/4.0/).

Article

# Adduct under Field—A Qualitative Approach to Account for Solvent Effect on Hydrogen Bonding

Ilya G. Shenderovich [1,*] and Gleb S. Denisov [2]

[1] Institute of Organic Chemistry, University of Regensburg, Universitaetstrasse 31, 93053 Regensburg, Germany
[2] Department of Physics, Saint-Petersburg State University, 198504 Saint-Petersburg, Russia; gldenisov@yandex.ru
* Correspondence: Ilya.Shenderovich@ur.de; Tel.:+49-941-9434027

Academic Editor: James Sherwood
Received: 1 January 2020; Accepted: 20 January 2020; Published: 21 January 2020

**Abstract:** The location of a mobile proton in acid-base complexes in aprotic solvents can be predicted using a simplified Adduct under Field (AuF) approach, where solute–solvent effects on the geometry of hydrogen bond are simulated using a fictitious external electric field. The parameters of the field have been estimated using experimental data on acid-base complexes in $CDF_3/CDClF_2$. With some limitations, they can be applied to the chemically similar $CHCl_3$ and $CH_2Cl_2$. The obtained data indicate that the solute–solvent effects are critically important regardless of the type of complexes. The temperature dependences of the strength and fluctuation rate of the field explain the behavior of experimentally measured parameters.

**Keywords:** solvent effect; hydrogen bond; NMR; condensed matter; polarizable continuum model; reaction field; external electric field; proton transfer

---

## 1. Introduction

Proton transfer represents the simplest possible chemical reaction [1] and is ubiquitous in chemistry [2,3], material science [4–6], and biology [7,8]. In the latter case, the complexity of the process can increase to a hydrogen atom transfer [9,10]. In condensed matter, the mechanism and the pathway of proton transfer depend on the local environment. As a result, the study of proton transfer processes in a given system can be used as a tool to study the local environment. In most cases, it will require a theoretical simulation of the proton transfer under question. Such simulations are still very challenging as they depend on a compromise between the size of the modeled molecular system and the quality of accounting for intermolecular interactions. The size should be large enough to include all relevant interactions; the quality should be good enough to estimate their effect correctly. One may prefer to simulate a given molecular system in condensed matter using oversimplified approaches, looking only for a qualitative description of the system. Often, such approaches are fully justified. The available theories of nonadiabatic [11] and adiabatic [12] proton transfer reactions provide a useful background for understanding experimental results as on reversible proton transfer in the Zundel cation [13–15] as well as on fast proton dynamics in general [16–18]. The precision of such analysis can be improved further [19]. However, the most challenging part is to account for the effect of fluctuating solute–solvent interactions [20–22].

Often, one needs to restrict proton mobility in order to stabilize individual structures. This is especially important for high-resolution nuclear magnetic resonance spectroscopy (NMR) whose characteristic time is of the order of $10^{-3}$ s. Basically, proton and molecular exchange can be suppressed by lowering the temperature. However, when studying intermolecular interactions in solution, one is strictly limited with the available temperature range. For aprotic polar solvents, the lowest possible temperature is about 100 K [23]. This temperature is not always low enough to affect proton dynamics [24,25]. Another problem is that the required temperature depends on intermolecular interactions in a complex way [26–28]. Even the structures of complexes with strong noncovalent interactions are affected by interactions with the environment [29–32]. In solution, this can be visualized by molecular dynamics simulations [33,34]. Thus, conventional gas-phase calculations can neither be used to predict at what temperature in a given molecular system proton exchange can be suppressed in a given solvent nor to simulate the mean structure of the system in solution.

Solvent effect can be divided into two parts: (i) fluctuating solute–solvent specific interactions and (ii) macroscopic electric field. The polarizable continuum model (PCM) includes only the latter effect [35]. As a result, this model is not sufficient to simulate the structure of noncovalently bound complexes in polar solvents [36]. The effects of specific solute–solvent interactions are to some extent implicitly included in the SMD (Solvation Model based on Density) model [37]. This model uses a number of solvent-specific parameters. In reality, the tabulated values of Abraham's hydrogen bond acidity and basicity, aromaticity, and electronegative halogenicity of solvents are not always the optimal choice for a given molecular system. The temperature dependence of these parameters is not known. Thus, given standard conditions, this model can be a good approximation and would fail otherwise. The problem can be overcome by using molecular dynamics approaches. However, they are computationally consuming and challenging when in non-aqueous solutions [38].

Alternatively, the effect of environment can be simulated using a fictitious external electric field [39–41]. The main advantage of this approach in relation to complexes with noncovalent interactions is that their experimental structures can be reproduced using only one parameter—the strength of the external electric field. The physical meaning of this field is illustrated in Figure 1. For the sake of simplicity, we consider a hydrogen-bonded (H-bond) complex in an aprotic polar solvent. The strongest intermolecular interaction in this complex is the acid-base H-bond. The geometry of this bond is affected by the macroscopic electric field generated by dipole moments of solvent molecules. This effect can be simulated by the PCM model. Besides that, there are weak yet multiple interactions with solvent molecules. They also cause changes in the electron density in the acid and base that affect the position of the mobile proton. The PCM model ignores this effect. The SMD model can include this effect through the empirical parameters. Using the external electric field model, we can estimate the relative amplitudes of both macroscopic and specific effects on the properties of the hydrogen bond under question when simulating a given experimental property of H-bond with the external field alone and in combination with the PCM model. The efficiency of this Adduct under Field (AuF) approach was demonstrated using a complex of hydrogen fluoride with pyridine [42].

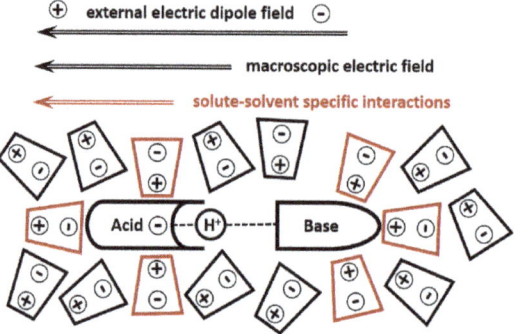

**Figure 1.** The direction of the external electric field that simulates the effect of solute–solvent interactions on the H-bond.

In this work, we use the AuF model in order to simulate experimentally observed solvent-driven proton transfer in a number of H-bonded complexes. The aim of this study is to formulate a simplified computational approach capable of predicting the temperature at which proton exchange will be suppressed in any given solute–solvent system. The model molecular systems are shown in Figure 2. These complexes have been experimentally studied in the past in a liquid $CDF_3/CDClF_2$ mixture, exhibiting a dielectric constant between 20 at 170 K and 38 at 103 K [23]. The proton-bound homodimer of pyridine (1) does not have chemically active sites exposed to the solvent while the carboxylic moiety in 2–6 can specifically interact with solvent molecules [34].

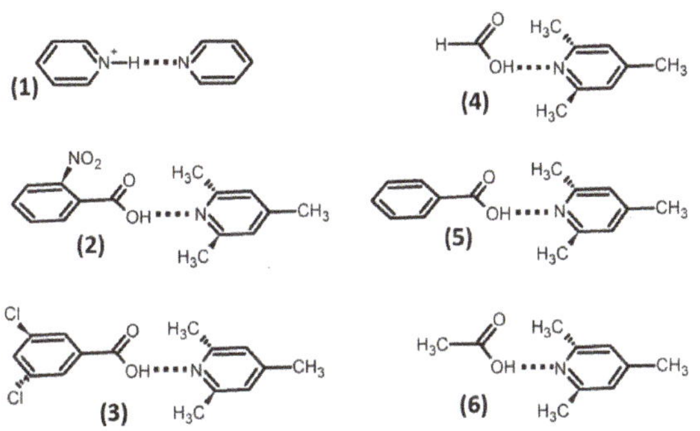

**Figure 2.** H-bonded complexes studied in this paper: proton-bound homodimer of pyridine (1) and complexes of 2,4,6-trimethylpyridine (collidine) with 2-nitrobenzoic acid (2), 3,5-dichlorobenzoic acid (3), formic acid (4), benzoic acid (5), and acetic acid (6).

Proton-bound homodimers can be of two types—symmetric, in which case the partners equally share the binding proton, and asymmetric, where the proton has a stronger bond to one of the partners at any given moment in time [43]. In the proton-bound homodimers of pyridine derivatives in $CDF_3/CDClF_2$ mixtures, the mobile proton jumps between the two bases faster than $10^3$ s$^{-1}$ down to 120 K [44] and slower than $10^{11}$ s$^{-1}$ up to 290 K [45].

In Table 1 $^1J(^{15}N^1H)$ scalar coupling constants in 2–6 in $CDF_3/CDClF_2$ solution are collected [46]. These constants were measured at different temperatures—the reason being that above these temperatures the solvent-driven exchange between O-H···N and O$^-$···[H-N]$^+$ forms of the complexes was fast on the NMR time scale. Proton tautomerism in such complexes has previously been studied in detail [34]. For our purpose, it is important that the process strongly depends on the $pK_a$ of the involved acid. As a result, the solute–solvent interactions can be analyzed in a large temperature range. We know that in the O$^-$···[H-N]$^+$ form $^1J(^{15}N^1H) \gtrsim 90$ Hz [47]. Thus, for some of these complexes, the tautomerism can be slow on the NMR time scale of chemical shifts and fast on the NMR time scale of scalar couplings. However, such aspects are beyond the precision of our qualitative model.

**Table 1.** Experimental $^1J(^{15}N^1H)$ scalar couplings in 2–6 in $CDF_3/CDClF_2$ solution [46].

| Complex | $^1J(^{15}N^1H)$, Hz | T, K | $pK_a$ [1] |
|---|---|---|---|
| 2 | −87.0 | 200 | 2.16 |
| 3 | −81.1 | 130 | 3.46 |
| 4 | −79.1 | 120 | 3.75 |
| 5 | −76.9 | 120 | 4.19 |
| 6 | −65.4 | 110 | 4.75 |

[1] The $pK_a$'s of the involved acids.

## 2. Results

### 2.1. Proton-Bound Homodimer of Pyridine

Figure 3a shows the potential energy curve of a non-adiabatic proton transfer in **1** under the PCM approximation at $\varepsilon = 29.3$. The minimum energy structures of the pyridines of **1** are not equal. Therefore, when the mobile proton is transferred from one pyridine to the other while all other atoms are fixed, the second minimum has a larger energy. In reality, this fictitious profile is not present and only shown to illustrate further changes. The ground vibrational level of the proton is higher than the energy of the transition state. The frequency of the stretching vibration ($v_{NHN}$) estimated under the harmonic approximation is 2486 cm$^{-1}$. While the potential is anharmonic, this value is a rough estimate and is given for illustrative purposes only [48]. The accuracy of the calculations can only be increased at the cost of making them very time-consuming [49,50]. The value of $\varepsilon$ can also be challenged; in CDF$_3$/CDClF$_2$ solution at about 130 K $\varepsilon \approx 30$ [23]. However, the non-adiabatic proton transfer depends on an optical dielectric constant of about 2 [16]. Under the gas phase harmonic approximation, $v_{NHN} = 2142$ cm$^{-1}$. Thus, a qualitatively similar potential surface will be observed for any value of $\varepsilon$. We are interested in the situation when this transfer is suppressed. What is important is that solvent polarization alone cannot cause this effect.

CDF$_3$/CDClF$_2$ solution cannot be simulated using the SMD approximation because its parameters are not known. Instead, chemically similar CH$_2$Cl$_2$ can be used. Although $v_{NHN}$ increases under this approximation to 2517 cm$^{-1}$, it is still higher than the energy of the transition state, Figure 3b—meaning that this model cannot reproduce the experimentally observed single-well location of the mobile proton in **1**.

The single-well location becomes possible in the presence of an external electric field. Under the PCM approximation ($\varepsilon = 29.3$) and the field of 0.001 a.u., the energy of the ground vibrational level of the mobile proton is very close to the energy of the transition state, Figure 3c. At 0.002 a.u., the former is lower than the latter, Figure 3d. This increase of the field is accompanied by an increase of $v_{NHN}$ from 2593 cm$^{-1}$ to 2684 cm$^{-1}$, Figure 3c,d. Thus, the experimentally observed proton jumps in **1** can be simulated under the PCM approximation and $\varepsilon \approx 30$ when the strength of the external field is above 0.001–0.002 a.u. There is no hard criteria for choosing the most appropriate value of the field. We can only state that the lower limit of its strength is 0.001 a.u. Within the gas phase approximation, this limits increases to at least 0.003 a.u., Figure 3e,f.

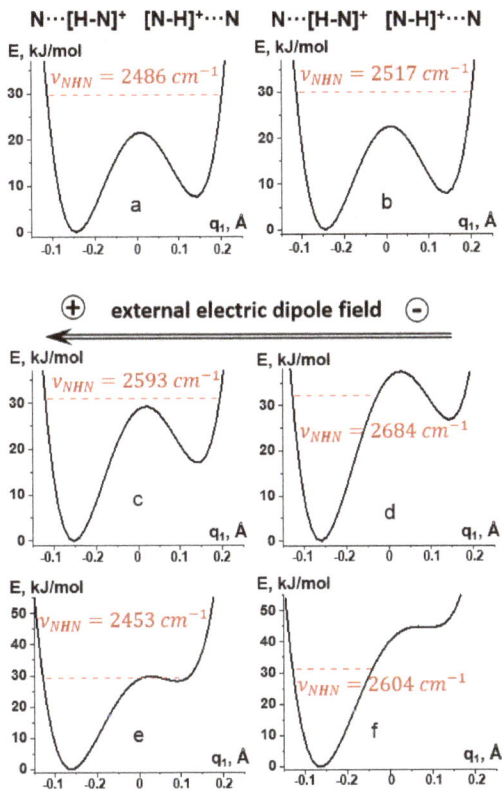

**Figure 3.** Potential energy curve of a proton transfer within **1** at different approximations: (**a**) PCM ($\varepsilon$=29.3), (**b**) SMD (CH$_2$Cl$_2$, $\varepsilon$=8.9), (**c**) PCM ($\varepsilon$=29.3) and the external electric field of 0.001 a.u., (**d**) PCM ($\varepsilon$=29.3) and the external electric field of 0.002 a.u., (**e**) the external electric field of 0.003 a.u., and (**f**) the external electric field of 0.005 a.u. Dashed lines indicate the energy of the ground state. $\nu_{NHN}$ are the frequencies of the mobile proton stretching vibration. q$_1$ corresponds to the distance of the mobile proton with respect to the H-bond center [51].

## 2.2. Complexes of Collidine with Acids

In Table 2, geometric parameters of H-bond in **4** under different approximations are reported. Although these parameters depend on the level of approximation, the mobile proton is located at the acid in all cases. Only at $\varepsilon$ > 29 do there appear higher energy local minima on the potential energy curve of proton transfer that correspond to the proton location at collidine. Taking into account a qualitative character of our analysis, we studied the effect of the external electric field on the location of the mobile proton in **2–6** at a computationally efficient $w$B97XD/def2svp approximation. We also restricted our analysis to the comparison of the difference between the energies of the two minima (proton at acid and proton at base) on the potential energy curve of proton transfer. The values of $\varepsilon$ under the PCM approximation were taken equal to 12.5 for **2** and 29.3 for **3–6**. These values are close to the dielectric constant of CDF$_3$/CDClF$_2$ solution at 200 K and 130 K, respectively [23]. There is no need to select $\varepsilon$ with a higher precision because Table 2 clearly demonstrates that, at $\varepsilon$ > 10, its effect on H-bond geometry remains rather constant.

Table 2. H-bond geometry of **4** under different DFT (Density Functional Theory) approximations.

| DFT Functional | Basis Set | PCM, ε | N···H, Å | N ... O, Å |
|---|---|---|---|---|
| wB97XD | def2svp | – | 1.689 | 2.696 |
| wB97XD | def2svpp | – | 1.655 | 2.674 |
| wB97XD | def2tzvp | – | 1.701 | 2.709 |
| wB97XD | def2tzvpp | – | 1.707 | 2.712 |
| B2PLYPD3, gd3 | def2svp | – | 1.692 | 2.697 |
| B2PLYPD3, gd3 | def2svpp | – | 1.677 | 2.694 |
| B2PLYPD3, gd3 | def2tzvp | – | 1.687 | 2.700 |
| B2PLYPD3, gd3 | def2tzvpp | – | 1.694 | 2.703 |
| wB97XD | def2svp | 12.5 | 1.625 | 2.650 |
| wB97XD | def2svp | 29.3 | 1.620 | 2.647 |
| wB97XD | def2svp | 46.8 | 1.619 | 2.646 |
| wB97XD | def2svp | 108.9 | 1.617 | 2.645 |

Figure 4 demonstrates the effect of the external electric field on the energy of the molecular (O-H···N) and ionic (O⁻···[H-N]⁺) forms of H-bonds in **2** (Figure 4a,b), **3** (Figure 4c,d), **4** (Figure 4e,f), **5** (Figure 4g,h), and **6** (Figure 4i,j) under the PCM and gas-phase approximations. For all complexes in both approximations, an increase of the field causes an energy decrease of both forms, although the favor is towards the ionic one. Upon this increase, the profile of a potential energy curve changes from a single-well (molecular) to a double-well to a single-well (ionic) one. The double-well potential interval is shown in Figure 4. For each complex, there is a unique value of the field strength for which the energy minima of the two forms are equal. $\Delta E$ corresponds to the energy of the complex with respect to the value at this field.

Strictly speaking, in order to find which value of the external field is the best approximation of experimental conditions, one needs (i) to estimate the molar fractions of the two forms from NMR spectra and (ii) to then find at what field the same ratio is be obtained in calculations. The former can be done using either the value of $^1J(^{15}N^1H)$ in Table 1 or the $^1$H-NMR chemical shift of the mobile proton [52]; the latter—by calculating the effect of the field on the free energy. However, both of these estimates are rough and are redundant in the case of the present qualitative analysis. Instead, the lower limit of the external electric field can be associated to the value at which the energy minima of the two forms are equal, Table 3.

Table 3. The external electric field at which the energy minima of the molecular (O-H···N) and ionic (O⁻···[H-N]⁺) forms of H-bonds in **2–6** are equal.

| Method | 2 | 3 | 4 | 5 | 6 |
|---|---|---|---|---|---|
| Field, a.u. & PCM ε = | 0.0004 12.5 | 0.0005 29.3 | 0.0015 29.3 | 0.0014 29.3 | 0.0027 29.3 |
| Field, a.u. (gas-phase) | 0.0044 | 0.0046 | 0.0075 | 0.0061 | 0.0082 |

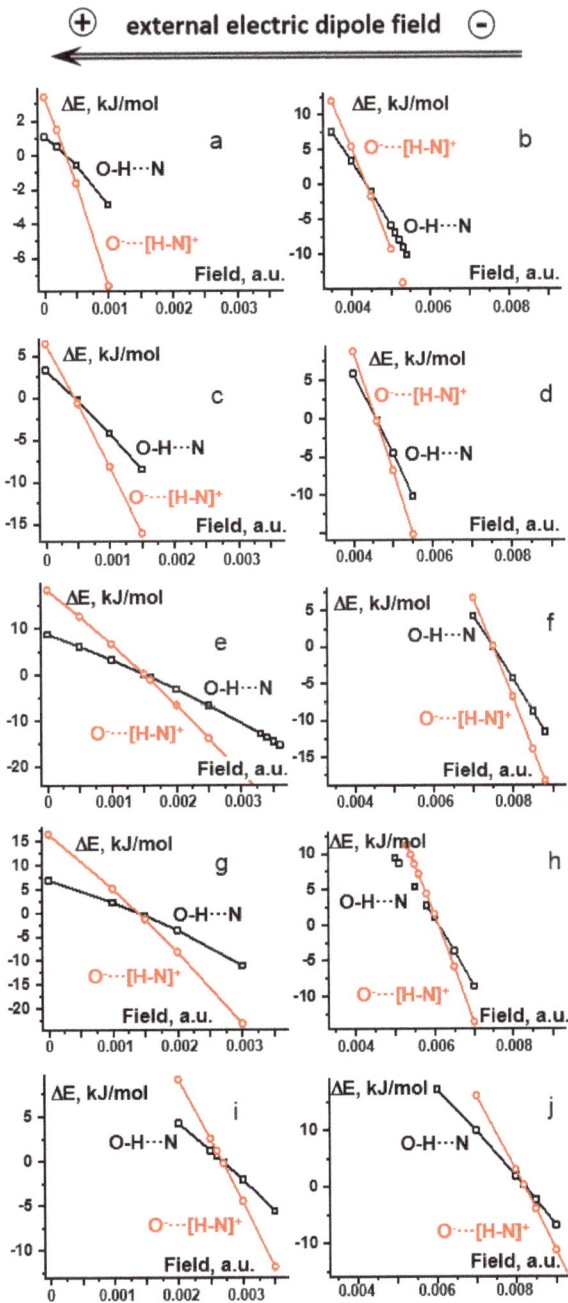

**Figure 4.** The effect of the external electric field on the energy of O-H···N (black □) and O⁻···[H-N]⁺ (red ○) forms of H-bonds in **2–6**. For each complex, there is a unique value of the field strength for which the energy minima of the two forms are equal. $\Delta E$ corresponds to the energy with respect to the value at this field. **2**: (**a**) PCM ($\varepsilon$=12.5), (**b**) gas-phase; **3**: (**c**) PCM ($\varepsilon$=29.3), (**d**) gas-phase; **4**: (**e**) PCM ($\varepsilon$=29.3), (**f**) gas-phase; **5**: (**g**) PCM ($\varepsilon$=29.3), (**h**) gas-phase; **6**: (**i**) PCM ($\varepsilon$=29.3), (**j**) gas-phase.

## 2.3. The Gas-Phase Proton Affinities

For the further discussion of the obtained results, we will use the values of the gas-phase proton affinities (PA). These values are listed in Table 4 for a number of selected proton acceptors.

Table 4. Gas-phase proton affinities of selected proton acceptors.

| Acceptor | PA, kJ/mol | Acceptor | PA, kJ/mol | Acceptor | PA, kJ/mol |
|---|---|---|---|---|---|
| pyridine | 936 | 2-nitrobenzoate | 1382 | Formate | 1431 |
| collidine | 988 | 3,5-dichlorobenzoate | 1379 | Acetate | 1447 |
| benzoate | 1421 | 4-nitrophenolate | 1354 | fluoride | 1547 |
|  |  | $F^-\cdots HCF_3$ | 1429 |  |  |

## 3. Discussion

Figure 5 shows lower limits of the external electric fields simulated the effect of $CDF_3/CDClF_2$ on **1-6** under the PCM (5a) and gas-phase (5b) approximations as a function of the $pK_a$ of the proton-donor. The $pK_a$ of pyridine is 5.32 [46]; other values are listed in Table 1; Table 3. For **2–6**, the strength of the field required to transfer the proton to the base correlates with the strength of the acids in both approaches. **1** deviates strongly from these correlations as it should. The energy minima of two tautomeric forms of **1** are equal at zero field. The values shown for **1** (Figure 5) correspond to the case when this double-well potential energy curve becomes a single-well one (Figure 3). However, in contrast to **2–6**, proton tautomerism in **1** remains fast on the NMR time scale. Thus, the physical meanings of the values reported here for **2–6** and **1** are different. What is important is that (i) the order of magnitude of the electric field simulated the effect of $CDF_3/CDClF_2$ on **1**, **2–6**, and pyridine$\cdots$HF$\cdots$(HCF$_3$)$_n$ [42] is the same and (ii) its effect on H-bond geometry correlates with the proton donating power of involved acids. The former means that the AuF approach is appropriate for a qualitative description of solute–solvent interactions. The latter suggests that it should be possible to predict the effect of a given solvent on the geometry of a given H-bonded complex. What is the most reliable representation of the correlation between the expected strength of the external electric field and the chemical properties of involved acids and bases?

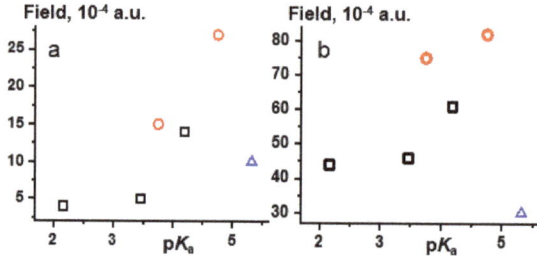

**Figure 5.** The lower limit of the external electric field simulated the effect of $CDF_3/CDClF_2$ on **2**, **3**, **5** (black □), **4**, **6** (red ○), and **1** (blue △) under the PCM (**a**) and gas-phase (**b**) approximations as a function of the $pK_a$ of the proton-donor.

The use of $pK_a$ as a measure of acid's proton-donating power in non-aqueous solutions introduces an error into the correlation. The reason is that the $pK_a$ depends on solvation in water that is very specific solvent [53–55]. The pKa's of ionizable groups in a non-aqueous environment can be estimated theoretically [56]. However, such calculations are quite demanding. Alternatively, they can be empirically corrected to a solvent under question [57]. The easiest way to estimate the proton-donating and proton-accepting powers is to calculate the gas-phase proton affinity (PA), Table 4 [58,59]. These values are very close to available experimental data for pyridine (930 kJ/mol) [60,61], collidine (980 kJ/mol) [61], benzoate (1422 kJ/mol) and 2-nitrobenzoate (1383 kJ/mol) [62], formate (1445 kJ/mol) and acetate (1456 kJ/mol) [63], and fluoride (1550 kJ/mol) [64].

Figure 6 demonstrates the lower limits of the external electric field simulated the effect of $CDF_3/CDClF_2$ on 2–6 under the PCM (Figure 6a) and gas-phase (Figure 6b) approximations as a function of the PA of the involved conjugate bases. We are aware that the use of the PCM approximation perturbs such correlations due to its dependence on the size of a molecular complex under study [58]. Therefore, we intend to use the gas-phase approximation. The analytical expression for the correlation shown in Figure 6b is:

$$F^{gas}[in\ a.u.] = \{(0.55 \pm 0.07) \cdot PA[in\ kJ/mol] - (700 \pm 100)\} \cdot 10^{-4}.$$

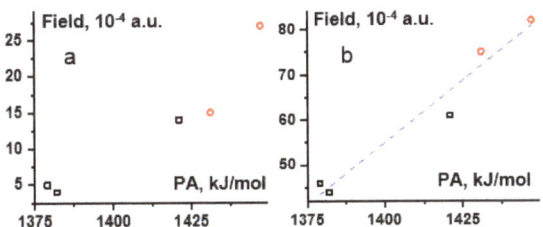

**Figure 6.** The lower limit of the external electric field simulated the effect of $CDF_3/CDClF_2$ on **2, 3, 5** (black □) and **4, 6** (red ○) under the PCM (**a**) and gas-phase (**b**) approximations as a function of the PA of the involved acids.

This correlation can be generalized by replacing the PA of the conjugate bases with a difference between the PA's of a proton donor (conjugate base) and an acceptor: $\Delta PA = PA^{donor} - PA^{acceptor}$. The analytical expression for this final correlation as shown in Figure 7 is:

$$F^{gas}[in\ a.u.] = 1.3 \cdot 10^{-4} \cdot \{\exp(0.009 \cdot \Delta PA[in\ kJ/mol]) - 1\}.$$

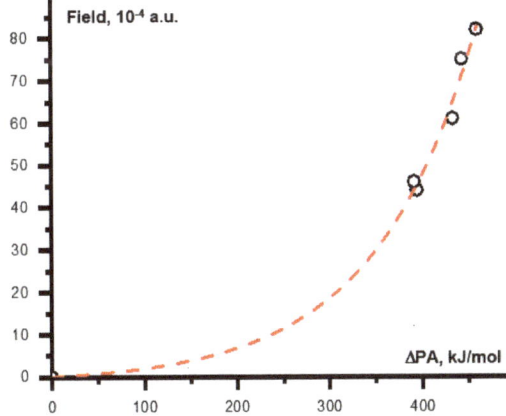

**Figure 7.** A functional dependence between the lower limit of the external electric field at which the energy minima of the Acid-H···Base and Acid⁻···[H-Base]⁺ forms of an H-bonded complex in $CDF_3/CDClF_2$ are equal and the difference between the PAs of a proton donor and an acceptor, $\Delta PA$.

Here, $F^{gas}$ tends to zero as $\Delta PA$ tends to zero that is physically correct. Results obtained for **1**, **2**, and **6** provide limiting values for the strengths of the external electric field simulated the effect of $CDF_3/CDClF_2$ on H-bond geometry at 300 K, 200 K, and 100 K, respectively. When the gas-phase approximation is used for the field-strength calculations, these values are about 0.003 a.u., 0.004 a.u., and 0.082 a.u., respectively. Only a part of this field can be associated to solvent polarization and accounted for in the frameworks of the PCM approach. The effect of this contribution on H-bond geometry is roughly constant and temperature independent. Another part of the field simulates the effect of solute–solvent interactions. Their impact is fluctuating and depends on temperature. Let us estimate the magnitudes of these two contributions.

For **1**, the lower limit of the external electric field estimated under the PCM approximations is about 0.001 a.u. This value can be associated to the effect of solute–solvent interactions. Notice that both pyridines of **1** are affected by these interactions—meaning that 0.001 a.u. reflects the difference between the effects of solvation on the protonated pyridine and the H-bonded one. This value fluctuates faster than $10^3$ s$^{-1}$ down to 120 K and slower than $10^{11}$ s$^{-1}$ up to 290 K [44,45]. As a result, proton exchange within **1** is fast on the NMR and slow on the IR time scales in this temperature range.

Proton tautomerism *acceptor*···*H-donor* ⇌ *[acceptor-H]$^+$···(donor)$^-$* in $CDF_3/CDClF_2$ is strongly shifted towards the *[acceptor-H]$^+$···(donor)$^-$* tautomer already at 200 K when the difference between the PAs of the proton donor and the acceptor is smaller than 400 kJ/mol, **2**. The larger the difference, the lower the temperature should be. In $CDF_3/CDClF_2$, at the lowest experimentally achievable temperature of 100 K, this tautomer dominates completely only when $\Delta PA$ is smaller than 500 kJ/mol, **6**. The lower limits of the external electric field estimated under the PCM approximations required to stabilize the *[acceptor-H]$^+$···(donor)$^-$* tautomers of **2–6** in $CDF_3/CDClF_2$ vary from 0.0004 a.u. to 0.0027 a.u., Table 3. However, in contrast to **1**, these values fluctuate slow on the NMR time scale. Tentatively, this field can mostly be associated to solvation of the carbonyl group. The lower the temperature, the more stable the interaction with solvent molecules. For a rough estimate, it can be assumed that the value of this slow fluctuating field increases from 0.0005 a.u. to 0.0030 a.u. in the temperature range from 200 K to 100 K.

For the chemically similar $CH_2Cl_2$, the lowest experimentally achievable temperature is about 170 K [27]. Thus, when $\Delta PA$ is smaller than 400 kJ/mol, the *[acceptor-H]$^+$···(donor)$^-$* tautomer should dominate. This estimate can be checked using a complex of 4-nitrophenol with acetate in $CD_2Cl_2$ [65]. In this complex, $\Delta PA = PA^{acetate} - PA^{phenolate} = 93$ kJ/mol, Table 4. At 173 K, the *phenol*···*(acetate)$^-$* form of the complex dominated. Ab initio molecular dynamics demonstrated that this form was stabilized by interactions of the carbonyl group with solvent molecules. This interaction is implicitly included in our correlation. Formally speaking, these results support our estimate. However, molecular dynamics showed that the location of a tetraalkylammonium anion was also critically important in this case. This interaction is not covered by our correlation. This effect is absent for charged H-bonded complexes only for very bulky anions [25]. Notice that solvation of the phenolate oxygen will reduce the effect of the carbonyl solvation.

The importance of specific interactions is extreme in the case of a complex of pyridine with hydrogen fluoride. This complex was studied by NMR [23,66] and model calculations [43,67]. The strength of the external electric field, at which the experimental geometry of the N···H and H-F bonds of pyridine···HF in $CDF_3/CDClF_2$ is reproduced, depends on the number of the solvent molecules coordinated to the fluorine in model adducts pyridine···HF···$(HCF_3)_n$. It is about 0.017 a.u. for pyridine···HF, 0.010 a.u. for pyridine···HF···$HCF_3$, and 0.006 a.u. for pyridine···HF···$(HCF_3)_2$. For the former adduct, $\Delta PA = 611$ kJ/mol, Table 4. For pyridine···HF···$HCF_3$, the structure of the proton donor is HF···$HCF_3$ and $\Delta PA = 493$ kJ/mol. It is not clear how to estimate the PA of F$^-$···$(HCF_3)_2$ because the structure of such composite donor critically depends on its protonation state. In any case, it should be smaller than 500 kJ/mol that explains a near central location of the mobile proton between the nitrogen and fluorine atoms as observed in experiments. Thus, also for this complex, our qualitative analysis agrees with a high-level molecular dynamics [33].

## 4. Materials and Methods

Gaussian 09.D.01 program package was used [68]. If not stated otherwise, geometry optimizations were done at the $w$B97XD/def2tzvpp and $w$B97XD/def2svp approximations for **1** and **2–6**, respectively [69,70]. The identity of minima was confirmed by the absence of imaginary vibrational frequencies. The default SCRF=PCM method has been used to construct the solute cavity. The parameters for SMD calculations were adapted from the Minnesota Solvent Descriptor Database: eps = 8.93, epsinf = 1.4242, H-bond acidity = 0.1, H-bond basicity = 0.05, surface tension at interface = 39.15, carbon aromaticity = 0.0, electronegative halogenicity = 0.667 [71]. Although the SMD model was parametrized for the Minnesota functionals family, for the qualitative analysis presented in this work, we decided to use the same functional for all types of calculations.

The external electric field was added to calculations using a keyword Field. The $C_2$ symmetry axis of pyridine or collidine was fixed along the direction of the field using a keyword Z-Matrix. The electric dipole field in Gaussian is directed from the negative to the positive potential that is opposite to the conventional direction of electric field.

The gas-phase proton affinities (PA) were calculated as follows:

$$PA = \Delta H^{298}(B) + 5RT/2 - \Delta H^{298}(BH).$$

Here, $\Delta H^{298}(B)$ and $\Delta H^{298}(BH)$ stand for the sums of the electronic and thermal enthalpies of a base and its conjugate acid or the conjugate base of an acid and the acid at 298 K. The enthalpies were estimated at the B3LYP/6-311++g(3df,2p) level. This level provides a reasonable description of the structure and harmonic frequencies of the neutral and charged H-bonded systems in the gas phase [72]. It is also sufficient to obtain correct values of enthalpies [73].

## 5. Conclusions

The gas-phase proton affinity (PA) of conjugate bases is larger than that of most neutral bases. Proton transfer in condensed matter requires either an H-bond network [74] or solvation [54,75]. In specific cases, small alterations can cause pronounced changes [76]. Therefore, neither gas-phase nor PCM calculations can reproduce the geometry of an acid-base complex in condensed matter, if its environment is ignored. In contrast, useful qualitative data can be obtained using the Adduct under Field (AuF) approach. The weak yet multiple interactions between the acid-base complex and solvent molecules influence the electron density in the acid and base that affects the position of the mobile proton. These changes can be simulated using a fictitious external electric field. The macroscopic electric field can be either accounted for by the PCM approach or included in the strength of the field. The strength of the solute–solvent interactions fluctuates and its effective magnitude depends on temperature. In this paper, we report estimates of the strength of the fictitious field that simulates solvation effect of $CDF_3/CDClF_2$ on homo- and heterogeneous acid-base complexes in the temperature range from 300 K to 100 K. With some limitations, the obtained results can be extended onto the chemically similar $CHCl_3$ and $CH_2Cl_2$. The computational simplicity of the AuF approach could lend itself to wide application including large molecular systems [77–80].

In the presence of the external electric field, the potential energy curve of a proton transfer within the proton-bound homodimers of pyridines changes from a symmetric double-well potential to an asymmetric single-well one. In the temperature range 120–290 K, the fluctuation rate of this field is between $10^3$ and $10^{11}$ s$^{-1}$ that defines the rate of proton exchange within the homodimers. The lower limits of this field are reported above. For [FHF]$^-$ [81,82] or [H$_{2n+1}$O$_n$]$^+$ [83] proton-bound homodimers, the same strength of the field can be an acceptable approximation only when several solvent molecules are explicitly included into calculations.

Below 200 K, solvent effects on heterogeneous acid-base complexes can be simulated using a quasi-constant fictitious field. For complexes of pyridine with carboxylic acids, the strength of this field and its temperature dependence are reported above. For complexes of pyridine with alcohols and phenols, the strength will be smaller because interaction of the carbonyl oxygen with solvent molecules increases the proton-donating power of the hydroxylic group of carboxylic acids. When a proton-donating or proton-accepting center is open for a strong interaction with solvent molecules, these molecules should be included into the model adduct. See, for example, pyridine$\cdots$HF$\cdots$HCF$_3$ and pyridine$\cdots$HF$\cdots$(HCF$_3$)$_2$ adducts.

The most important conclusion of this study is that solute–solvent interactions remarkably affect the geometry of acid-base complexes in aprotic solvents even if the active sites of these complexes are not accessible for solvent molecules. As a result, these complexes exhibit proton tautomerism *acceptor$\cdots$H-donor* $\rightleftharpoons$ *[acceptor-H]$^+\cdots$(donor)$^-$* in a large temperature range. The rate of this process is often slow on the time scales of electronic excitations and molecular vibrations while fast on the time scale of NMR. Therefore, both tautomers can be observed in the former cases while exchange averaged parameters will be obtained in the latter. Only in the presence of moderately strong solvation effects, for example, when solvent molecules interact with the proton-donating group, can the lifetime of the *[acceptor-H]$^+\cdots$(donor)$^-$* tautomer become long on the NMR time scale in the temperature range from 200 K to 100 K.

**Author Contributions:** Conceptualization, I.G.S.; methodology, I.G.S. and G.S.D.; data curation, G.S.D.; writing—original draft preparation, I.G.S.; writing—review and editing, G.S.D.; visualization, I.G.S.; supervision, I.G.S. All authors have read and agreed to the published version of the manuscript.

**Funding:** This research was funded by the Russian Foundation of Basic Research (Project 20-03-00231). APC was sponsored by MDPI.

**Acknowledgments:** The authors gratefully acknowledge the Gauss Centre for Supercomputing e.V. (www.gauss-centre.eu) for funding this project by providing computing time on the GCS Supercomputer SuperMUC at Leibniz Supercomputing Centre (LRZ, www.lrz.de).

**Conflicts of Interest:** The authors declare no conflict of interest.

## References

1. Hynes, J.T.; Klinman, J.P.; Limbach, H.H.; Schowen, R.L. (Eds.) *Hydrogen-Transfer Reactions*; WILEY-VCH: Weinheim, Germany, 2007. [CrossRef]
2. Berendsen, H.J.C.; Mavri, J. Quantum Simulation of Reaction Dynamics by Density-Matrix Evolution. *J. Phys. Chem.* **1993**, *97*, 13464–13468. [CrossRef]
3. Bekcioglu, G.; Allolio, C.; Sebastiani, D. Water Wires in Aqueous Solutions from First-Principles Calculations. *J. Phys. Chem. B* **2015**, *119*, 4053–4060. [CrossRef] [PubMed]
4. Ciacka, P.; Fita, P.; Listkowski, A.; Radzewicz, C.; Waluk, J. Evidence for Dominant Role of Tunneling in Condensed Phases and at High Temperatures: Double Hydrogen Transfer in Porphycenes. *J. Phys. Chem. Lett.* **2016**, *7*, 283–288. [CrossRef] [PubMed]
5. Kabbe, G.; Dressler, C.; Sebastiani, D. Toward Realistic Transfer Rates within the Coupled Molecular Dynamics/Lattice Monte Carlo Approach. *J. Phys. Chem. C* **2016**, *120*, 19905–19912. [CrossRef]
6. Piwonski, H.; Sokolowski, A.; Kijak, M.; Nonell, S.; Waluk, J. Arresting Tautomerization in a Single Molecule by the Surrounding Polymer: 2,7,12,17-Tetraphenyl Porphycene. *J. Phys. Chem. Lett.* **2013**, *4*, 3967–3971. [CrossRef]
7. Karerlin, S.C.L.; Mavri, J.; Warshel, A. Examining the Case for the Effect of Barrier Compression on Tunneling, Vibrationally Enhanced Catalysis, Catalytic Entropy and Related Issues. *FEBS Lett.* **2010**, *584*, 2759–2766. [CrossRef]
8. Chan-Huot, M.; Dos, A.; Zander, R.; Sharif, S.; Tolstoy, P.M.; Compton, S.; Fogle, E.; Toney, M.D.; Shenderovich, I.; Denisov, G.S.; et al. NMR Studies of Protonation and Hydrogen Bond States of Internal Aldimines of Pyridoxal 5′-Phosphate Acid-Base in Alanine Racemase, Aspartate Aminotransferase, and PolyLlysine. *J. Am. Chem. Soc.* **2013**, *135*, 18160–18175. [CrossRef]
9. Prah, A.; Franciskovic, E.; Mavri, J.; Stare, J. Electrostatics as the Driving Force Behind the Catalytic Function of the Monoamine Oxidase A Enzyme Confirmed by Quantum Computations. *ACS Catal.* **2019**, *9*, 1231–1240. [CrossRef]

10. Pregelj, D.; Jug, U.; Mavri, J.; Stare, J. Why Does the Y326I Mutant of Monoamine Oxidase B Decompose an Endogenous Amphetamine at a Slower Rate Than the Wild Type Enzyme? Reaction Step Elucidated by Multiscale Molecular Simulations. *Phys. Chem. Chem. Phys.* **2018**, *20*, 4181–4188. [CrossRef]
11. Borgis, D.C.; Lee, S.; Hynes, J.T. A dynamical theory of nonadiabatic proton and hydrogen atom transfer reaction rates in solution. *Chem. Phys. Lett.* **1989**, *162*, 19–26. [CrossRef]
12. Kiefer, P.M.; Hynes, J.T. Nonlinear Free Energy Relations for Adiabatic Proton Transfer Reactions in a Polar Environment. II. Inclusion of the Hydrogen Bond Vibration. *J. Phys. Chem. A* **2002**, *106*, 1850–1861. [CrossRef]
13. Dahms, F.; Fingerhut, B.P.; Nibbering, E.T.J.; Pines, E.; Elsaesser, T. Large-amplitude Transfer Motion of Hydrated Excess Protons Mapped by Ultrafast 2D IR Spectroscopy. *Science* **2017**, *357*, 491–495. [CrossRef] [PubMed]
14. Kundu, A.; Dahms, F.; Fingerhut, B.P.; Nibbering, E.T.J.; Pines, E.; Elsaesser, T. Ultrafast Vibrational Relaxation and Energy Dissipation of Hydrated Excess Protons in Polar Solvents. *Chem. Phys. Lett.* **2018**, *713*, 111–116. [CrossRef]
15. Dahms, F.; Costard, R.; Pines, E.; Fingerhut, B.P.; Nibbering, E.T.J.; Elsaesser, T. The Hydrated Excess Proton in the Zundel Cation $H_5O_2^+$: The Role of Ultrafast Solvent Fluctuations. *Angew. Chem. Int. Ed.* **2016**, *55*, 10600–10605. [CrossRef]
16. Kiefer, P.M.; Pines, E.; Pines, D.; Hynes, J.T. Solvent-Induced Red-Shifts for the Proton Stretch Vibrational Frequency in a Hydrogen-Bonded Complex. 1. A Valence Bond-Based Theoretical Approach. *J. Phys. Chem. B* **2014**, *118*, 8330–8351. [CrossRef]
17. Keinan, S.; Pines, D.; Kiefer, P.M.; Hynes, J.T.; Pines, E. Solvent-Induced O–H Vibration Red-Shifts of Oxygen-Acids in Hydrogen-Bonded O–H⋯Base Complexes. *J. Phys. Chem. B* **2015**, *119*, 679–692. [CrossRef]
18. Ditkovich, J.; Mukra, T.; Pines, D.; Huppert, D.; Pines, E. Bifunctional Photoacids: Remote Protonation Affecting Chemical Reactivity. *J. Phys. Chem. B* **2015**, *119*, 2690–2701. [CrossRef]
19. Scherrer, A.; Agostini, F.; Sebastiani, D.; Gross, E.K.U.; Vuilleumier, R. On the Mass of Atoms in Molecules: Beyond the Born-Oppenheimer Approximation. *Phys. Rev. X* **2017**, *7*, 031035. [CrossRef]
20. Perrin, C.L. Symmetry of Hydrogen Bonds in Solution. *Pure Appl. Chem.* **2009**, *81*, 571–583. [CrossRef]
21. Perrin, C.L.; Karri, P.; Moore, C.; Rheingold, A.L. Hydrogen-Bond Symmetry in Difluoromaleate Monoanion. *J. Am. Chem. Soc.* **2012**, *134*, 7766–7772. [CrossRef]
22. Dopieralski, P.; Perrin, C.L.; Latajka, Z. On the Intramolecular Hydrogen Bond in Solution: Car-Parrinello and Path Integral Molecular Dynamics Perspective. *J. Chem. Theory Comput.* **2011**, *7*, 3505–3513. [CrossRef] [PubMed]
23. Shenderovich, I.G.; Burtsev, A.P.; Denisov, G.S.; Golubev, N.S.; Limbach, H.-H. Influence of the temperature-dependent dielectric constant on the H/D isotope effects on the NMR chemical shifts and the hydrogen bond geometry of the collidine-HF complex in $CDF_3/CDClF_2$ solution. *Magn. Reson. Chem.* **2001**, *39*, S91–S99. [CrossRef]
24. Shenderovich, I.G.; Tolstoy, P.M.; Golubev, N.S.; Smirnov, S.N.; Denisov, G.S.; Limbach, H.-H. Low-Temperature NMR Studies of the Structure and Dynamics of a Novel Series of Acid—;Base Complexes of HF with Collidine Exhibiting Scalar Couplings Across Hydrogen Bonds. *J. Am. Chem. Soc.* **2003**, *125*, 11710–11720. [CrossRef] [PubMed]
25. Lesnichin, S.B.; Tolstoy, P.M.; Limbach, H.-H.; Shenderovich, I.G. Counteranion-Dependent Mechanisms of Intramolecular Proton Transfer in Aprotic Solution. *Phys. Chem. Chem. Phys.* **2010**, *12*, 10373–10379. [CrossRef] [PubMed]
26. Pollice, R.; Fleckenstein, F.; Shenderovich, I.; Chen, P. Compensation of London Dispersion in the Gas Phase and in Aprotic Solvents. *Angew. Chem. Int. Ed.* **2019**, *58*, 14281–14288. [CrossRef]
27. Pollice, R.; Bot, M.; Kobylianskii, I.J.; Shenderovich, I.; Chen, P. Attenuation of London Dispersion in Dichloromethane Solutions. *J. Am. Chem. Soc.* **2017**, *139*, 13126–13140. [CrossRef]
28. Vyalikh, A.; Emmler, T.; Shenderovich, I.; Zeng, Y.; Findenegg, G.H.; Buntkowsky, G. H-2-solid state NMR and DSC study of isobutyric acid in mesoporous silica materials. *Phys. Chem. Chem. Phys.* **2007**, *9*, 2249–2257. [CrossRef]
29. Grabowski, S.J. Analysis of Hydrogen Bonds in Crystals. *Crystals* **2016**, *6*, 59. [CrossRef]
30. Grabowski, S.J. [FHF]⁻—The Strongest Hydrogen Bond under the Influence of External Interactions. *Crystals* **2016**, *6*, 3. [CrossRef]

31. Vener, M.V.; Chernyshov, I.Y.; Rykounov, A.A.; Filarowski, A. Structural and spectroscopic features of proton hydrates in the crystalline state. Solid-state DFT study on HCl and triflic acid hydrates. *Mol. Phys.* **2018**, *116*, 251–262. [CrossRef]
32. Mukhopadhyay, M.; Banerjee, D.; Koll, A.; Mandal, A.; Filarowski, A.; Fitzmaurice, D.; Das, R.; Mukherjee, S. Excited state intermolecular proton transfer and caging of salicylidine-3,4,7-methyl amine in cyclodextrins. *J. Photochem. Photobiol. A Chem.* **2005**, *175*, 94–99. [CrossRef]
33. Pylaeva, S.A.; Elgabarty, H.; Sebastiani, D.; Tolstoy, P.M. Symmetry and Dynamics of FHF$^-$ Anion in Vacuum, in $CD_2Cl_2$ and in $CCl_4$. Ab Initio MD Study of Fluctuating Solvent–Solute Hydrogen and Halogen Bonds. *Phys. Chem. Chem. Phys.* **2017**, *19*, 26107–26120. [CrossRef] [PubMed]
34. Koeppe, B.; Pylaeva, S.A.; Allolio, C.; Sebastiani, D.; Nibbering, E.T.J.; Denisov, G.S.; Limbach, H.-H.; Tolstoy, P.M. Polar solvent fluctuations drive proton transfer in hydrogen bonded complexes of carboxylic acid with pyridines: NMR, IR and ab initio MD study. *Phys. Chem. Chem. Phys.* **2017**, *19*, 1010–1028. [CrossRef] [PubMed]
35. Tomasi, J.; Mennucci, B.; Cammi, R. Quantum mechanical continuum solvation models. *Chem. Rev.* **2005**, *105*, 2999–3093. [CrossRef]
36. Shenderovich, I.G. Simplified calculation approaches designed to reproduce the geometry of hydrogen bonds in molecular complexes in aprotic solvents. *J. Chem. Phys.* **2018**, *148*, 124313. [CrossRef]
37. Marenich, A.V.; Cramer, C.J.; Truhlar, D.G. Universal Solvation Model Based on Solute Electron Density and on a Continuum Model of the Solvent Defined by the Bulk Dielectric Constant and Atomic Surface Tensions. *J. Phys. Chem. B* **2009**, *113*, 6378–6396. [CrossRef]
38. Mori, Y.; Takano, K. Location of protons in N–H$\cdots$N hydrogen-bonded systems: A theoretical study on intramolecular pyridine–dihydropyridine and pyridine–pyridinium pairs. *Phys. Chem. Chem. Phys.* **2012**, *14*, 11090–11098. [CrossRef]
39. Hofmeister, C.; Coto, P.B.; Thoss, M. Controlling the Conductance of Molecular Junctions Using Proton Transfer Reactions: A Theoretical Model Study. *J. Chem. Phys.* **2017**, *146*, 092317. [CrossRef]
40. Liang, H.; Chai, B.; Chen, G.; Chen, W.; Chen, S.; Xiao, H.; Lin, S. Electric Field-driven Acid-base Transformation: Proton Transfer from Acid (HBr/HF) to Base ($NH_3/H_2O$). *Chem. Res. Chin. Univ.* **2015**, *31*, 418–426. [CrossRef]
41. Dominikowska, J.; Palusiak, M. Tuning Aromaticity of para-Substituted Benzene Derivatives with an External Electric Field. *ChemPhysChem* **2018**, *19*, 590–595. [CrossRef]
42. Shenderovich, I.G.; Denisov, G.S. Solvent effects on acid-base complexes. What is more important: A macroscopic reaction field or solute-solvent interactions? *J. Chem. Phys.* **2019**, *150*, 204505. [CrossRef] [PubMed]
43. Chan, B.; Del Bene, J.E.; Radom, L. What Factors Determine Whether a Proton-Bound Homodimer Has a Symmetric or an Asymmetric Hydrogen Bond? *Mol. Phys.* **2009**, *107*, 1095–1105. [CrossRef]
44. Kong, S.; Borissova, A.O.; Lesnichin, S.B.; Hartl, M.; Daemen, L.L.; Eckert, J.; Antipin, M.Y.; Shenderovich, I.G. Geometry and Spectral Properties of the Protonated Homodimer of Pyridine in the Liquid and Solid States. A Combined NMR, X-ray Diffraction and Inelastic Neutron Scattering Study. *J. Phys. Chem. A* **2011**, *115*, 8041–8048. [CrossRef] [PubMed]
45. Melikova, S.M.; Rutkowski, K.S.; Gurinov, A.A.; Denisov, G.S.; Rospenk, M.; Shenderovich, I.G. FTIR Study of the Hydrogen Bond Symmetry in Protonated Homodimers of Pyridine and Collidine in Solution. *J. Mol. Struct.* **2012**, *1018*, 39–44. [CrossRef]
46. Tolstoy, P.M.; Smirnov, S.N.; Shenderovich, I.G.; Golubev, N.S.; Denisov, G.S.; Limbach, H.-H. NMR studies of solid state—Solvent and H/D isotope effects on hydrogen bond geometries of 1:1 complexes of collidine with carboxylic acids. *J. Mol. Struct.* **2004**, *700*, 19–27. [CrossRef]
47. Andreeva, D.V.; Ip, I.; Gurinov, A.A.; Tolstoy, P.M.; Denisov, G.S.; Shenderovich, I.G.; Limbach, H.-H. Geometrical Features of Hydrogen Bonded Complexes Involving Sterically Hindered Pyridines. *J. Phys. Chem. A* **2006**, *110*, 10872–10879. [CrossRef]
48. Lankau, T.; Yu, C.-H. A quantum description of the proton movement in an idealized NHN$^+$ bridge. *Phys. Chem. Chem. Phys.* **2011**, *13*, 12758–12769. [CrossRef]
49. Asmis, K.R.; Yang, Y.; Santambrogio, G.; Brümmer, M.; Roscioli, J.R.; McCunn, L.R.; Johnson, M.A.; Kühn, O. Gas-Phase Infrared Spectroscopy and Multidimensional Quantum Calculations of the Protonated Ammonia Dimer $N_2H_7^+$. *Angew. Chem. Int. Ed.* **2007**, *46*, 8691–8694. [CrossRef]

50. Giese, K.; Petković, M.; Naundorf, H.; Kühn, O. Multidimensional quantum dynamics and infrared spectroscopy of hydrogen bonds. *Phys. Rep.* **2006**, *430*, 211–276. [CrossRef]
51. Limbach, H.-H.; Tolstoy, P.M.; Pérez-Hernández, N.; Guo, J.; Shenderovich, I.G.; Denisov, G.S. OHO Hydrogen Bond Geometries and NMR Chemical Shifts: From Equilibrium Structures to Geometric H/D Isotope Effects, with Applications for Water, Protonated Water, and Compressed Ice. *Israel J. Chem.* **2009**, *49*, 199–216. [CrossRef]
52. Sharif, S.; Shenderovich, I.G.; González, L.; Denisov, G.S.; Silverman, D.H.; Limbach, H.-H. Nuclear Magnetic Resonance and ab Initio Studies of Small Complexes Formed between Water and Pyridine Derivatives in Solid and Liquid Phases. *J. Phys. Chem. A* **2007**, *111*, 6084–6093. [CrossRef] [PubMed]
53. Shenderovich, I.G. The Partner Does Matter: The Structure of Heteroaggregates of Acridine Orange in Water. *Molecules* **2019**, *24*, 2816. [CrossRef] [PubMed]
54. Gurinov, A.A.; Mauder, D.; Akcakayiran, D.; Findenegg, G.H.; Shenderovich, I.G. Does Water Affect the Acidity of Surfaces? The Proton-Donating Ability of Silanol and Carboxylic Acid Groups at Mesoporous Silica. *Chem. Phys. Chem.* **2012**, *13*, 2282–2285. [CrossRef] [PubMed]
55. Kunz, W. Specific ion effects in colloidal and biological systems. *Curr. Opin. Colloid Interface Sci.* **2010**, *15*, 34–39. [CrossRef]
56. Sham, Y.Y.; Chu, Z.T.; Warshel, A. Consistent Calculations of pKa's of Ionizable Residues in Proteins: Semi-microscopic and Microscopic Approaches. *J. Phys. Chem. B* **1997**, *101*, 4458–4472. [CrossRef]
57. Rossini, E.; Bochevarov, A.D.; Knapp, E.W. Empirical Conversion of pKa Values between Different Solvents and Interpretation of the Parameters: Application to Water, Acetonitrile, Dimethyl Sulfoxide, and Methanol. *ACS Omega* **2018**, *3*, 1653–1662. [CrossRef]
58. Gurinov, A.A.; Denisov, G.S.; Borissova, A.O.; Goloveshkin, A.S.; Greindl, J.; Limbach, H.-H.; Shenderovich, I.G. NMR Study of Solvation Effect on the Geometry of Proton-Bound Homodimers of Increasing Size. *J. Phys. Chem. A* **2017**, *121*, 8697–8705. [CrossRef]
59. Gurinov, A.A.; Lesnichin, S.B.; Limbach, H.-H.; Shenderovich, I.G. How Short is the Strongest Hydrogen Bond in the Proton-Bound Homodimers of Pyridine Derivatives? *J. Phys. Chem. A* **2014**, *118*, 10804–10812. [CrossRef]
60. Hunter, E.P.L.; Lias, S.G. Evaluated Gas Phase Basicities and Proton Affinities of Molecules: An Update. *J. Phys. Chem. Ref. Data* **1998**, *27*, 413–656. [CrossRef]
61. Bräuer, P.; Situmorang, O.; Ng, P.L.; D'Agostino, C. Effect of Al content on the strength of terminal silanol species in ZSM-5 zeolite catalysts: A quantitative DRIFTS study without the use of molar extinction coefficients. *Phys. Chem. Chem. Phys.* **2018**, *20*, 4250–4262. [CrossRef]
62. Błaziak, K.; Sendys, P.; Danikiewicz, W. Experimental versus Calculated Proton Affinities for Aromatic Carboxylic Acid Anions and Related Phenide Ions. *ChemPhysChem* **2016**, *17*, 850–858. [CrossRef] [PubMed]
63. Bieńkowski, P.; Świder, P.; Błaziak, K.; Danikiewicz, W. Proton affinities of the anions of aromatic carboxylic acids measured by kinetic method. *Int. J. Mass Spectrom.* **2014**, *357*, 29–33. [CrossRef]
64. Ervin, K.M. Experimental Techniques in Gas-Phase Ion Thermochemistry. *Chem. Rev.* **2001**, *101*, 391–444. [CrossRef] [PubMed]
65. Pylaeva, S.; Allolio, C.; Koeppe, B.; Denisov, G.S.; Limbach, H.-H.; Sebastiani, D.; Tolstoy, P.M. Proton transfer in a short hydrogen bond caused by solvation shell fluctuations: An ab initio MD and NMR/UV study of an (OHO)$^-$ bonded system. *Phys. Chem. Chem. Phys.* **2015**, *17*, 4634–4644. [CrossRef]
66. Golubev, N.S.; Shenderovich, I.G.; Smirnov, S.N.; Denisov, G.S.; Limbach, H.-H. Nuclear Scalar Spin-Spin Coupling Reveals Novel Properties of Low-Barrier Hydrogen Bonds in a Polar Environment. *Chem. Eur. J.* **1999**, *5*, 492–497. [CrossRef]
67. Del Bene, J.E.; Bartlett, R.J.; Elguero, J. Interpreting $^{2h}J(F,N)$, $^{1h}J(H,N)$ and $^{1}J(F,H)$ in the hydrogen-bonded FH–collidine complex. *Magn. Reson. Chem.* **2002**, *40*, 767–771. [CrossRef]
68. Frisch, M.J.; Trucks, G.W.; Schlegel, H.B.; Scuseria, G.E.; Robb, M.A.; Cheeseman, J.R.; Scalmani, G.; Barone, V.; Mennucci, B.; Petersson, G.A.; et al. *Gaussian 09, Revision D.01*; Gaussian, Inc.: Wallingford, CT, USA, 2013.
69. Chai, J.-D.; Head-Gordon, M. Long-range corrected hybrid density functionals with damped atom-atom dispersion corrections. *Phys. Chem. Chem. Phys.* **2008**, *10*, 6615–6620. [CrossRef]
70. Weigend, F.; Ahlrichs, R. Balanced basis sets of split valence, triple zeta valence and quadruple zeta valence quality for H to Rn: Design and assessment of accuracy. *Phys. Chem. Chem. Phys.* **2005**, *7*, 3297–3305. [CrossRef]

71. Winget, P.; Dolney, D.M.; Giesen, D.J.; Cramer, C.J.; Truhlar, D.G. Minnesota Solvent Descriptor Database. Available online: http://comp.chem.umn.edu/solvation/mnsddb.pdf (accessed on 20 January 2020).
72. Zhu, H.; Blom, M.; Compagnon, I.; Rijs, A.M.; Roy, S.; von Helden, G.; Schmidt, B. Conformations and Vibrational Spectra of a Model Tripeptide: Change of Secondary Structure upon Micro-Solvation. *Phys. Chem. Chem. Phys.* **2010**, *12*, 3415–3425. [CrossRef]
73. Moser, A.; Range, K.; York, D.M. Accurate Proton Affinity and Gas-Phase Basicity Values for Molecules Important in Biocatalysis. *J. Phys. Chem. B* **2010**, *114*, 13911–13921. [CrossRef]
74. Manriquez, R.; Lopez-Dellamary, F.A.; Frydel, J.; Emmler, T.; Breitzke, H.; Buntkowsky, G.; Limbach, H.-H.; Shenderovich, I.G. Solid-State NMR Studies of Aminocarboxylic Salt Bridges in L-Lysine Modified Cellulose. *J. Phys. Chem. B* **2009**, *113*, 934–940. [CrossRef] [PubMed]
75. Mauder, D.; Akcakayiran, D.; Lesnichin, S.B.; Findenegg, G.H.; Shenderovich, I.G. Acidity of Sulfonic and Phosphonic AcidFunctionalized SBA-15 under Almost WaterFree Conditions. *J. Phys. Chem. C* **2009**, *113*, 19185–19192. [CrossRef]
76. Ip, B.C.K.; Shenderovich, I.G.; Tolstoy, P.M.; Frydel, J.; Denisov, G.S.; Buntkowsky, G.; Limbach, H.-H. NMR Studies of Solid Pentachlorophenol-4-Methylpyridine Complexes Exhibiting Strong OHN Hydrogen Bonds: Geometric H/D Isotope Effects and Hydrogen Bond Coupling Cause Isotopic Polymorphism. *J. Phys. Chem. A* **2012**, *116*, 11370–11387. [CrossRef] [PubMed]
77. Levina, V.A.; Filippov, O.A.; Gutsul, E.I.; Belkova, N.V.; Epstein, L.M.; Lledós, A.; Shubina, E.S. Neutral Transition Metal Hydrides as Acids in Hydrogen Bonding and Proton Transfer: Media Polarity and Specific Solvation Effects. *J. Am. Chem. Soc.* **2010**, *132*, 11234–11246. [CrossRef]
78. Dub, P.A.; Baya, M.; Houghton, J.; Belkova, N.V.; Daran, J.-C.; Poli, R.; Epstein, L.M.; Shubina, E.S. Solvent Control in the Protonation of [Cp*Mo(dppe)H$_3$] by CF$_3$COOH. *Eur. J. Inorg. Chem.* **2007**, *2007*, 2813–2826. [CrossRef]
79. Belkova, N.V.; Gribanova, T.N.; Gutsul, E.I.; Minyaev, R.M.; Bianchini, C.; Peruzzini, M.; Zanobini, F.; Shubina, E.S.; Epstein, L.M. Specific and non-specific influence of the environment on dihydrogen bonding and proton transfer to [RuH$_2${P(CH$_2$CH$_2$PPh$_2$)$_3$}]. *J. Mol. Struct.* **2007**, *844–845*, 115–131. [CrossRef]
80. Belkova, N.V.; Besora, M.; Epstein, L.M.; Lledós, A.; Maseras, F.; Shubina, E.S. Influence of Media and Homoconjugate Pairing on Transition Metal Hydride Protonation. An IR and DFT Study on Proton Transfer to CpRuH(CO)(PCy3). *J. Am. Chem. Soc.* **2003**, *125*, 7715–7725. [CrossRef]
81. Shenderovich, I.G.; Smirnov, S.N.; Denisov, G.S.; Gindin, V.A.; Golubev, N.S.; Dunger, A.; Reibke, R.; Kirpekar, S.; Malkina, O.L.; Limbach, H.-H. Nuclear Magnetic Resonance of Hydrogen Bonded Clusters between F$^-$ and (HF)$_n$: Experiment and Theory. *Ber. Bunsenges.* **1998**, *102*, 422–428. [CrossRef]
82. Golubev, N.S.; Melikova, S.M.; Shchepkin, D.N.; Shenderovich, I.G.; Tolstoy, P.M.; Denisov, G.S. nterpretation of hydrogen/deuterium isotope effects on NMR chemical shifts of [FHF]$^-$ ion based on calculations of nuclear magnetic shielding tensor surface. *Z. Phys. Chem.* **2003**, *217*, 1549–1563. [CrossRef]
83. Vener, M.V.; Kong, S.; Levina, A.A.; Shenderovich, I.G. Spectroscopic Signatures of [H$_9$O$_4$]$^+$ and [H$_{13}$O$_6$]$^+$ Ions in a Polar Aprotic Environment Revealed under DFT-PCM Approximation. *Acta Chim. Slov.* **2011**, *58*, 402–410.

**Sample Availability:** Not available.

© 2020 by the authors. Licensee MDPI, Basel, Switzerland. This article is an open access article distributed under the terms and conditions of the Creative Commons Attribution (CC BY) license (http://creativecommons.org/licenses/by/4.0/).

Review

# Hydrogen Bond and Other Lewis Acid–Lewis Base Interactions as Preliminary Stages of Chemical Reactions

Sławomir J. Grabowski [1,2]

[1] Polimero eta Material Aurreratuak: Fisika, Kimika eta Teknologia, Kimika Fakultatea, Euskal Herriko Unibertsitatea UPV/EHU & Donostia International Physics Center (DIPC) PK 1072, 20080 Donostia, Euskadi, Spain; s.grabowski@ikerbasque.org; Tel.: +34-943-018-187
[2] IKERBASQUE, Basque Foundation for Science, 48011 Bilbao, Spain

Academic Editor: Ilya G. Shenderovich
Received: 23 September 2020; Accepted: 10 October 2020; Published: 13 October 2020

**Abstract:** Various Lewis acid–Lewis base interactions are discussed as initiating chemical reactions and processes. For example, the hydrogen bond is often a preliminary stage of the proton transfer process or the tetrel and pnicogen bonds lead sometimes to the $S_N2$ reactions. There are numerous characteristics of interactions being first stages of reactions; one can observe a meaningful electron charge transfer from the Lewis base unit to the Lewis acid; such interactions possess at least partly covalent character, one can mention other features. The results of different methods and approaches that are applied in numerous studies to describe the character of interactions are presented here. These are, for example, the results of the Quantum Theory of Atoms in Molecules, of the decomposition of the energy of interaction or of the structure-correlation method.

**Keywords:** Lewis acid–Lewis base interactions; hydrogen bond; tetrel bond; pnicogen bond; triel bond; electron charge shifts

## 1. Introduction

It is well known that the hydrogen bond plays a crucial role in numerous chemical, physical and biological processes [1,2]. However, other interactions are also important in various processes, particularly biochemical ones [3,4]. It is worth mentioning that such terms as interaction and reaction are even used interchangeably sometimes since interactions often lead to corresponding reactions or at least they are initiative steps of chemical reactions. The aim of this study is to display dependencies between the latter phenomena. The interrelations between interactions and reactions as well as between them and other phenomena and processes, or reasons for the lack of such relations, were discussed in numerous studies, even in very early ones. For example, Lewis has described that "at the recent conference of the Faraday Society (July, 1923) all of those who participated seemed agreed that the average organic molecule is very little polarized, but there were some who believed that polarisation and indeed ionisation precede every reaction" [5].

More recent studies indicate an important role of electron charge shifts (related to the polarisation) in processes corresponding to interactions and then to chemical reactions. For example, Rauk has stated that "all reactions of organic compounds are treated within the framework of generalized Lewis acid - Lewis base theory, their reactivity being governed by the characteristics of the frontier orbitals of the two reactants. All compounds have occupied molecular orbitals and so can donate electrons, that is, act as bases in the Lewis sense. All compounds have empty molecular orbitals and so can accept electrons, that is, act as acids in the Lewis sense" [6].

It was also indicated that the term "noncovalent interactions" is not a proper one since it concerns hydrogen bond and halogen bond as well as other interactions that often possess characteristics of covalent bonds [7]. The covalent character is often related to the occurrence of polarization and charge transfer processes [8,9]. On the other hand, for very weak interactions ruled mainly by dispersion forces, the electron charge shifts are negligible, if there are any, also electrostatic interactions are not important there. This is why the term Lewis acid–Lewis base interactions seems to be the proper one since it is related to those interactions where the electron charge shifts occur between Lewis acid and Lewis base units. It seems also that this term excludes very weak interactions, as those that occur between methane molecules, for example, where dispersion forces are the most important ones and where it is difficult to indicate the Lewis acid and Lewis base centres [7].

Kaplan has concluded in his monograph that "intermolecular interactions are involved in the formation of complicated chemical complexes, such as charge-transfer and hydrogen-bond complexes. Study of the mechanism of elementary chemical reactions is impossible without knowledge of the exchange processes between the translational and electron-vibration energies, which depend on the interaction of particles under collisions. Knowledge of the potential surface, characterizing the mutual trajectories of the reactants, is necessary to obtain the rates of chemical reactions" [10].

This is why the aim of this review is to point out relationships between interactions and reactions since the former phenomena may lead to the latter ones thus they initiate numerous structural changes and electron charge redistributions in species being in contact. This is of particular interest, which characteristics possess interactions that lead to chemical reactions. It was discussed in one of recent reviews that the stronger electrostatic interactions lead to the stronger Pauli repulsion that implies the greater electron charge shifts, mainly from the Lewis base unit to the Lewis acid; these phenomena may lead to the chemical reaction [11]. In this review, few types of interactions are discussed and it is analysed which conditions should be fulfilled for them that they lead to chemical reactions.

## 2. The Hydrogen Bond as a Preliminary Stage of the Proton Transfer Process

It has been described in early studies that a fragment of a crystal structure may be treated as "a frozen stage" of a considered chemical reaction. The related fragments that differ by geometry and that are taken from various crystal structures may correspond to the analysed chemical reaction since they reflect structural changes accompanying this reaction [12–14]. This approach is known as the structure-correlation method. It was applied to analyse different reactions such as the nucleophilic addition to a carbonyl group, nucleophilic substitution at tetrahedral coordinated atoms ($S_N1$ and $S_N2$ reactions); electrocyclic ring closure of polyenes and other chemical processes [12–15].

It is important that the proton transfer, PT, process related to the hydrogen bond, HB, may be also discussed in terms of the structure-correlation method. For example, PT in O-H···O hydrogen bonds was analysed since -C=O···H-O-C- fragments taken from different crystal structures were compared to reconstruct the corresponding reaction path [9,11,16]. For these analyses, the high-precision neutron diffraction geometries were taken from the Cambridge Structural Database, CSD [17,18]. The recent CSD release (CSD updates up to May 2020) was applied here to search the above-mentioned -C=O···H-O-C- fragment with the following search criteria; accurate crystal structures with e.s.d's ≤ 0.005 Å, R ≤ 7.5%, error free structures, without disorder, no polymers and no powder diffraction results. Only neutron diffraction results were taken into account here since they are characterised by precisely determined positions of H-atoms [19] in contrast to the X-ray results, where the refinement of crystal structures is usually based on the spherical approximation of the atomic electron densities that results in the spherical symmetry of atomic scattering factors [20]. One may say that the sample of fragments described above and corresponding to different crystal structures of organic and organometallic compounds may reflect the reaction path of the following PT process; -C=O···H-O-C- ⇔ -C-O-H···O=C-. In some of structures H-atom is situated in the mid-point of the O...O distance or near to this point therefore -C=O···H$^+$···O=C- fragments are also included in the sample. The search has led to a finding of 56 geometrical fragments corresponding to the above-mentioned PT process. The similar search with the same criteria for accuracy of results was performed for the similar fragments where the H-atom is replaced by the deuterium, i.e., the -C=O...D-O-C- fragments; in this case only four structures were found.

Figure 1 presents the PT reaction path based on two above described CSD searches; this is the relationship between the Δr parameter and the O···O distance. The same relationships were discussed before [9,11,16] but they were based on earlier CSD updates. The Δr parameter is the distance of the H-atom of the O-H...O bridge from the O···O mid-point. For the linear O-H(D)···O systems the O···O distance may be expressed as the $r_{O\text{-}H} + r_{H\ldots O}$ sum while the Δr parameter as the $(r_{H\ldots O} - r_{OH})/2$ term. The $r_{OH}$ and $r_{H\ldots O}$ values correspond to the O-H bond length and the H...O distance, respectively. The points of Figure 1, which correspond to fragments of crystal structures, may be considered as positions of the proton in PT process. The results of this figure are symmetrised around Δr = 0; this symmetrisation corresponds to the equivalency of systems during PT reaction because the homonuclear O-H···O hydrogen bond is discussed here. The "points" in the middle of O···O distance may be considered as the transition state of the proton transfer reaction. These are strongly elongated O-H bonds and they are observed for the O···O distances amounting about ~2.4–2.5 Å. For long O···O distances the H-atoms are located far from their mid-points, they are situated close to one of oxygen centres rather.

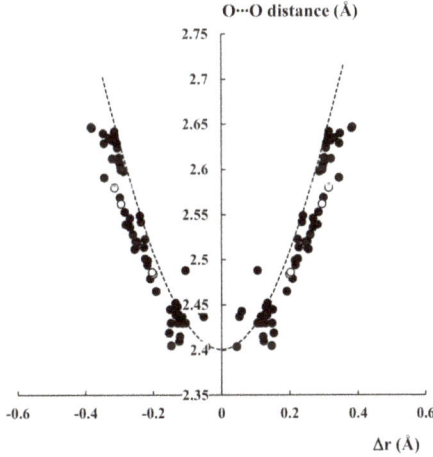

**Figure 1.** The dependence between the Δr (Å) - the displacement of the proton (or deuter) position from the O···O mid-point and the O···O distance (Å), for the O-H···O systems (black circles) and the O-D···O ones (white circles). The broken line corresponds to the bond number conservation rule (Equation (1)).

The broken line of Figure 1 corresponds to the relationship expressing the bond order (number) conservation rule (Equation (1)) [12,13,21].

$$\exp\left(\frac{\Delta r_i}{c}\right) + \exp\left(\frac{\Delta r_j}{c}\right) = 1 \tag{1}$$

The $\Delta r_i$ and $\Delta r_j$ terms in the above equation correspond to the $(r_0 - r_{OH})$ and $(r_0 - r_{H\ldots O})$ differences; $r_0$ is the typical single O-H bond length not perturbed by any interaction. The bond length of water in the gas phase equal to 0.957 Å was chosen here and in other studies [9]. The exponential terms of Equation (1), $\exp(\Delta r_i/c)$, may be treated as the definition of the bond order. However, other names for this term as well as for similar expressions are often applied in various studies; the bond valence [21] or the bond number [22], for example. The constant c for the O-H···O hydrogen bonds is determined from the above exponential expression assuming that for the O-H···O linear system with the H-atom located in the O···O mid-point, and the O···O distance of 2.4 Å, two equivalent H···O distances possess the bond order equal to 0.5; in other words; c = (0.957 − 1.2)/ln(0.5).

The bond order and the bond order conservation rule ideas for the O-H···O hydrogen bonds may be described in the following way. For the single O-H bond in the gas phase, the bond order is equal to unity. If this bond is involved in the hydrogen bond thus it is elongated and its bond order decreases. However, this decrease is compensated by the H···O contact. Hence the sum of bond orders (Equation (1)) of the O-H bond and the H···O contact is equal to unity. The greater is the O-H bond elongation for the stronger hydrogen bond thus the greater is the bond order of the H···O contact that is shorter accordingly; the latter is also accompanied by the decrease of the O···O distance (Figure 1). The extreme cases of very short O···O distances with the H-atom location at the O···O mid-point for very strong hydrogen bonds correspond to the transition states of the proton transfer reaction.

One can see that the broken line of Figure 1 that was derived from the bond order conservation rule is only approximately in agreement with the neutron diffraction results. Similar disagreement concerning the relationship between the O-H bond length and the H···O distance was explained by the influence of electrostatic forces that are not properly included in the bond order conservation concept [23]. Hence the corrected reference single O-H bond length of 0.925 Å was proposed to take into account the electrostatic contribution and to have better agreement between experimental results and theoretical evaluations [23]. Figure 1 contains also the O-D···O systems; it is pointed out in several studies that the deuteration of the O-H···O systems results in shortening of the O-D bond and lengthening of the O···O and D···O distances in comparison with their non-deuterated counterparts; it is known as the Ubbelohde effect [24]. Figure 1 shows that the deuterated O-D···O systems are approximately in agreement with the broken curve derived from Equation (1). One may conclude that the reaction path presented in Figure 1 shows that the hydrogen bond, especially for the strong O-H···O interactions, may be treated as the initial stage of PT process.

It is worth mentioning that similar relationships to this one presented in Figure 1 were analysed for other hydrogen bonded systems. For example, it was found that the dependence between $q_1$ and $q_2$ parameters for the N-H···N hydrogen bond geometries taken from experimental NMR and crystal structures' results is in agreement with the bond order conservation rule expressed by an equation similar to that one presented above here (Equation (1)) [25]. The $q_1$ and $q_2$ parameters are equal to ($r_{H···N} - r_{N-H}$)/2 and $r_{H···N} + r_{N-H}$, respectively; $r_{H···N}$ is the H···N distance and $r_{N-H}$ is the N-H bond length in the N-H···N system.

In another study the proton bound water dimer, $H_5O_2^+$ (Zundel cation) was discussed [26]. The relationships between molecular structures of this cation and the $^1$H-NMR chemical shifts were presented; low-temperature neutron diffraction results were used for these relationships. The dependence between $q_1 = (r_{H···O} - r_{O-H})/2$ and $q_2 = r_{H···O} + r_{O-H}$ parameters that is very similar to that one presented in Figure 1 has been also discussed [26]. The chemical shifts and other NMR parameters were also analysed for the N-H···N [25,27], F-H···N [28] and F-H···F$^-$ [28] hydrogen bonded systems. The bond order conservation rule [9,12] was compared there with the experimental results and with the theoretical calculations [25,27,28].

One can see that the geometries of hydrogen bonded systems represent various stages of the proton transfer reaction. It is discussed above here for different types of hydrogen bonds. As a consequence, the question arises, what are the characteristics of hydrogen bonds that may be treated as the initial stages of the PT process? It was discussed in earlier studies [9,11,29] that the strong hydrogen bonds may lead to the proton transfer reactions. It is worth recalling here effects that accompany the A-H···B hydrogen bond formation. It is the Lewis acid–Lewis base interaction thus the noticeable electron charge shift from the base unit to the acidic one occurs here [7,11]. This is connected with the $n_B \rightarrow \sigma_{AH}^*$ orbital-orbital interaction [8], where $n_B$ is the lone electron pair orbital of the B-centre while $\sigma_{AH}^*$ is the antibonding orbital of the A-H σ-bond. If we exclude from our discussion the blue shift hydrogen bonds [30] which do not lead to the PT process rather thus the hydrogen bond formation is connected with the increase of the polarization of the A-H bond, its elongation and consequently with its weakening. In extreme cases of the strong hydrogen bonds possessing covalent or at least partly covalent character [9] the PT reaction occurs. The terms resulting from the decomposition of the energy of interaction and related to the electron charge shifts are described as those expressing the covalency. The following terms occur in different decomposition schemes: charge transfer, polarization, delocalization or induction term. The name of term depends on the decomposition scheme applied. The interaction energy contributions differ by physical meaning in different schemes although they similarly express processes related to the electron charge shifts.

The covalent character may be also detected by the Quantum Theory of Atoms in Molecules (QTAIM) approach [31,32]. The value of the electron density at the H$\cdots$B bond critical point (BCP), $\rho_{BCP}$, of the order of 0.1 u and more, and the negative Laplacian of this electron density, $\nabla^2\rho_{BCP}$, inform of the covalent character of the hydrogen bond. However, it is assumed in numerous studies that even for $\nabla^2\rho_{BCP} > 0$, the negative value of the total electron energy density at BCP, $H_{BCP}$, shows the partly covalent character of interaction [9,33–36].

Figure 2 presents the relationship between the hydrogen$\cdots$Lewis base distance and the $H_{BCP}$ value at the corresponding BCP, for the sample of hydrogen bonds analysed in earlier study [37]. These results are based on the MP2/6-311++G(d,p) calculations. The following complexes linked by hydrogen bonds were discussed there. The complexes connected by charge assisted hydrogen bonds (CAHBs): (FHF)$^-$, $H_2O\cdots H_3O^+$, $H_3O^+\cdots HCN$, $OH^-\cdots H_2O$, $NH_3\cdots NH_4^+$, $C_5H_5^-\cdots HF$ and $C_5H_5^-\cdots C_2H_2$. The complexes with $\pi$-electrons acting as the acceptor of proton, these are two latter CAHB systems as well as the following species; $C_2H_2\cdots HF$, $C_2H_4\cdots HF$, $C_6H_6\cdots HF$, $(C_2H_2)_2$ (T-shaped dimer), $C_2H_4\cdots C_2H_2$, $C_6H_6\cdots C_2H_2$, $C_6H_6\cdots CH_4$, $C_6H_6\cdots CHCl_3$, $C_2H_2\cdots CH_4$ and $C_2H_2\cdots CHCl_3$. There is the sub-sample of complexes linked by the C-H$\cdots$B hydrogen bonds: $F_3CH\cdots NCCH_3$, $H_3CH\cdots NCCH_3$, $HCCH\cdots NCCH_3$, $F_3CH\cdots OCH_2$, $H_3CH\cdots OCH_2$, $HCCH\cdots OCH_2$, $H_3CH\cdots SH_2$, $HCCH\cdots SH_2$ and $HCCH\cdots S(CH_3)_2$. The other hydrogen bonds analysed in the above-mentioned study [37] may be classified as moderate or strong ones, these are interactions in the following complexes; $(C_6H_5COOH)_2$, $(CH_3COOH)_2$, $(HCOOH)_2$, $(HCONH)_2$, $(HCSNH)_2$, $(H_2O)_2$ (trans-linear dimer), $H_2O\cdots HF$ and $H_2CO\cdots HF$. The hydrogen bonds are divided into three groups in Figure 2, C-H$\cdots\pi$ (open circles, the $C_5H_5^-\cdots HF$ complex with the F-H proton donating bond is also included there), C-H$\cdots$B (black squares) and remaining ones, among them CAHB systems (black circles). The hydrogen$\cdots$Lewis base distance is understood here in the following way: it is the H$\cdots$B distance for the 3c–4e (three centre–four electron [8]) A-H$\cdots$B hydrogen bonds. For the C-H$\cdots\pi$ interactions this is the distance between the H-atom of Lewis acid unit and the carbon atom or the bond critical point of CC bond of the Lewis base [37]. The latter depends on the kind of bond path characterizing the intermolecular link [37]. Figure 2 shows that for stronger interactions characterised by shorter distances between Lewis acid-base units the covalent character is revealed that is expressed by the negative $H_{BCP}$ values, the corresponding systems may be treated as the potentially preliminary stages of PT processes. The weaker hydrogen bonds are characterised by longer distances between these units, these are mainly the C-H$\cdots\pi$ and C-H$\cdots$B systems.

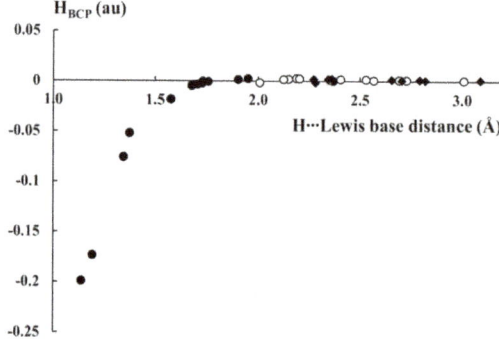

**Figure 2.** The dependence between the H⋯Lewis base distance (between H-atom of the proton donor and the centre of Lewis base unit, in Å) and the total electron energy density at BCP, $H_{BCP}$ for hydrogen bonded systems. The Lewis base centre is an atom for 3c–4e A-H⋯B hydrogen bonds (black squares for C-H⋯B hydrogen bonds and black circles for other 3c–4e systems) and it is BCP of CC bond or C-atom in Lewis base unit for the A-H⋯π hydrogen bonds (while circles).

## 3. The Case of Halogen Bonds

There are several various interactions that are treated in numerous studies as the hydrogen bond counterparts, particularly it concerns so-called σ-hole and π-hole bonds [38–42]. The σ-hole is a region of electron charge depletion at a centre considered approximately in the direction from an atom bonded to this centre, in the elongation of this bond [38,39]. On the other hand, the π-hole concerns a centre in a planar molecule or a planar molecular fragment [41,42], the triel centres such as boron one in trihalides and trihydrides are examples of such a situation [43,44]. The electron charge depletion at the σ-holes and π-holes often leads to the positive electrostatic potential, EP, at these regions; hence they often act as the Lewis acid centres.

The halogen bond, XB, is a case of an interaction where the σ-hole at halogen centre, X (F, Cl, Br, I or At) i.e., the Lewis acid site, interacts with the Lewis base centre [38,39]. This is why this interaction is often considered as the hydrogen bond counterpart, in numerous studies the comparison of these interactions is performed [11,40,45–47]. However, there is the question if XB, similarly as the hydrogen bond, may be considered as a preliminary stage of the chemical reaction. The PT process follows sometimes the hydrogen bond formation. Does a similar transfer of a halogen atom occur? Let us discuss shortly studies, particularly the recent ones, on the entities with halonium cation that links two Lewis base centres. The latter topic is revealed in recent studies by two challenges. The first one concerns finding of Lewis base–halonium ion–Lewis base arrangements with asymmetrically located halonium cation [48]. They are designated later here as [LbXLb]$^+$, where in further discussions Lb is replaced by the specific atomic centre while X marks the halogen. The second challenge concerns finding of the [LbXLb]$^+$ arrangement with fluorine, X = F, situated between the Lewis base sides since rather such systems with heavier halogens are known [49]. As concerns the latter challenge, the generation of a symmetrical fluoronium ion in solution was discussed, its existence was evidenced indirectly [50]. These experimental studies were supported by the calculations performed at various levels, all levels applied confirmed this ion existence, for example, the B3LYP/6-311++G(d,p) results show equal F⋯C distances amounting 1.6 Å in the [CFC]$^+$ arrangement [50]. It was proved in numerous earlier and recent studies that the heavier halogen atom (Cl, Br, I) may be engaged in the hypervalent link in solution possessing formally positive charge. Such a link was usually doubted for the fluorine centre; the studies of Lectka and co-workers discussed here [49,50] are related to this challenge. However, it was evidenced only indirectly the occurrence of the fluoronium ion as a short-lived reaction intermediate [50]. It was discussed also that this ion is formed in a solvolysis process from a precursor molecule probably according to the $S_N1$ reaction through the fluorine centre [50,51]. The crystal structure of the fluoronium ion precursor was determined [50]; the structure of this ion taken from the Cambridge Structural Database, CSD [17,18] is presented in Figure 3. The direct observation of the similar symmetrical [CFC]$^+$ ion was described recently [49], this is a stable species in solution and it was characterised by $^{19}$F, $^1$H, and $^{13}$C NMR.

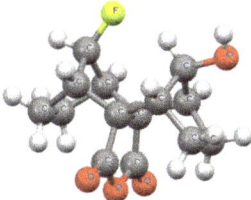

**Figure 3.** The fluoronium ion precursor from the crystal structure [50], BEXNOJ refcode.

Another challenge announced here concerns asymmetric [NXN]$^+$ systems since the symmetrical systems containing the halogen, X, are known rather from various studies [48]. For example, the experimental NMR spectroscopic and crystal structure studies as well as the theoretical calculations were performed recently on the systems containing halogen cation as well as other cations between nitrogen centres, i.e., the [NZN]$^+$ systems where Z$^+$ = H$^+$, Li$^+$, Na$^+$, F$^+$, Cl$^+$, Br$^+$, I$^+$, Ag$^+$ and Au$^+$ [52]. Two series, one of bis(pyridine) entities and the second one containing the (1,2-bis(pyridin-2-ylethynyl)benzene) structure were considered [52]. In all cases the symmetrical arrangements are observed, only for Z$^+$ = H$^+$ and F$^+$ the asymmetric systems occur; this concerns both above-mentioned series. However, in a case of the fluorine species there is no experimental confirmation of such asymmetry, only theoretical calculations reveal such arrangements. Figures 4 and 5 show fragments of bis(pyridine) crystal structures. In two cases (Figure 4); silver [53] and iodine [53] cations are located in the centre of the [NAgN]$^+$ and [NIN]$^+$ systems, respectively. In a case of the bis(pyridine)proton structure the asymmetric [NHN]$^+$ arrangements are observed [54] (see Figure 5).

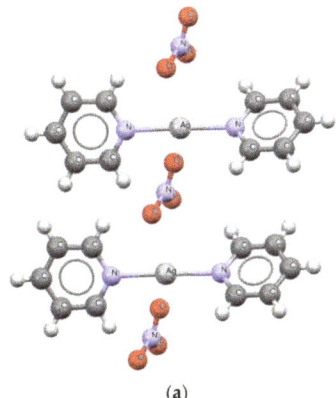

(a)

**Figure 4.** *Cont.*

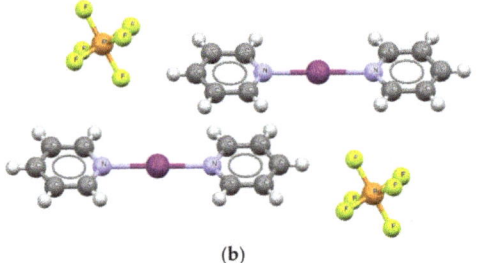

(b)

**Figure 4.** The fragment of the crystal structure of bis(pyridine)-silver(i) nitrate monohydrate, AGPYNO02 refcode (**a**) and di(pyridin-1-yl)iodonium hexafluorophosphate, CICQIQ03 refcode (**b**), reference [53].

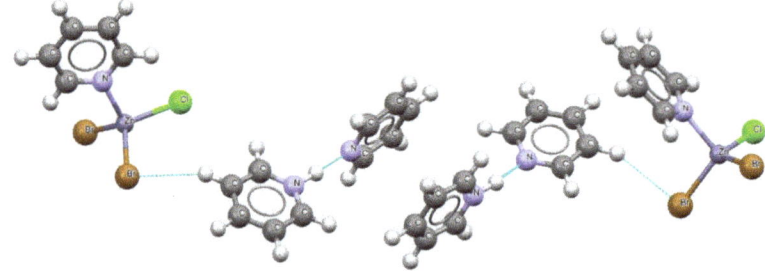

**Figure 5.** The fragment of the crystal structure of pyridinium dibromo-chloro-pyridine-zinc(ii) pyridine solvate, PYCBZN01 refcode, reference [54].

It is worth mentioning that the potential energy curves were analysed theoretically for the bis(pyridine) series of complexes [52], the displacement of the $Z^+$ cation from the N···N mid-point was plotted versus the electronic energy of the system. The single well potential energy curve is observed for all systems except of $Z^+ = H^+$, $F^+$ where the double minimum symmetrical curve occurs [52]. Since the similar situation occurs for the series of 1,2-bis(pyridin-2-ylethynyl)benzene structures thus one can expect that the potential energy curve shape depends on the cation. However, this shape is a result of numerous factors; not only a kind of a cation. It depends on the type of a complex, the environment that results from the arrangement of molecules in crystals or from the type of solvent in liquids. For example, the first asymmetric linear silver complexes and the first asymmetric halonium complexes characterised by the [NAgN]$^+$ and [NIN]$^+$ arrangements, respectively, were analysed and they were confirmed by $^1$H and $^1$H-$^{15}$N HMBC NMR spectroscopy, and by X-ray diffraction results in crystals [55]. Thus one may expect that the halonium ion transfer and the transfer of other cations as Ag$^+$, for example, is possible for some systems. It is worth to mention that Zundel compared the hydrogen bond with analogues interactions where the proton is replaced by the lithium, sodium or potassium cations; the transfers similar to the proton transfer were analysed [56].

Let us compare systems containing $Z^+$ cations that were discussed above with the hydrogen bonds. The examples presented earlier here indicate the asymmetric position of proton in [NHN]$^+$ arrangement for both 1,2-bis(pyridin-2-ylethynyl)benzene and bis(pyridine) structures. It seems that the asymmetry of the H-atom position results, at least partly, from the asymmetry of environment in crystals. In another crystal structure containing protonated homodimer of pyridine, i.e., pyridinium tetrakis(3,5-bis(trifluoromethyl)phenyl)borate pyridine solvate, the asymmetry of the [NHN]$^+$ arrangement is also observed; however the asymmetric [NHN]$^+$ bridges occur also in solution [57]. The NMR signals in solution show the fast reversible proton transfer. All results concerning crystal as well as solution indicate the asymmetry of the [NHN]$^+$ contact, in spite the

N-H⋯N hydrogen bond is very strong and covalent in nature and the weak anion coordination is observed. Other strong N-H⋯N hydrogen bonds in derivatives of proton-bound homodimer of pyridine were analysed in solution and it was also found they are asymmetric in spite of these being very strong interactions [58].

It is worth to note however that hydrogen bonds characterised by a single symmetrical minimum of potential energy curve are also well known. The [FHF]⁻ anion is an example, the symmetrical species is known in the gas phase [59] although in crystal structure this anion is disturbed by packing forces since the movement of proton from the central position is often observed [60]. Similarly, the OH⋯O hydrogen bonds were analysed as it was discussed which conditions have to be fulfilled for the central position of proton [61]. Figure 1 presented earlier here shows geometries of numerous OH⋯O systems, the asymmetric ones as well as those where the proton is very close to the O⋯O mid-point. Various studies on hydrogen bonds, and particularly on the O-H⋯O systems indicate the complex character of PT process [62], for example, the potential barrier height should be taken into account in the analysis of this process. The same probably concerns the transfer of other cations, among them of halonium ones. There is no sufficient experimental and theoretical results however, and there is room for more extended investigations in this matter.

## 4. The Dihydrogen Bond as a Stage of the Molecular Hydrogen Uptake

The dihydrogen bond, DHB, is a special type of the hydrogen bond where the Lewis base centre is the negatively charged hydrogen atom [63–65]. In other words, it is the link between hydrogen atoms of the opposite charge, $H^{+\delta}\cdots^{-\delta}H$. The question arises if the proton transfer process A-H⋯B ⇔ A⁻⋯⁺H-B known for the A-H⋯B hydrogen bonded systems [62] occurs for dihydrogen bonds. This is discussed in detail in the monograph of Bakhmutov where numerous studies related to this topic are discussed [65]. It was generalised that the protonation of the hydride species characterised by the negatively charged hydrogen centre may be reversible or not, the latter phenomenon is connected with the molecular hydrogen elimination that may be expressed by the following transformations.

$$A-H^{+\delta} + {}^{-\delta}H - B \rightarrow [B(H_2)]^+ A^- \rightarrow A^- + H_2 + B^+ \quad (2)$$

It is worth mentioning that the B centre connected with the hydric hydrogen is most often the transition metal, the numerous moieties containing molecular hydrogen attached to the transition metal centre are known, such systems often occur in crystal structures [66,67]. The model $F-H^{+\delta}\cdots^{-\delta}H-Li$ complex corresponding to the phenomena expressed by Equation (2) was analysed theoretically early at the HF/6-31G(d,p) level [68]. It was found that the presence of the external electric field may lead to the formation and elimination of the molecular hydrogen since there are the following products; $F^- + H_2 + Li^+$, of the proton transfer reaction. The reverse process is observed for the system containing dihydrogen molecule inserted between Lewis acid and Lewis base units, $H_3B\cdots H_2\cdots NH_3$ since the DHB system is formed in the external electric field; $H_3B^- -H\cdots H-^+NH_3$ [68]. Bakhmutov, in his monograph [65], gives numerous examples of PT process in DHBs systems. The PT reaction and the dihydrogen elimination were discussed in the X-ray crystal structure of N-[2-(6-aminopyridyl)]acetamidine cyanoborohydride [69], the solid state transformation from the dihydrogen bonded LiBH₄-triethanolamine system to the covalent bonded material was described in another study [70].

There is an early example of the theoretical study [71] where the following reaction that leads to the molecular hydrogen uptake through the dihydrogen bond formation and next the proton transfer was analysed for a series of complexes of the $AlH_4^-$ anion.

$$[AlH_4]^- + HX \rightarrow AlH_3 - H^- \cdots HX \rightarrow [AlH_3X]^- + H_2 \quad (3)$$

The MP2/6-311++G(d,p) calculations were performed for this process (Equation (3)) with the HF, HCl, and H₂O species acting as the proton donors. The transition states of the process of transformation from the DHB systems to the products containing dihydrogen molecule were also calculated indicating that the PT reactions are energetically possible here, for example the potential barrier height for the $AlH_3\text{-}H^-\cdots HCl \rightarrow [AlH_3Cl]^- + H_2$ reaction is equal to 6.7 kcal/mol.

In one of more recent studies [72] the reverse process of the molecular hydrogen cleavage at the metal centre that leads to the dihydrogen bonded complex formation was analysed; $FM + H_2 \rightarrow FH\cdots HM$ (M = Cu, Ag, Au). The calculations were performed with the use of the aug-cc-pVQZ basis set (aug-cc-pVQZ-PP for metal atoms) and the following methods, MP2, CCSD(T) and CASSCF/CASPT2. The systems with dihydrogen molecule attached to the metal centre correspond to global minima, while the dihydrogen bonded complexes are characterised by higher energies (local energetic minima). Only for complexes containing the gold centre the DHB system is lower in energy than the system with the $H_2$ molecule for CASSCF/CASPT2 and CCSD(T) calculations. For example, for the CCSD(T)/aug-cc-pVQZ(aug-cc-pVQZ-PP) level of calculations for the $FAu + H_2 \rightarrow FH\cdots HAu$ reaction the potential barrier height amounts 30.9 kcal/mol and the products of reaction are lower in energy than reactants by 1.5 kcal/mol.

## 5. The Change of Trigonal Planar Triel Configuration into the Tetrahedral One—Triel Bonds

The triel bond is an interaction between the 13th group element acting as the Lewis acid centre and the electron-rich region of another or the same species [43,44,73]. It is worth to note that the acidic properties of triel atoms are related to the so-called $\pi$-holes [41,42] that are regions in planar molecules or in planar fragments of molecules at centres characterised by the depletion of the electron charge. For example, the trivalent triel atoms in trihydrides and trihalides possess formally empty p-orbital situated perpendicularly to the planes of molecules and being capable to act as the electron acceptor; numerous early [74–77] and more recent studies [43,44,73] related to interactions of the above-mentioned species with Lewis base units are known. The special attention should be paid to the studies of Phillips and co-workers who analysed such complexes from the theoretical and experimental points of view since the high level ab initio calculations were performed as well as the corresponding crystal structures were discussed [78–82]; particularly the potential energy as a function of the distance between the Lewis acid-base units was described [78]. It was found that the double local minimum occurs for some of complexes, deeper one corresponding to shorter distances and the flat, shallow energy minimum for the longer distances. The former minimum corresponds to strong and partly covalent interactions while the latter local minimum to weak interactions characterised mostly by dispersion forces. These findings are in line with other more recent studies [73,83].

It is worth to mention that for the triel bonds, similarly as for the hydrogen bonds, different sub-classes of interactions exist [84]. There are triel bonds with the one-centre electron donor, A-T ... B (where T is the triel atom, B is the electron donor and A is any atom connected with the triel centre), the interactions with $\pi$-electrons acting as the electron donor, A-T$\cdots\pi$ where the alkenes and alkines were considered as the Lewis base units [85,86] as well as benzene entity [87]. Another sub-class concerns links where the $\sigma$-bonds' electrons play a role of the electron donors, A-T$\cdots\sigma$ interactions; the complexes of molecular hydrogen acting as the Lewis base unit with the boron trihydride and boron trihalides were analysed in early study [85] and more recently the complexes of molecular hydrogen with boron and aluminium trihydrides and trifluorides were analysed theoretically [88]. Finally, one can mention intramolecular and bifurcated triel bonds that occur in the crystal structures [89]. It is well known that such types of hydrogen bonds are common in crystals [1,2,90].

The most interesting is which structural changes follow the formation of triel bonds? For example, the trivalent triel species described shortly above here change their planar structure if they interact with Lewis bases. The stronger interactions result in greater changes in the triel species geometry; in a case of extremely strong interactions, for example with anions, the triel centre conformation becomes to be close to tetrahedral structures. Even it acquires an ideal $T_d$ symmetry like in the $BF_4^-$ anions which occur very often in crystal structures [91]. Scheme 1 presents an example of the $AlCl_3\cdots NH_3$ complex linked by the Al$\cdots$N triel bond. The $\alpha$-angle which may be treated as the parameter of the deformation resulting from complexation is defined in this scheme. It is the angle between the Al···N line corresponding to the triel bond and one of the Al-Cl bonds. The angle parameter may be applied to other simple triel species complexes that informs the Lewis acid unit deformation resulting from the triel bond formation. This angle is equal to 90° if the interaction is very weak and the molecular deformations are not detected, it means that the triel Lewis acid species is still planar, or at least it is close to planarity. The increase of the α-angle is observed with the increase of the strength of interaction, for the $BF_4^-$ anion being the result of the $BF_3 \ldots F^-$ interaction it amounts 109.5° like for methane and other species possessing $T_d$ symmetry.

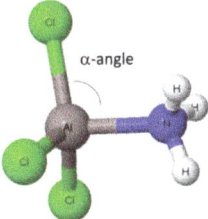

**Scheme 1.** The $AlCl_3 \cdots NH_3$ complex linked by the Al···N triel bond.

Figure 6 presents the relationship between the triel centre - Lewis base centre distance and the α-angle for simple complexes of boron and aluminium trihydrides and trichlorides. This dependence is based on the MP2/aug-cc-pVTZ results that come from one of recent studies [73]. The following Lewis base units were taken into account there; HCN, $NH_3$, $N_2$ and $Cl^-$. In such a way three sub-classes of complexes are presented in Figure 6; the ionic species being complexes with chloride anion and the remaining complexes are divided into two groups: complexes of aluminium Lewis acid units and complexes of boron compounds. The excellent linear correlation is observed for the neutral complexes of aluminium. For the remaining complexes, only tendency is observed that for stronger interactions (shorter Lewis acid-base distances) the greater α-angle occurs. In two cases of ionic systems, $BCl_4^-$ and $AlCl_4^-$ the ideal $T_d$ symmetry (α-angle equal to 109.5°) is observed certainly.

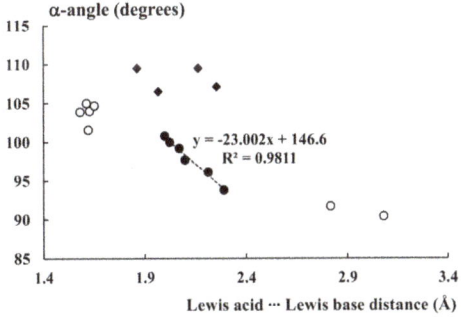

**Figure 6.** The dependence between the Lewis acid···Lewis base distance, T···B, (Å) and the α-parameter (see Scheme 1) in A-T···B triel bonds (T-triel centre). Black and white circles correspond to the neutral complexes, aluminium and boron species, respectively. Squares correspond to anion complexes.

Thus one can see that the relationship presented in Figure 6 may be treated as the reaction path for the process of the change of trigonal and planar configuration into the tetrahedral structure; few complexes linked by the triel bond are stable structures characterised by large dissociation energies. It is interesting that these boron adducts characterised by the structures close to tetrahedral ones may further interact with the electron rich species according to the $S_N2$ reaction mechanism.

The corresponding studies related to this topic appear from time to time, for example the reactions of borine carbonyl with trimethylamine and triethylamine were analysed experimentally early [92,93], or in more recent study, the $S_N2$ reaction for the $Cl^-\cdots CH_3Cl$ and $NH_3\cdots H_3BNH_3$ complexes were analysed theoretically [94], the latter one concerns the tetrahedral structure of boron.

## 6. Tetrel Bonds and the $S_N2$ Reaction

The tetrel bond is an interaction between the tetrel centre (14th group element) of the Lewis acid unit and the electron rich region of the moiety playing a role of the Lewis base [41,42,95–98]. The tetrel centre region that is responsible for the acidic (electron accepting) properties is classified as the σ-hole [95] but there are also cases of the tetrel centres that possess π-hole regions [98]. If the tetrel centre interacts with the Lewis base through the σ-hole thus the corresponding tetrel bond is often a preliminary stage of the $S_N2$ reaction [97]. The tetrel bond as a type of the σ-hole bond leads to the geometrical changes of interacting units that the complex structure is close to the trigonal bipyramid, especially in a case of strong interactions [97]. Scheme 2 presents an example of the species connected by the σ-hole tetrel bond. It is the $SiFH_3\cdots Cl^-$ complex with the $Si\cdots Cl$ intermolecular link and the $SiH_3$ group in the central part of the complex that tends to the planarity. The latter situation is similar to those observed for transition states of $S_N2$ reactions. Hence the similar parameter to that one discussed earlier here for the triel bonds (Scheme 1) may be introduced for tetrel σ-hole bonds. It is the α-angle (Scheme 2) between the tetrel···Lewis base link and one of bonds of the tetrel species. For the example shown in Scheme 2 it is the angle between $Si\cdots Cl$ line and the Si-H bond. For very strong interactions this angle is equal to 90° or nearly so. On the contrary, for extremely weak interactions that do not disturb the geometry of the Lewis acid unit the α-angle should be equal to ~70.5° that corresponds to the ideal tetrahedral structure of the tetrel species that is not perturbed by external forces.

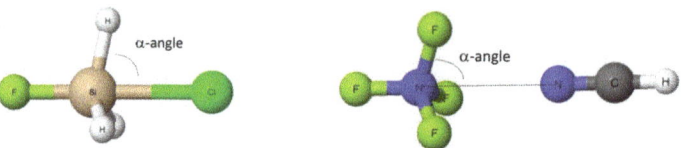

**Scheme 2.** The $SiFH_3\cdots Cl^-$ and $NF_4^+\cdots NCH$ complexes linked by the tetrel bond and pnicogen bond, respectively.

This is interesting that similar characteristics are often observed for the charge assisted pnicogen species, which interact with the Lewis base units [99]. The complexes of $ZH_4^+$, $ZF_4^+$ and $ZFH_3^+$ cations (Z = N, P, As) were analysed theoretically and the Z···Lewis base intermolecular links were observed [99] that may be classified as the type of σ-hole bonds, i.e., pnicogen bonds. These charge assisted pnicogen bonds also lead to the structural changes similar to those occurring for tetrel bonds, and they may be also classified as the preliminary stages of the $S_N2$ reactions [99]. Scheme 2 presents the $NF_4^+\cdots NCH$ complex connected by the N···N pnicogen bond. One of the nitrogen centres plays here a role of the electron acceptor while another N-centre is the electron donor. This is a very similar situation to that one occurring for the σ-hole tetrel bonds discussed here (see the same Scheme 2 with an example of the tetrel bonded complex).

Figure 7 presents the relationships between the tetrel/pnicogen centre···Lewis base centre distance and the α-angle. Two sub-samples of complexes of tetrel species and one sub-sample of pnicogen species complexes are presented here. These geometrical results were taken from earlier studies [97,99] where the MP2/aug-cc-pVTZ calculations were performed on complexes presented in this figure. Let us describe the sub-samples presented in Figure 7. For that one with units linked by tetrel bond, the $ZH_4$, $ZFH_3$ and $ZF_4$ (Z = C, Si, Ge) species interact with the HCN and LiCN acting as Lewis bases (nitrogen atom is the Lewis base centre here). For the complexes with negatively charge assisted tetrel bonds the same Lewis acid units as for the former sub-sample interact with chloride anion. And for complexes linked by the pnicogen bond the $ZH_4^+$, $ZFH_3^+$ and $ZF_4^+$ species (Z = N, P and As) interact with HCN and LiCN through the Z and nitrogen centres.

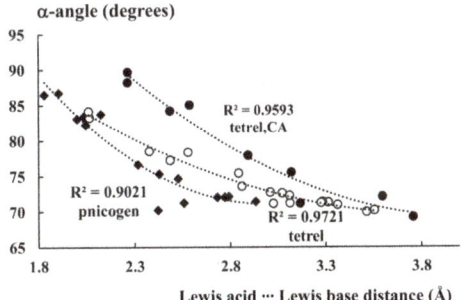

**Figure 7.** The dependence between the Lewis acid⋯Lewis base distance, Z⋯B, (Å) and the α-parameter (see Scheme 2) in A-Z⋯B tetrel and pnicogen bonds (Z-tetrel or pnicogen centre). Black and white circles correspond to the tetrel complexes linked by charge assisted tetrel bonds and neutral tetrel bonds, respectively. Squares correspond to complexes linked by the pnicogen bond.

Figure 7 shows that for shorter Lewis acid-base distances corresponding to stronger interactions, the α-parameter (Scheme 2) tends to 90°. It means that the central part of the complex becomes to be planar as it is observed for the transition state of the $S_N2$ reaction. The second order polynomial dependences between the distance and the α-angle are observed here with high values of the correlation coefficients. These dependencies may be treated as the corresponding reaction paths of the above-mentioned $S_N2$ reactions. This is very interesting that the pnicogen and tetrel σ-bonds considered here (Figure 7) do not practically differ between themselves. For both interactions the Lewis acid units possess the tetrahedral structure, for both interactions, the complexation leads to the geometrical changes towards the structure of trigonal bipyramid. And finally, both interactions may be considered as the preliminary stages of the $S_N2$ reactions.

One can also mention C-H⋯M (M designates metal) contacts that often play an important role in catalysis [100]. These interactions are usually classified as attractive ones and they are named as agostic interactions in terms of the Dewar-Chatt-Duncanson model [101,102]. However, the term anagostic interactions was introduced [103] for such contacts that are sterically enforced in square-planar transition metal $d^8$ complexes to distinguish them from attractive agostic interactions. The orbital interaction schemes were presented by Scherer and co-workers for various types of the C-H⋯M interactions [104]; these are: the above-mentioned repulsive anagostic 3c-4e interaction being the contact between hydric hydrogen and the filled $M-dz^2$ orbital, the attractive 3c-4e hydrogen bond being the interaction between the protic H-atom and the filled $M-dz^2$ orbital, preagostic attractive 3c-2e interaction (π-back donation) and the σ-agostic 3c-2e attractive interaction (hydric hydrogen–empty $M-dz^2$ orbital). On the other hand, in another study it was justified that the formation of numerous C-H ... M structures is mainly driven by the dispersion forces thus the models based on the orbital-orbital interactions schemes are not sufficient to describe the nature of these structures [105]. It seems that both explanations, the "orbital based" explanation as well as that one considering dispersion forces, are valid. For example, in one of recent studies on the C-H ... Ni contacts in $Ni^{II}$ planar isomers the occurrence of the covalent type charge delocalisation and of the London dispersion forces was justified [106]. The latter study is based on experimental results that are supported by theoretical analyses where various approaches were applied [106]. However, it is worth noting that these C-H ... M interactions, regardless of the mechanism of their formation, lead to the metal centre coordination change. In the case of square-planar structures with two additional C-H ... M contacts the metal structure tends to the octahedral one. Hence the agostic interaction may be considered as a preliminary stage of the process of structural reconstruction.

## 7. Summary

Different interactions are discussed in this article; the hydrogen bond and its special type—the dihydrogen bond, the pnicogen, and tetrel bonds as representatives of the σ-hole bond, and the triel bond as an example of the π-hole bond. These interactions may be treated as preliminary stages of various reactions and processes; the proton transfer, the release of the molecular hydrogen, or the $S_N2$ reaction.

However, it is also very important that intra- and intermolecular connections lead to numerous structural changes of the interacting units. The triel planar and trigonal species interacting with Lewis bases tend to achieve the tetrahedral geometry. On the other hand the tetrahedral tetrel moieties as well as the tetrahedral pnicogen cations, both characterised by the lack of the lone electron pairs, tend to attain the structure of the trigonal bipyramid. It is worth to mention that different structural changes take place for different species of the same element, like in a case of the σ-hole bonds on one hand and in a case of the π-hole bonds on the other hand. In general, there is a variety of numerous structural changes accompanying the processes of complexation.

Finally it is worth noting here that other interactions not discussed here may initiate various chemical reactions, there is room for numerous future studies related to this topic. One of the most important topics to be discussed in next studies is the application of the structure-correlation method [12–14]. This method may be generalised as the analysis of geometrical changes during the chemical reactions. One can see that Figures 1 and 2 presented in this study express the changes following the proton transfer process, Figure 6 shows such changes related to the transformation of the trigonal planar system into the tetrahedral structure while Figure 7 shows the reaction paths of the $S_N2$ reactions.

**Funding:** This research was funded by the Spanish Government MINECO/FEDER, grant number PID2019-109555GB-I00 and Eusko Jaurlaritza, grant number IT-1254-19.

**Acknowledgments:** Technical and human support provided by Informatikako Zerbitzu Orokora - Servicio General de Informática de la Universidad del País Vasco (SGI/IZO-SGIker UPV/EHU), Ministerio de Ciencia e Innovación (MICINN), Gobierno Vasco Eusko Jaurlanitza (GV/EJ), European Social Fund (ESF) is gratefully acknowledged.

**Conflicts of Interest:** The authors declare no conflict of interest.

## References

1. Jeffrey, G.A.; Saenger, W. *Hydrogen Bonding in Biological Structures*; Springer: Berlin, Germany, 1991.
2. Jeffrey, G.A. *An Introduction to Hydrogen Bonding*; Oxford University Press: New York, NY, USA, 1997.
3. *Intermolecular Interactions in Crystals: Fundamentals of Crystal Engineering*; Novoa, J.J. Ed.; The Royal Society of Chemistry: London, UK, 2018.
4. Politzer, P.; Murray, J.S. Halogen Bonding: An Interim Discussion. *Chem. Phys. Chem.* **2013**, *14*, 278–294. [CrossRef] [PubMed]
5. Lewis, G.N. *Valence and the Structure of Atoms and Molecules*; American Chemical Society Monograph Series; The Chemical Catalog Company, Inc.: New York, NY, USA, 1923; p. 146.
6. Rauk, A. *Orbital Interaction Theory of Organic Chemistry*, 2nd ed.; John Wiley & Sons, Inc.: New York, NY, USA, 2001; p. XIII.
7. Grabowski, S.J. Hydrogen Bond and Other Lewis acid–Lewis Base Interactions–Mechanisms of Formation. In *Practical Aspects of Computational Chemistry IV*; Leszczynski, J., Shukla, M.K., Eds.; Springer Science: New York, NY, USA, 2016; Chapter 9; pp. 245–278.
8. Weinhold, F.; Landis, C. *Valency and Bonding, a Natural Bond Orbital Donor—Acceptor Perspective*; Cambridge University Press: Cambridge, UK, 2005.
9. Grabowski, S.J. What is the Covalency of Hydrogen Bonding? *Chem. Rev.* **2011**, *11*, 2597–2625. [CrossRef] [PubMed]
10. Kaplan, I.G. *Intermolecular Interactions: Physical Picture, Computational Methods and Model Potentials*; John Wiley & Sons, Ltd.: Chichester, UK, 2006; p. 1.

11. Grabowski, S.J. Hydrogen bonds, and σ-hole and π-hole bonds–mechanisms protecting doublet and octet electron structures. *Phys. Chem. Chem. Phys.* **2017**, *19*, 29742–29759. [CrossRef] [PubMed]
12. Dunitz, J.D. *Analysis of the Structure of Organic Molecules*; Cornell University Press: Ithaca, NY, USA, 1979.
13. Bürgi, H.B. Stereochemistry and Reaction Paths as Determined from Crystal Structure Data—A Relationship between Structure and Energy. *Angew. Chem. Int. Ed.* **1975**, *14*, 460–473. [CrossRef]
14. Bürgi, H.B.; Dunitz, J.D. From Crystal Statics to chemical dynamics. *Acc. Chem. Res.* **1983**, *16*, 153–161. [CrossRef]
15. Bürgi, H.B.; Dunitz, J.D.; Shefter, E. Geometrical reaction coordinates. II. Nucleophilic addition to a carbonyl group. *J. Am. Chem. Soc.* **1973**, *95*, 5065–5067. [CrossRef]
16. Grabowski, S.J.; Krygowski, T.M. The proton transfer path for C=O . . . H-O systems modelled from crystal structure data. *Chem. Phys. Lett.* **1999**, *305*, 247–250. [CrossRef]
17. Groom, C.R.; Bruno, I.J.; Lightfoot, M.P.; Ward, S.C. The Cambridge structural database. *Acta Cryst.* **2016**, *B72*, 171–179. [CrossRef]
18. Wong, R.; Allen, F.H.; Willett, P. The scientific impact of the Cambridge Structural Database: A citation-based study. *J. Appl. Cryst.* **2010**, *43*, 811–824. [CrossRef]
19. Wilson, C.C. *Single Crystal Neutron Diffraction from Molecular Materials*; World Scientific Publishing Co. Pre. Ltd.: Singapore, 2000.
20. Luger, P. *Modern X-Ray Analysis on Single Crystals*, 2nd ed.; Walter de Gruyter: Berlin, Germany, 2014.
21. Brown, I.D. Bond valences—A simple structural model for inorganic chemistry. *Chem. Soc. Rev.* **1978**, *7*, 359–376. [CrossRef]
22. Pauling, L. Atomic radii and interatomic distances in metals. *J. Am. Chem. Soc.* **1947**, *69*, 542–553. [CrossRef]
23. Gilli, P.; Bertolasi, V.; Ferretti, V.; Gilli, G. Covalent nature of the strong homonuclear hydrogen-bond–study of the OH . . . O system by crystal-structure correlation methods. *J. Am. Chem. Soc.* **1994**, *116*, 909–915. [CrossRef]
24. Robertson, J.M.; Ubbelohde, A.R. Structure and thermal properties associated with some hydrogen bonds in crystals. I. The isotope effect. *Proc. R. Soc. Lond. Ser. A* **1939**, *170*, 222–240.
25. Benedict, H.; Limbach, H.-H.; Wehlan, M.; Fehlhammer, W.-P.; Golubev, N.S.; Janoschek, R. Solid State $^{15}$N-NMR and Theoretical Studies on Primary and Secondary Geometric H/D Isotope Effects on Low-Barrier NHN-Hydrogen Bonds. *J. Am. Chem. Soc.* **1998**, *120*, 2939–2950. [CrossRef]
26. Limbach, H.-H.; Tolstoy, P.M.; Pérez-Hernández, N.; Guo, J.; Shenderovich, I.G.; Denisov, G.S. OHO hydrogen bond geometries and NMR chemical shifts: From equilibrium structures to geometric H/D isotopic effects, with applications for water, protonated water, and compresses ice. *Isr. J. Chem.* **2009**, *49*, 199–216. [CrossRef]
27. Benedict, H.; Shenderovich, I.G.; Malkina, O.L.; Malkin, V.G.; Denisov, G.S.; Golubev, N.S.; Limbach, H.-H. Nuclear Scalar Spin-Spin Couplings and Geometries of Hydrogen Bonds. *J. Am. Chem. Soc.* **2000**, *122*, 1979–1988. [CrossRef]
28. Shenderovich, I.G.; Tolstoy, P.M.; Golubev, N.S.; Smirnov, S.N.; Denisov, G.S.; Limbach, H.-H. Low-Temperature NMR Studies of the Structure and Dynamics of a Novel Series of Acid–Base Complexes of HF with Collidine Exhibiting Scalar Couplings across Hydrogen Bonds. *J. Am. Chem. Soc.* **2003**, *125*, 11710–11720. [CrossRef] [PubMed]
29. Grabowski, S.J. Hydrogen bonds and other interactions as a response to protect doublet/octet electron structure. *J. Mol. Model.* **2018**, *24*, 38. [CrossRef] [PubMed]
30. Kryachko, E.S. Neutral Blue-Shifting and Blue-Shifted Hydrogen Bonds. In *Hydrogen Bonding—New Insights*; Grabowski, S.J., Ed.; Springer: Berlin/Heidelberg, Germany, 2006; Chapter 8; pp. 293–336.
31. Bader, R.F.W. *Atoms in Molecules, a Quantum Theory*; Oxford University Press: Oxford, UK, 1990.
32. Matta, C.; Boyd, R.J. (Eds.) *Quantum Theory of Atoms in Molecules: From Solid State to DNA and Drug Design*; Wiley-VCH Verlag GmbH & Co. KGaA: Weinheim, Germany, 2007.
33. Cremer, D.; Kraka, E. A Description of the Chemical Bond in Terms of Local Properties of Electron Density and Energy. *Croat. Chem. Acta* **1984**, *57*, 1259–1281.
34. Jenkins, S.; Morrison, I. The chemical character of the intermolecular bonds of seven phases of ice as revealed by AB initio calculation of electron densities. *Chem. Phys. Lett.* **2000**, *317*, 97–102. [CrossRef]
35. Arnold, W.D.; Oldfield, E. The Chemical Nature of Hydrogen Bonding in Proteins via NMR: J-Couplings, Chemical Shifts, and AIM Theory. *J. Am. Chem. Soc.* **2000**, *122*, 12835–12841. [CrossRef]

36. Rozas, I.; Alkorta, I.; Elguero, J. Behavior of ylides containing N, O, and C atoms as hydrogen bond acceptors. *J. Am. Chem. Soc.* **2000**, *122*, 1154–11161. [CrossRef]
37. Grabowski, S.J.; Lipkowski, P. Characteristics of XH . . . π Interactions: Ab Initio and QTAIM Studies. *J. Phys. Chem. A* **2011**, *115*, 4765–4773. [CrossRef] [PubMed]
38. Clark, T.; Hennemann, M.; Murray, J.S.; Politzer, P. Halogen bonding: The σ-hole. *J. Mol. Model.* **2007**, *13*, 291–296. [CrossRef] [PubMed]
39. Politzer, P.; Lane, P.; Concha, M.C.; Ma, Y.; Murray, J.S. An overview of halogen bonding. *J. Mol. Model.* **2007**, *13*, 305–311. [CrossRef]
40. Politzer, P.; Riley, K.E.; Bulat, F.A.; Murray, J.S. Perspectives on halogen bonding and other σ-hole interactions: Lex parsimoniae (Occam's Razor). *Comput. Theor. Chem.* **2012**, *998*, 2–8. [CrossRef]
41. Politzer, P.; Murray, J.S.; Clark, T. Halogen bonding: An electrostatically-driven highly directional noncovalent interaction. *Phys. Chem. Chem. Phys.* **2010**, *12*, 7748–7758. [CrossRef]
42. Politzer, P.; Murray, J.S.; Clark, T. Halogen bonding and other σ-hole interactions: A perspective. *Phys. Chem. Chem. Phys.* **2013**, *15*, 11178–11189. [CrossRef]
43. Grabowski, S.J. Boron and other triel Lewis acid centers: From hypovalency to hypervalency. *Chem. Phys. Chem.* **2014**, *15*, 2985–2993. [CrossRef]
44. Grabowski, S.J. π-hole bonds: Boron and aluminium Lewis acid centers. *Chem. Phys. Chem.* **2015**, *16*, 1470–1479. [CrossRef]
45. Scheiner, S. Detailed Comparison of the Pnicogen Bond with Chalcogen, Halogen, and Hydrogen Bonds. *Int. J. Quantum Chem.* **2013**, *113*, 1609–1620. [CrossRef]
46. Scheiner, S. The Pnicogen Bond: Its Relation to Hydrogen, Halogen, and Other Noncovalent Bonds. *Acc. Chem. Res.* **2013**, *46*, 280–288. [CrossRef] [PubMed]
47. Grabowski, S.J. Hydrogen and halogen bonds are ruled by the same mechanisms. *Phys. Chem. Chem. Phys.* **2013**, *15*, 7249–7259. [CrossRef] [PubMed]
48. Turunen, L.; Erdélyi, M. Halogen Bonds and Halonium Ions. *Chem. Soc. Rev.* **2020**, *49*, 2688–2700. [CrossRef]
49. Holl, M.G.; Pitts, C.R.; Lectka, T. Quest for a Symmetric [C–F–C]+ Fluoronium Ion in Solution: A Winding Path to Ultimate Success. *Acc. Chem. Res.* **2020**, *53*, 265–275. [CrossRef]
50. Struble, M.D.; Scerba, M.T.; Siegler, M.; Lectka, T. Evidence for a Symmetrical Fluoronium Ion in Solution. *Science* **2013**, *340*, 57–60. [CrossRef]
51. Hennecke, U. Revealing the Positive Side of Fluorine. *Science* **2013**, *340*, 41–42. [CrossRef]
52. Reiersølmoen, A.C.; Battaglia, S.; Øien-Ødegaard, S.; Gupta, A.K.; Fiksdahl, A.; Lindh, R.; Erdélyi, M. Symmetry of three-center, four-electron bonds. *Chem. Sci.* **2020**, *11*, 7979–7990. [CrossRef]
53. Bedin, N.; Karim, A.; Reitti, M.; Carlsson, A.C.C.; Topić, F.; Cetina, M.; Pan, F.; Havel, V.; Al-Ameri, F.; Sindelar, V.; et al. Counterion influence on the N-I-N halogen bond. *Chem. Sci.* **2015**, *6*, 3746–3756. [CrossRef]
54. Villarreal-Salinas, B.E.; Schlemper, E.O. Crystal structure of a salt of the pyridinium-pyridine ion by X-ray and neutron diffraction. *J. Cryst. Mol. Struct.* **1978**, *8*, 217–237. [CrossRef]
55. Ward, J.S.; Fiorini, G.; Frontera, A.; Rissanen, K. Asymmetric [N-I-N]+ halonium complexes. *Chem. Commun.* **2020**, *56*, 8428–8431. [CrossRef]
56. Zundel, G. *Series of Ten Lectures On: Proton Polarizability of Hydrogen Bonds and Proton Transfer Processes, Their Role in Electrochemistry and Biology*; Institut für Physikalische Chemie der Universität München: München, Germany, 1997.
57. Kong, S.; Borissova, A.O.; Lesnichin, S.B.; Hartl, M.; Daemen, L.L.; Eckert, J.; Antipin, M.Y.; Shenderovich, I.G. Geometry and Spectral Properties of the Protonated Homodimer of Pyridine in the Liquid and Solid States. A Combined NMR, X-ray Diffraction and Inelastic Neutron Scattering Study. *J. Phys. Chem. A* **2011**, *115*, 8041–8048. [CrossRef]
58. Gurinov, A.A.; Lesnichin, S.B.; Limbach, H.-H.; Shenderovich, I.G. How Short is the Strongest Hydrogen Bond in the Proton-Bound Homodimers of Pyridine Derivatives? *J. Phys. Chem. A* **2014**, *118*, 10804–10812. [CrossRef] [PubMed]
59. Wenthold, P.G.; Squires, R.R. Bond Dissociation Energies of $F_2^-$ and $HF_2^-$. A Gas-Phase Experimental and G2 Theoretical Study. *J. Phys. Chem.* **1995**, *99*, 2002–2005. [CrossRef]
60. Grabowski, S.J. [FHF]−-The Strongest Hydrogen Bond under the Influence of External Interactions. *Crystals* **2016**, *6*, 3. [CrossRef]

61. Grabowski, S.J.; Ugalde, J.M. High-level ab initio calculations on low barrier hydrogen bonds and proton bound homodimers. *Chem. Phys. Lett.* **2010**, *493*, 37–44. [CrossRef]
62. Sobczyk, L.; Grabowski, S.J.; Krygowski, T.M. Interrelation between H-Bond and Pi-Electron Delocalization. *Chem. Rev.* **2005**, *105*, 3513–3560. [CrossRef]
63. Richardson, T.B.; de Gala, S.; Crabtree, R.H. Unconventional Hydrogen Bonds: Intermolecular B-H ... H-N Interactions. *J. Am. Chem. Soc.* **1995**, *117*, 12875–12876. [CrossRef]
64. Epstein, L.M.; Shubina, E.S. New types of hydrogen bonding in organometallic chemistry. *Coord. Chem. Rev.* **2002**, *231*, 165–181. [CrossRef]
65. Bakhmutov, V.I. *Dihydrogen Bonds, Principles, Experiments, and Applications*; John Wiley & Sons: Hoboken, NJ, USA, 2008.
66. Kubas, G.J. *Metal Dihydrogen and σ-Bond Complexes–Structure, Theory, and Reactivity*; Kluwer Academic/Plenum Publishers: New York, NY, USA, 2001.
67. Crabtree, R.H. *The Organometallic Chemistry of the Transition Metals*; John Wiley & Sons, Inc.: Hoboken, NJ, USA, 2005.
68. Rozas, I.; Alkorta, I.; Elguero, J. Field effects on dihydrogen bonded systems. *Chem. Phys. Lett.* **1997**, *275*, 423–428. [CrossRef]
69. Custelcean, R.; Jackson, J.E. Topochemical Control of Covalent Bond Formation by Dihydrogen Bonding. *J. Am. Chem. Soc.* **1998**, *120*, 12935–12941. [CrossRef]
70. Custelcean, R.; Jackson, J.E. Topochemical Dihydrogen to Covalent Bonding Transformation in $LiBH_4 \cdot TEA$: A Mechanistic Study. *J. Am. Chem. Soc.* **2000**, *122*, 5251–5257. [CrossRef]
71. Marincean, S.; Jackson, J.E. Quest for IR-pumped reactions in dihydrogen-bonded complexes. *J. Phys. Chem. A* **2004**, *108*, 5521–5526. [CrossRef]
72. Grabowski, S.J.; Ruipérez, F. Dihydrogen bond interactions as a result of $H_2$ cleavage at Cu, Ag and Au centres. *Phys. Chem. Chem. Phys.* **2016**, *18*, 12810–12818. [CrossRef]
73. Grabowski, S.J. The Nature of Triel Bonds, a Case of B and Al Centres Bonded with Electron Rich Sites. *Molecules* **2020**, *25*, 2703. [CrossRef]
74. Brinck, T.; Murray, J.S.; Politzer, P. A computational analysis of the bonding in boron trifluoride and boron trichloride and their complexes with ammonia. *Inorg. Chem.* **1993**, *32*, 2622–2625. [CrossRef]
75. Bessac, F.; Frenking, G. Why Is $BCl_3$ a Stronger Lewis Acid with Respect to String Bases than $BF_3$? *Inorg. Chem.* **2003**, *42*, 7990–7994. [CrossRef]
76. Jonas, V.; Frenking, G.; Reetz, M.T. Comparative Theoretical Study of Lewis Acid-Base Complexes of $BH_3$, $BF_3$, $BCl_3$, $AlCl_3$, and $SO_2$. *J. Am. Chem. Soc.* **1994**, *116*, 8741–8753. [CrossRef]
77. Van der Veken, B.J.; Sluyts, E.J. Reversed Lewis Acidity of Mixed Boron Halides: An Infrared Study of the Van der Waals Complexes of $BF_xCl_y$ with $CH_3F$ in Cryosolution. *J. Am. Chem. Soc.* **1997**, *119*, 11516–11522. [CrossRef]
78. Phillips, J.A. Structural and energetic properties of nitrile–$BX_3$ complexes: Substituent effects and their impact on condensed-phase sensitivity. *Theor. Chem. Acc.* **2017**, *136*, 16. [CrossRef]
79. Giesen, D.J.; Phillips, J.A. Structure, Bonding, and Vibrational Frequencies of $CH_3CN$-$BF_3$: New Insight into Medium Effects and the Discrepancy between the Experimental and Theoretical Geometries. *J. Phys. Chem. A* **2003**, *107*, 4009–4018. [CrossRef]
80. Phillips, J.A.; Cramer, C.J. B-N Distance Potential of $CH_3CN$-$BF_3$ Revisited: Resolving the Experiment-Theory Structure Discrepancy and Modeling the Effects of Low-Dielectric Environments. *J. Phys. Chem. B* **2007**, *111*, 1408–1415. [CrossRef] [PubMed]
81. Wrass, J.P.; Sadowsky, D.; Bloomgren, K.M.; Cramer, C.J.; Phillips, J.A. Quantum chemical and matrix-IR characterization of $CH_3CN$-$BCl_3$: A complex with two distinct minima along the B-N bond potential. *Phys. Chem. Chem. Phys.* **2014**, *16*, 16480–16491. [CrossRef] [PubMed]
82. Helminiak, H.M.; Knauf, R.R.; Danforth, S.J.; Phillips, J.A. Structural and Energetic Properties of Acetonitrile-Group IV (A & B) Halide Complexes. *J. Phys. Chem. A* **2014**, *118*, 4266–4277.
83. Grabowski, S.J. Two Faces of Triel Bonds in Boron Trihalide Complexes. *J. Comput. Chem.* **2018**, *39*, 472–480. [CrossRef]
84. Grabowski, S.J. Triel bond and coordination of triel centres–Comparison with hydrogen bond interaction. *Coord. Chem. Rev.* **2020**, *407*, 213171. [CrossRef]

85. Fau, S.; Frenking, G. Theoretical investigation of the weakly bonded donor—Acceptor complexes $X_3B$—$H_2$, $X_3B$—$C_2H_4$, and $X_3B$—$C_2H_2$ (X= H, F, Cl). *Mol. Phys.* **1999**, *96*, 519–527.
86. Grabowski, S.J. Triel Bonds, π-Hole-π-Electrons Interactions in Complexes of Boron and Aluminium Trihalides and Trihydrides with Acetylene and Ethylene. *Molecules* **2015**, *20*, 11297–11316. [CrossRef]
87. Grabowski, S.J. Triel bonds-complexes of boron and aluminum trihalides and trihydrides with benzene. *Struct. Chem.* **2017**, *28*, 1163–1171. [CrossRef]
88. Grabowski, S.J. Molecular Hydrogen as a Lewis Base in Hydrogen Bonds and Other Interactions. *Molecules* **2020**, *25*, 3294. [CrossRef]
89. Grabowski, S.J. Bifurcated Triel Bonds—Hydrides and Halides of 1,2-Bis(Dichloroboryl)Benzene and 1,8-Bis(Dichloroboryl)Naphthalene. *Crystals* **2019**, *9*, 503. [CrossRef]
90. Etter, M.C. Encoding and Decoding Hydrogen-Bond Patterns of Organic Compounds. *Acc. Chem. Res.* **1990**, *23*, 120–126. [CrossRef]
91. Grabowski, S.J. Hydrogen bonds with $BF_4^-$ anion as a proton acceptor. *Crystals* **2020**, *10*, 460. [CrossRef]
92. Grotewold, J.; Lisi, E.A.; Villa, A.E. Reactions of Co-ordinated Boron Compounds in the Gas Phase. Part I. Borine Carbonyl and Trimethylamine. *J. Chem. Soc. A* **1966**, 1034–1037. [CrossRef]
93. Grotewold, J.; Lisi, E.A.; Villa, A.E. Reactions of Co-ordinated Boron Compounds in the Gas Phase. Part II. Triethylamine as Scavenger of Borine. *J. Chem. Soc. A* **1966**, 1038–1041. [CrossRef]
94. Yang, S.-Y.; Fleurat-Lessard, P.; Hristov, I.; Ziegler, T. Free Energy Profiles for the Identity $S_N2$ Reactions $Cl^-$ + $CH_3Cl$ and $NH_3 + H_3BNH_3$: A Constraint Ab Initio Molecular Dynamics Study. *J. Phys. Chem. A* **2004**, *108*, 9461–9468. [CrossRef]
95. Bundhun, A.; Ramasami, P.; Murray, J.S.; Politzer, P. Trends in σ-hole Strengths and Interactions of $F_3MX$ Molecules (M = C, Si, Ge and X = F, Cl, Br, I). *J. Mol. Model.* **2013**, *19*, 2739–2746. [CrossRef]
96. Bauzá, A.; Mooibroek, T.J.; Frontera, A. Tetrel-Bonding Interaction: Rediscovered Supramolecular Force? *Angew. Chem. Int. Ed.* **2013**, *52*, 12317–12321. [CrossRef]
97. Grabowski, S.J. Tetrel bond-σ-hole bond as a preliminary stage of the $S_N2$ reaction. *Phys. Chem. Chem. Phys.* **2014**, *16*, 1824–1834. [CrossRef]
98. Zierkiewicz, W.; Michalczyk, M.; Scheiner, S. Comparison between Tetrel Bonded Complexes Stabilized by σ and π Hole Interactions. *Molecules* **2018**, *23*, 1416. [CrossRef]
99. Grabowski, S.J. σ-Hole Bond versus Hydrogen Bond: From Tetravalent to Pentavalent, N, P and As Atoms. *Chem. Eur. J.* **2013**, *19*, 14600–14611. [CrossRef] [PubMed]
100. Scherer, W.; McGrady, G.S. Agostic Interactions in $d^0$ Metal Alkyl Complexes. *Angew. Chem. Int. Ed.* **2004**, *43*, 1782–1806. [CrossRef] [PubMed]
101. Dewar, M.J.S. A review of the π-complex theory. *Bull. Soc. Chim. Fr.* **1951**, *18*, C71–C79.
102. Chatt, J.; Duncanson, L.A. Olefin Co-ordination Compounds. Part III. Infra-red Spectra and Structure: Attempted Preparation of Acetylene Complexes. *J. Chem. Soc.* **1953**, 2939–2947. [CrossRef]
103. Sundquist, W.I.; Bancroft, D.P.; Lippard, S.J. Synthesis, characterization, and biological activity of cis-diammineplatinum(II) complexes of the DNA intercalators 9-aminoacridine and chloroquine. *J. Am. Chem. Soc.* **1990**, *112*, 1590–1596. [CrossRef]
104. Scherer, W.; Dunbar, A.C.; Barquera-Lozada, J.E.; Schmitz, D.; Eickerling, G.; Kratzert, D.; Stalke, D.; Lanza, A.; Macchi, P.; Casati, N.P.M.; et al. Anagostic Interactions under Pressure: Attractive or Repulsive? *Angew. Chem. Int. Ed.* **2015**, *54*, 2505–2509. [CrossRef]
105. Lu, Q.; Neese, F.; Bistoni, G. Formation of Agostic Structures Driven by London Dispersion. *Angew. Chem. Int. Ed.* **2018**, *57*, 4760–4764. [CrossRef]
106. Mitoraj, M.P.; Babashkina, M.G.; Robeyns, K.; Sagan, F.; Szczepanik, D.W.; Seredina, Y.V.; Garcia, Y.; Safin, D.A. Chameleon-like Nature of Anagostic Interactions and Its Impact on Metalloaromaticity in Square-Planar Nickel Complexes. *Organometallics* **2019**, *38*, 1973–1981. [CrossRef]

© 2020 by the author. Licensee MDPI, Basel, Switzerland. This article is an open access article distributed under the terms and conditions of the Creative Commons Attribution (CC BY) license (http://creativecommons.org/licenses/by/4.0/).

*Review*

# Small Molecules, Non-Covalent Interactions, and Confinement

Gerd Buntkowsky [1,*,†] and Michael Vogel [2,*,†]

1   Institut für Physikalische Chemie, Technische Universität Darmstadt, 64287 Darmstadt, Germany
2   Institut für Festkörperphysik, Technische Universität Darmstadt, 64295 Darmstadt, Germany
*   Correspondence: gerd.buntkowsky@chemie.tu-darmstadt.de (G.B.); michael.vogel@physik.tu-darmstadt.de (M.V.)
†   These authors contributed equally to this work.

Academic Editor: Ilya Shenderovich
Received: 3 June 2020; Accepted: 15 July 2020; Published: 21 July 2020

**Abstract:** This review gives an overview of current trends in the investigation of small guest molecules, confined in neat and functionalized mesoporous silica materials by a combination of solid-state NMR and relaxometry with other physico-chemical techniques. The reported guest molecules are water, small alcohols, and carbonic acids, small aromatic and heteroaromatic molecules, ionic liquids, and surfactants. They are taken as characteristic role-models, which are representatives for the typical classes of organic molecules. It is shown that this combination delivers unique insights into the structure, arrangement, dynamics, guest-host interactions, and the binding sites in these confined systems, and is probably the most powerful analytical technique to probe these systems.

**Keywords:** confinement; solid-state NMR; molecular dynamics; interfaces and surfaces

---

## 1. Introduction

Porous silicates and alumosilicates include such diverse materials as the well-known microporous zeolites, a group of crystalline alumosilicates present in daily life, over mesoporous silica materials, such as the original ordered periodical mesoporous silica (PMS) [1,2] to controlled porous glasses and aerogels. They span a diameter range from fractions of a nanometer to ca. 50 nm and above. Owing to this large dispersion of well-defined diameters these systems are among the most versatile solid host-systems for studies of molecules in confinement.

In the present review an overview of confinement studies in PMS materials with a focus on neat and surface modified MCM-41 (Mobil Composition of Matter No. 41) [3] and SBA-15 (Santa Barbara Amorphous) [4,5] as hosts is given. Both MCM-41 and SBA-15 are characterized by well-ordered hexagonal pore-arrays, however, with different pore diameter distribution. Owing to their ordered structures, high porosity, high intrinsic surface area, low density, thermal stability, tunable pore sizes, and functional surface groups, PMS were successfully employed in a large variety of applications ranging from gas-storage and separation over heterogeneous catalysis to drug delivery (see, e.g., [6–22]), and many more.

PMS-type materials are ideal host systems for confinement studies since they combine narrow pore-diameter distributions and large specific surface with good chemical stability, easy handling, and chemical functionalization. They were employed in studies on the structural and dynamic properties of many different confined molecules, including water, alcohols, carbonic acids, protein solutions, and on thermophysical processes, such as freezing and melting points, glass transitions [23–29], or an electrochemical study of local pKa in confinement [30].

Of particular importance here are the periodic mesoporous silica materials MCM-41 [3] and SBA-15 [4] and their many derivates, which exhibit well-defined hexagonally arranged mesopore structures and three dimensional sponge like structures in porous glasses such as Vycor [31] or CPG-10-75 [32,33]. These materials opened up new research fields, as they allowed the introduction of larger molecular entities into well-defined pores. The prosperity of this field can be seen from the fact that, currently (March 2020), there are nearly 10,000 references in the Web of Science which have MCM-41 or SBA-15 in their title.

A large part of the interest in these materials stems from the fact that these materials are chemically very stable, because of the strength of the covalent Si-O-Si bonds, and that their surface silanol (Si-OH) groups are very potent reactive groups for chemical modification or functionalization of their surfaces. Thus, it is relatively easy to introduce the necessary functional groups by linker molecules, which serve, e.g., as possible binding sites for the chemical function of interest, such as a catalytic center. Such linker molecules can change the polarity or hydrophilicity of the surface, modify the hydrogen bonding properties of the surface, or add chemical functions, e.g., amide or carboxy functions [11] by binding functional groups such as amino, amide, carboxyl, phosphate, chloride, or peptide functions by post-synthetic grafting. Alternatively, it is also possible to add some of these functions directly during synthesis by co-condensation with appropriate additives [12,34].

A salient prerequisite for all these applications is a deep understanding of how the pores and pore surfaces interact with the confined molecules, which are, e.g., substrates of a catalytic reaction, a drug to be delivered or a fluid mixture which should be separated into its components.

This understanding is only obtainable by a combination of various complementary spectroscopic, thermodynamic, computational, and general physico-chemical characterization techniques, including multi-nuclear variable-temperature solid-state NMR (SSNMR), differential scanning calorimetry (DSC), powder X-ray diffraction (PXRD), small angle scattering (SAXS and SANS), thermogravimetric analysis (TGA) [35], and molecular dynamics (MD) simulations, as is shown by a number of recent papers (see, e.g., [36–47]). While X-ray diffraction techniques like XRD or SAXS reveal the ordered structures of these materials [48–50], nitrogen adsorption is employed to study the specific surface areas and pore diameters [51,52], DSC or TGA are used for the investigation of phase transitions inside the pores, NMR provides insights into the local ordering and dynamics on the molecular level, and computation interprets these results [35,53].

The purpose of this short review is to collect and report recent results on the investigation of guest molecules confined in mesoporous host systems with a particular emphasis on examples, where NMR techniques are prominently employed. As the description and theory of the NMR experiments employed for these investigations and their physical background are found in the literature, they are not described here in detail. Instead the reader is directed to a number of recent reviews on these experiments and references therein [54–57]. The same is true for the background of melting and glass transitions in confinement, where the reader is referred to papers [27,41,58,59] and references cited therein. The main advantage in the application of SSNMR techniques, which makes them complementary to diffraction techniques, is the fact that SSNMR works well on disordered systems or materials strongly affected by local impurities or multi-phase materials, on the one hand [39], and that it is able to analyze not only structural but also dynamical processes and in particular phase transitions on the other [35,53]. The drawback of NMR, in general, and SSNMR techniques, in particular, is their low sensitivity. For this reason microporous materials like zeolites [54,60–63] or mesoporous materials like MCM-41 and SBA-15 derivatives with high specific surfaces are commonly employed as host materials for NMR-confinement studies [64–67]. To battle this drawback, indirect detection methods under MAS, where the X-nucleus of interest is detected via the far more sensitive protons, a technique originally developed by Ishii and Tycko [68], were successfully applied to porous systems by the Pruski group to achieve remarkably sensitivity enhancements [69–72].

A recent alternative to this pure SSNMR technique for sensitivity enhancement is the application of hyperpolarization techniques like Dynamic Nuclear Polarization (DNP) enhanced SSNMR [73–76], which boost the sensitivity of SSNMR by several orders of magnitude [77–84], and in particular its variant SENS (Surface Enhanced NMR Spectroscopy) [85–95], or parahydrogen-induced polarization [96–98]

(PHIP), whose potential for surface studies was demonstrated by Hunger et al. [99,100] and Stepanov and coworkers [101], or spin-exchange optical pumping (SEOP) [102,103].

In particular for functionalized systems, SSNMR techniques provided unprecedented details about the interaction of the linker molecules and the surface and its wetness [104–106]. Motokura et al. [107] employed $^{13}$C CP MAS NMR to investigate the catalytic transformation of epoxides under $CO_2$ atmosphere on silica-supported aminopyridinium halides. Gath et al. [108] ascertained the properties of silylated amorphous silica materials. Wang et al. [109] studied a series of different linker molecules tethered on MCM-41 or SBA-15 by $^2$H MAS. Kandel et al. [110,111] investigated inhibitory processes in aldol reactions employing amine-functionalized silica supports. Jayanthi et al. [112–114] combined $^2$H and $^{29}$Si MAS NMR with MD simulations to study the dependence of the linker-surface interaction on the water concentration and on the temperature for $N$-(2-(triethoxysilyl)propyl)acetamide-$d_3$ grafted onto MCM-41 and $N$-(2-aminoethyl-$d_4$)propanamide grafted onto SBA-15. The Bluemel group pioneered in a series of seminal papers the application of CPMAS, in general and in particular, of HRMAS (high-resolution magic-angle spinning [115]), a solid-state NMR experiment, which employs the partial motional averaging of anisotropies for the investigation of physisorbed or chemisorbed molecules on surfaces, the characterization of novel porous catalysts [104,105,116–125]. Important contributions by the Coperet group were studies of various supported organometallic catalysts by SSNMR (see [86,87,90,92,126–132]) and by the Scott group [133–138], who developed a series of novel porous catalytic materials and investigated in detail the factors determining their adsorption and reactivity properties and by the Pruski [69–72,139–145] and Buntkowsky [146–152] groups, who employed conventional and DNP-enhanced SSNMR for the characterization of immobilized molecules. Another important aspect is that these materials are potential carriers for bioactive molecules, such as amino acids, peptides, or drugs [19,153–155]. Klimavicius et al. investigated silica confined ionic liquids by CPMAS NMR [156].

While the focus of this review is devoted to results obtained in the DFG special research unit FOR1583, there are also short reports about important contributions from outside this consortium. The rest of this review is organized as follows: Section 2 gives an introduction into the preparation and surface modification of the mesoporous host materials. Section 3 discusses the behavior of simple systems and Section 4 the behavior of complex systems, such as binary liquids or crowded solutions inside the confinement. The review is finished by a summary and an outlook into possible future developments of the field.

## 2. Porous Host Materials

*2.1. Microporous Materials as Hosts*

Owing to their immense importance both in daily life and in technology, zeolites are most probably the best characterized class of porous materials. They are well-ordered framework silicates with the composition $(A^+, E^{2+}_{0.5})_x (AlO_2)_x (SiO_2)_y \cdot (H_2O)_z$ ($A^+$ = $Na^+$, $K^+$ and $E^{2+}$ = $Mg^{2+}$, $Ca^{2+}$) and belong to the family of tectosilicates. A well-known example, found in nature, is faujasite ($Na_2Ca[Al_4Si O_{10}O_{28}] \cdot 20\ H_2O$) [157]. Zeolites find applications e.g., as molecular sieves [158,159], in heterogeneous catalysis [160,161], for gas storage [162], and as ion exchange resins [157]. An extensive recent overview about this fascinating class of materials is given in the recent special issue on zeolite chemistry

Owing to their narrow pore diameters and importance as catalysts for organic chemistry, most confinement studies employing zeolites use small molecules [163,164]. Typical recent examples are the deuterium NMR studies by Nishchenko et al. [165] and Lalowicz et al. [166,167] who analyzed the dynamics of tert.-butyl alcohol-$d_9$ and methanol-$d_4$ inside zeolites, respectively. Moreover, NMR field-gradient approaches yielded valuable insights into the diffusion of various small molecules in zeolites, including information about diffusion anisotropy, transport resistance at crystal surfaces, and pore connectivities [168–174].

## 2.2. Mesoporous Silica Materials as Hosts

While microporous zeolites are clearly still the technically most important class of porous host materials, their applications are limited to fairly small molecular sizes. For this reason the development of new classes of materials [3,6–10,16–21,106] with larger and adjustable pore sizes, such as mesoporous silica and mesoporous carbon materials, gained in importance. Depending on the material, they are characterized by adjustable pore sizes between ca. 10 and 1000 Å and, thus, close the gap between the microporous and the macroporous regime. Their combination of large specific volume and specific surface areas with high thermal stability and low specific weight creates a large application potential in physics, chemistry, pharmacy, polymer science, and related fields. Characteristic examples include applications in gas storage, in catalysis, in separation techniques, as additives to rubbers for tires media, and many more [11–15]. In confinement studies, SBA-15-type materials have the advantage of larger pore diameters, but MCM-41-type materials are better suited as models with narrow confinement. Additional merits of the latter are their generally better surface homogeneity and smoother inner surfaces.

## 2.3. Preparation and Chemical Functionalization of Mesoporous Silica; NMR Characterization

As mentioned above these mesoporous silica materials permit the comparatively simple synthesis of surface functionalized host systems with well-defined tunable narrow pore diameters [175–178], employing a synthesis protocol, which is based on Grünberg et al. [106] and Grün et al. [179]. Details of the synthesis and characterization are given in [176,180] and will not be repeated here. The changes of porosity, specific surfaces and the modification of the surface sites of the material can be monitored by the combination of nitrogen adsorption (BET and BJH) analysis and $^{29}$Si SSNMR spectroscopy. A typical example of such a synthesis is shown below (Figure 1), which displays the APTES ((3-aminopropyl)triethoxysilan) functionalization of MCM-41 materials. For this sample the BET measurements revealed a specific pore volume of 0.77 cm$^3$/g, a specific surface area of 1000 m$^2$/g and a specific pore diameter of 3.6 nm. From the $^{29}$Si SSNMR spectra the changes of the silanol groups during functionalization are determined by the change from Q-groups to T-groups.

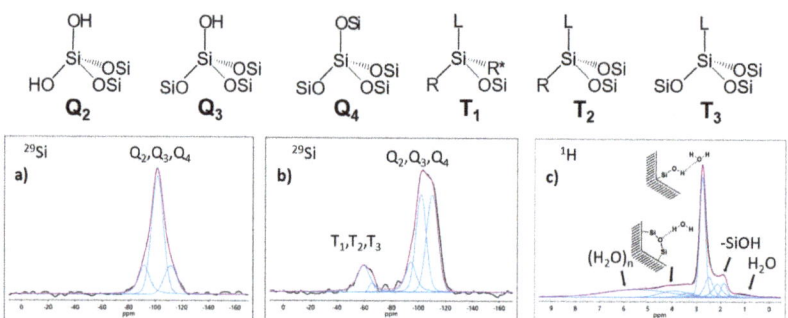

**Figure 1.** Upper panel: Different types of silica sites. Lower panel: $^{29}$Si CPMAS (10 kHz) spectra of (a) neat and (b) functionalized silica, showing the appearance of T$_n$-groups by the functionalization. (adapted from Weigler et al. [181]). (c): $^1$H SSNMR spectrum of a non-dried MCM-41 at 10 kHz (black). Deconvolution (blue), sum of deconvolution (magenta) and assignment of water species to the peaks [36].

At this point it is important to note that the freshly prepared samples in general contain a substantial amount of surface bound water molecules [36,39,182]. Since the latter can strongly influence the outcome of confinement studies, it is in generally necessary to check the hydration state of the sample by $^1$H MAS NMR measurements (the lower right panel of Figure 1 shows a typical example) and to employ special drying protocols for the preparation of "water-free" silica samples (for details see Brodrecht et al. [183]).

Since naturally-occurring porous hybrid materials like the skeleton of diatoms are based on modified silica materials consisting of silica and sillafins (polyamines) [184–187], the functionalization of mesoporous silica with peptides and peptoides [44,47,113,175–178,188–194] creates controllable well-defined model systems for the in-vitro study of, e.g., biomineralization.

There are two different strategies for obtaining such peptide-functionalized silica materials, namely an activation of the silica by virtue of a linker group, followed by a grafting of the previously synthesized peptide as shown, e.g., in [47,112] or the direct synthesis of the desired peptide inside the pores employing a modification of the standard SPPS protocol [176,177,194].

Figure 2 displays examples of both strategies and, in particular, how the success of the synthesis can be monitored by $^{13}$C CPMAS SSNMR spectroscopy. In the first example (Figure 2a), the collagen-model nonapeptide H-(Gly-Pro-Hyp)$_3$-OH is grafted to silica [47]. The intensity reduction of the succinimidyl signal at ca. 15 ppm, which is visible by comparison of a.ii and a.iv, and the peaks for the nonapeptide in the carbonyl region and also in the aliphatic region indicate the successful immobilization of the peptide, which was proven by DNP enhanced natural abundance $^{15}$N spectra (not shown). The second example displays the steps of the SPPS inside mesoporous silica for the addition of one amino acid residue to an N-terminus (for details see Brodrecht et al. [175]).

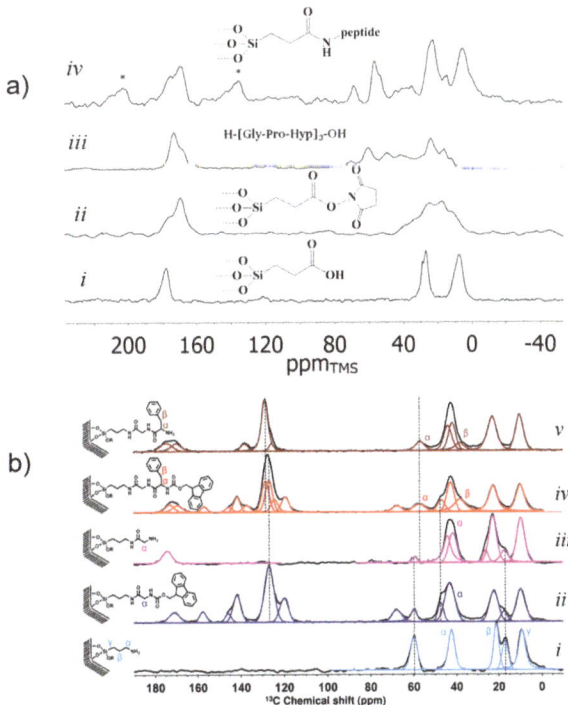

**Figure 2.** (a) $^{13}$C CPMAS NMR of the immobilization via the grafting approach of the nonapeptide H-[Gly-Pro-Hyp]$_3$-OH on carboxyl functionalized mesoporous silica: (*i*) neat carboxyl functionalized silica support; (*ii*) TSTU (N,N,N′,N′-tetramethyl-O-(N-succinimidyl)uroniumtetrafluorborat) pre-activated silica; (*iii*) free nonapeptide H-[Gly-Pro-Hyp]$_3$-OH; (*iv*) nonapeptide grafted on silica. Note: Signals marked with * refer to spinning sidebands. (adapted from [47]). (b) $^{13}$C CP MAS NMR of the steps of the in-pore SPPS (solid phase peptide synthesis) of functionalized SBA-15 (*i*), Fmoc-glycine functionalized species (*ii*), glycine functionalized species (*iii*), Fmoc-phenylalanine-glycine functionalized species (*iv*), phenylalanine-glycine functionalized species (*v*) (adapted from Brodrecht et al. [175]).

Although mesoporous silica materials offer a larger range of pore diameters they are still limited with respect to the accessible confinement sizes. In the case that functionalized pores with larger diameters are desired, they can be created in a hierarchical three-step process, which is sketched in Figure 3a. In the first step a membrane, such as a polycarbonate foil, is irradiated in a heavy ion accelerator, creating an ion-track. The carbonate material inside this iron track is then removed by etching, creating a channel through the polycarbonate foil. By selecting the ion dose, the irradiation directions, and the etching time, the number of channels, their dimensionality, and the channel diameters are selected [195]. In the second step these etched ion channels are coated with silica by atomic layer deposition (ALD). In the third step they can be functionalized by grafting of linkers, such as APDMS (3-aminopropyldimethyl silane) or APTES, to the silica inside the channels. Owing to their lower specific surfaces the detailed chemical characterization and monitoring of the surface functionalization by SSNMR is only feasible by means of DNP enhancement, which boosts the SSNMR sensitivity. As a typical example of these experiments, Figure 3b displays the DNP enhanced $^{29}$Si SSNMR spectra of the material. The broad low-field lines around ca. 60–75 ppm proves the formation of the characteristic $T_n$-groups, which result from the binding of APTES to the silica inside the channels.

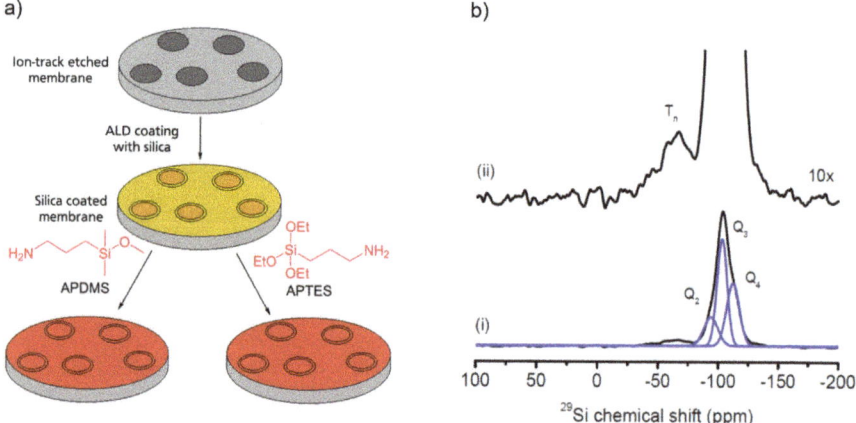

Figure 3. (a): Sketch of the three-step process to synthesize amine functionalized silica coated porous polycarbonate-membranes (see text for details). (b): DNP enhanced $^{29}$Si CPMAS spectra (i) revealing the characteristic $T_n$-groups in the tenfold magnefication (ii) (adapted from [196])).

## 3. Simple Liquids in Confinement

The simplest case of a confined system is a single component liquid confined inside the pores of a host material, such as silica. In this case the behavior of the liquid is governed by the competition of the interactions between the liquid's molecules and that of the host surface with the liquid's molecules. In addition to the typical interactions between the liquid molecules, such as hydrogen bonding, hydroaffinity, polarity, and aromaticity, there are also steric effects, which influence the dynamics. In the remainder of this section, some characteristic examples of single component liquids, such as small polar, nonpolar, and aromatic molecules, confined in mesoporous silica are discussed. The competition of these interactions leads to pronounced changes of their phase behavior, in particular when confined in narrow pores, where a large percentage of the molecules is close to, or in contact with, the pore surface. The confinement in a pore causes in general a depression of the melting/freezing point of the confined molecules, respectively prevents melting at all if the pore diameter is too low, causing a glass-transition instead. As a consequence, many molecules, which are a solid in their bulk phase at a given temperature, become a liquid when confined inside pores.

The situation becomes even more complicated, when molecules are employed as solvents, e.g., of a chemical reagent such as a drug or in filtering processes. In this situation there will be a competition of solvent-surface and solute-surface interactions with solvent-solute and solvent-solvent interactions. In order to be able to understand these complicated systems, it is a prerequisite to understand the behavior and properties of confined simple liquids.

To obtain this understanding, various analytical methods such as DSC, TGA, near-infrared spectroscopy (n-IR), and variable-temperature XRD are combined with manifold NMR methods, including one- and two-dimensional spectroscopy, spin-lattice relaxometry, and correlation function analysis, as well as broadband dielectric spectroscopy (BDS).

### 3.1. Water inside Mesoporous Silica

Polar molecules such as water [36,40,197–202], alcohols like methanol [166,167], tert.-butyl alcohol [165] or octanol [203,204] and carboxylic acids, such as isobutyric acid [38,42], can form hydrogen bonds among themselves and also with the surface's silanol groups (Si–OH).

Owing to its ubiquitous presence, its importance as a solvent and its importance for the life-sciences, water is the most interesting molecule for confinement studies. It is commonly used as a green solvent, employed both in technical processes and in medical applications. Moreover, due to its ability to build hydrogen-bonding networks with itself and also with surface-silanol groups and its rich phase-diagram, it is also a fascinating subject for basic scientific investigations.

Various aspects of water confined in mesoporous silica were investigated in a number of studies. An important outcome of these investigations was that the morphology of water inside the pores depends strongly on the pore diameter. For narrow pores a coexistence of two different water phases (a surface layer and nano-droplets or water-clusters) and for larger diameters a single water phase were detected by $^1$H MAS NMR [36,40] and $^2$H SSNMR [197,198]. It was feasible to assign different water species confined in mesoporous MCM-41 by virtue of combined $^1$H and $^2$H SSNMR experiments [199,200]. The exchange of the two spin species as a function of the hydration level was studied for MCM-41 filled with $D_2O$ [201] employing 2D selective soft-hard inversion recovery experiments [205,206] and the results were interpreted using a three site exchange model. In this model the highest exchange rate of 300 s$^{-1}$ is found between single hydroxyl protons and water protons. Moreover, as they did not observe any coalescence for the lines corresponding to the surface-water chemical exchange rate, they could provide an upper boundary (<1000 s$^{-1}$) for this rate.

When the confinement size is reduced, the melting temperature of water decreases, as described by the Gibbs–Thomson relation, until crystallization is fully suppressed [27,65,207]. Thus, severe geometrical restriction provides access to the properties of deeply cooled liquid water, which are of fundamental importance for an understanding of the anomalies of this liquid, but masked by rapid crystallization in the bulk [208]. In particular, it was proposed that the anomalies of water originate in a liquid-liquid critical point in the supercooled regime, which terminates a phase transition between high-density (HDL) and low-density (LDL) liquid forms [209]. Moreover, it was argued that the associated structural modifications have also a dynamical signature, explicitly, that there is a change in the temperature dependence of the structural $\alpha$ relaxation from a non-Arrhenius behavior characteristic for HDL to an Arrhenius behavior typical of LDL. While a number of studies reported such dynamical crossover of confined water, it remains a subject of controversial discussion whether the phenomenon is indicative of a HDL-LDL transition [210–212]. For example, alternative explanations based on confinement effects were given.

To tackle this problem, $^2$H NMR was used to investigate reorientation dynamics of unfreezable water ($D_2O$) in MCM-41 and SBA-15 pores over wide temperature ranges towards the glass transition [45,176,181,183,213–219]. In particular, spin-lattice relaxation, line-shape analysis, and stimulated-echo experiments were combined to ensure broad dynamic ranges and the pore-size was systematically varied to study possible finite-size effects. Figure 4 shows temperature-dependent correlation times $\tau$ obtained from $^2$H NMR approaches to water reorientation in mesoporous silica.

Clearly, there is a dynamic crossover near 220 K. Combining this NMR analyses with DSC and BDS studies [215], it turned out that, for a pore diameter of 2.8 nm, the dynamic crossover occurs near the melting temperature $T_m$ so that liquid and crystalline water fractions with, respectively, faster and slower rotational dynamics coexist inside the pores below this temperature (see Figure 4a). It was concluded that partial crystallization causes the effect, explicitly, that the dynamics of water changes when ice forms and further restricts the accessible pore volume. To test this hypothesis, later work exploited that the melting temperature $T_m$ can be altered when the pore diameter of the MCM-41 and SBA-15 material is varied [213]. It was found that the dynamic crossover occurs near 220 K, independent of the pore diameter (see Figure 4b). Hence, partial crystallization is not the reason of the change in the temperature dependence in the general case. One may be tempted to argue that the lacking pore-size dependence also excludes finite-size effects as possible origin. However, this argument does not hold because liquid water forms an interfacial layer, the thickness of which is largely independent of the pore diameter and, hence, the available space between the silica walls and the ice crystallites remains unaltered.

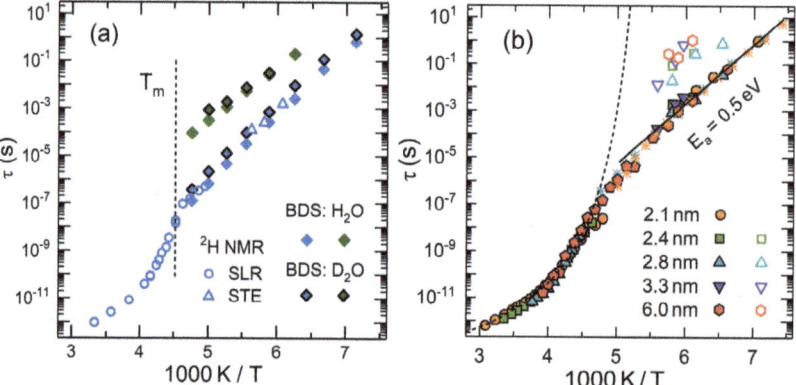

**Figure 4.** Correlation times of water reorientation in MCM-41 and SBA-15: (**a**) Results for $H_2O$ and $D_2O$ in MCM-41 pores with a diameter of 2.8 nm from BDS, $^2$H spin-lattice relaxation (SLR), and $^2$H stimulated-echo (STE) experiments [215]. The dashed line marks the melting temperature $T_m$ of water in these confinements, as obtained from DSC (adapted from [215]). (**b**) Results for $H_2O$ and $D_2O$ in MCM-41 pores with the indicated diameters from $^2$H NMR [213] and BDS (stars) [212,215]. The dashed line is an interpolation of the high-temperature data with a Vogel-Fulcher-Tammann relation. The solid line is an Arrhenius fit of the low-temperature results, yielding an activation energy of 0.5 eV (adapted from [213]).

To study the role of water-host interactions, analogous $^2$H NMR studies were performed for $D_2O$ in MCM-41 pores functionalized with APTES (see Section 2.3). It was found that water reorientation in native and functionalized MCM-41 pores is similar (see Figure 5a) [181]. In particular, the temperature-dependent correlation times $\tau$ show a crossover from non-Arrhenius behavior above ca. 220 K to an Arrhenius behavior below this temperature in both types of confinements. Moreover, a common activation energy of $E_a \cong 0.5$ eV was observed in the low-temperature regime for water in MCM-41 pores with and without APTES functionalization, but also for water in many other confinements, e.g., at protein surfaces [220–223]. Therefore, the term 'universal water relaxation' was coined for the low-temperature process. However, there are ongoing vigorous discussions whether this dynamic process can still be identified with the structural $\alpha$ relaxation of water or to a secondary $\beta$ relaxation, which has severe consequences for the value of the glass transition temperature $T_g$ of confined water and, possibly, also bulk water (see below) [210,212].

To obtain information about the nature of the low-temperature dynamics of water, it was utilized that $^2$H stimulated-echo experiments provide access to not only the rates but also the mechanisms of molecular reorientation dynamics [224–226]. In particular, it can be exploited that the angular resolution of the experiment is determined by the length of the evolution time $t_e$ in the stimulated-echo sequence.

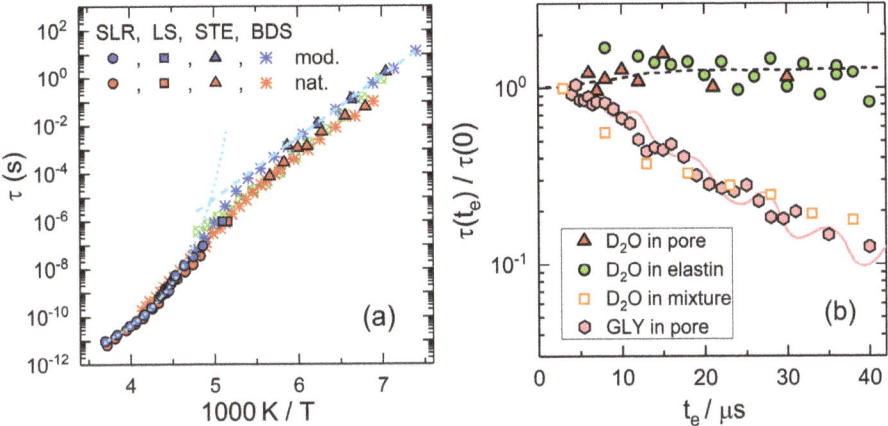

**Figure 5.** (a) Temperature-dependent correlation times of H$_2$O and D$_2$O reorientation in native MCM-41 (diameter 2.0 nm, red symbols) [214] and in APTES modified MCM-41 (diameter 2.2 nm, blue symbols) [181]. Results from $^2$H spin-lattice relaxation, line-shape analysis, and stimulated-echo experiments on D$_2$O dynamics and from BDS on H$_2$O dynamics in native MCM-41 [227] and D$_2$O dynamics in modified MCM-41 [181] are shown. For comparison, data from combined NMR and BDS studies on water reorientation in an elastin matrix are included (green crossed circles) [222]. The dotted line is an interpolation of the high-temperature data with a Vogel-Fulcher-Tammann relation. The dashed line is an Arrhenius fit of the low-temperature results, yielding an activation energy of 0.5 eV. (b) Evolution-time dependent normalized correlation times $\tau(t_e)/\tau(t_e \to 0)$ of D$_2$O reorientation in an APTES modified MCM-41 (pore) [181], in an elastin matrix [220], and in a 2:1 molar D$_2$O/DMSO mixture [228], and of glycerol (GLY) reorientation in MCM-41 (diameter 2.2 nm). The lines are expectations obtained from computer simulations [229]: (dashed) distorted tetrahedral jumps [220] and (solid) isotropic reorientation comprised of rotational jumps about angles of 2° (98%) and 30° (2%) [230].

It was reported that the observations for water in both silica and protein confinements are largely independent of the evolution time [45,181,214,219,220]. For example, the correlation times τ change in neither of these confinements when $t_e$ is extended and, thus, the angular resolution is enhanced (see Figure 5b). This ineffectiveness of geometrical filtering indicated that water reorientation results from jumps about large angles of the order of the tetrahedral angle. Closer analysis implied that the universal low-temperature water reorientation can be described as distorted tetrahedral jumps or, similarly, as quasi-isotropic large-angle jumps [45,181,214,219,220]. However, it remains elusive whether or not the observed rotational motion is coupled to translational diffusion. The existence of such coupling is a prerequisite for the interpretation of the dynamic crossover in terms of altered structural α relaxation in response to a HDL-LDL transition. By contrast, an absence of such coupling implies interpretations based on a crossover from structural α relaxation to localized β relaxation. Moreover, this aspect has major consequences for the nature of the much-debated glass transition of water at $T_g \cong 136$ K [231–233]. As the correlation times of the universal low-temperature water dynamics meet expectations for a glassy arrest at this temperature, a diffusive nature of this process entails a structural glass transition, whereas a localized nature implies an orientational glass transition, which is restricted to the rotational degrees of freedom.

## 3.2. Glycerol inside Mesoporous Silica

When investigating effects of geometrical restriction on liquid dynamics, it is desirable to compare behaviors of confined and bulk molecules over broad temperature ranges. In that respect, work on water has the drawback that its high tendency for crystallization hampers comparisons in the supercooled regime. On the other hand, use of good glass formers allows one to study molecular dynamics of both confined and bulk liquids in wide time windows. Major NMR contributions to this research field are discussed in a previous review article [67]. Therefore, we restrict ourselves to the case of the archetypal glass former glycerol in this contribution.

Figure 6 shows $^2$H NMR correlation times of bulk and confined glycerol. It is evident that the time scale of glycerol reorientation is unaffected even in narrow MCM-41 pores with diameters of 2–3 nm and at low temperatures near the glass transition. The temperature dependence is well described by the Vogel-Fulcher-Tammann relation typical of molecular glass formers. To arrive at these results, it is, however, necessary to avoid contamination with water by careful drying of the precursor materials [183,217]. Possibly, the hygroscopic nature of mesoporous silica and the high sensitivity of glycerol dynamics to water admixtures offer an explanation for different conclusions relating to confinement effects on the glass transition of glycerol in BDS studies [234,235]. Moreover, $^2$H NMR stimulated-echo studies showed that MCM-41 confinement does not alter the mechanism for glycerol reorientation (see Figure 5b). Specifically, the evolution-time dependence of the correlation times $\tau(t_e)$ of confined glycerol resembles that of bulk glycerol [230] but differs from that of confined water (see Section 3.1). The observed decrease of $\tau(t_e)$ indicates that the reorientation process of glycerol is composed of consecutive small-angle jumps, e.g., the data for the bulk liquid were successfully described by an isotropic reorientation model, which assumes that 98% of the rotational jumps occur about an angle of 2° and only 2% of them involve an angle of 30° [230]. On the other hand, $^2$H NMR correlation functions were found to be more stretched for confined glycerol than for bulk glycerol [217]. Hence, confinement results in higher dynamical heterogeneity, which, most probably, reflects mobility gradients across the pores with slower dynamics at the pore walls than in the pore centers, as commonly observed in simulation studies on confined liquids [236]. Consistent with these results for silica confinement, it was reported that protein matrices leave the rate of glycerol reorientation unaltered but increase its heterogeneity [237,238].

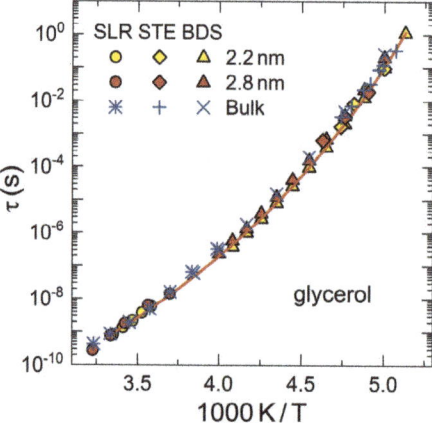

**Figure 6.** Temperature-dependent correlation times of glycerol-$d_5$ in the bulk liquid and in MCM-41 pores with the indicated diameters. Results from $^2$H NMR (spin-lattice relaxation and stimulated-echo experiments) and BDS are compared. The solid line is an interpolation with a Vogel-Fulcher-Tammann relation.

*3.3. Benzene, Biphenyl, and Naphthalene inside Mesoporous Silica*

While polar molecules like water exhibit strong hydrogen bonding interactions with the confinement, aromatic molecules such as benzene [23,37], biphenyl [43] or naphthalene [239] only weakly interact with the surface due to their hydrophobicity and strong π-π-stacking interactions among themselves. Employing a combination of $^2$H SSNMR and DSC the phase behavior of benzene confined in mesoporous silica was studied [23,37]. These experiments revealed a drastic lowering of the transition temperatures of the rotor and translational phases of the confined benzene. In particular for narrow pores a glass-like benzene phase with a broad distribution of activation energies was elucidated [23]. In this glass-like phase the rotational degrees of freedom were strongly decoupled from the translational ones. The comparison of these results with investigations employing a larger-diameter host material revealed that these glass-like phases are formed by roughly three outer molecular layers and that inside a "normal" crystalline benzene phase is formed, which behaves like bulk benzene [37].

This interesting behavior prompted the study of the next larger homologues of benzene, namely of biphenyl confined in inside narrow (nominal 2.5 nm and 2.9 nm diameter) silylated and non-silylated MCM-41 pores [43] and of naphthalene confined inside narrow (nominal 3.3 nm pore diameter) MCM-41 pores [239]. With respect to confinement studies, a major difference between these two molecules is that biphenyl has an internal rotational degree of freedom, which naphthalene is lacking. The confinement of biphenyl caused a depression of the melting point by ca. 110–120 K from the bulk value of 342.6 K down to 222 K to 229 K (depending on the pore diameter). Moreover, a careful line-shape analysis of the $^2$H NMR solid-echo spectra measured just below the melting points elucidated indications for the presence of a pre-melting process in the form of isotropic motions of a fraction of the biphenyl molecules (Figure 7). The best simulation of the spectra was obtained by a two-phase model, with a broad distribution of rotational correlation times, resulting from a broad distribution of activation energies for the rotational motion. For the confined naphthalene an even stronger reduction of the melting point (152 K), compared to the bulk material was found. For the detailed line shape analysis of the $^2$H SSNMR spectra, two different models were employed, namely on the one hand a two-phase model with a broad distribution of activation energies, which is similar for benzene and biphenyl, and on the other hand a crystal-like jump model employing an octahedral jump geometry. Both models revealed a narrow melting point distribution of the confined naphthalene, indicating a relatively well ordered structure of the confined naphthalene molecules [239]. These results were interpreted such that the confined naphthalene molecules most probably form a plastically crystalline phase, similar to naphthalene in ball-milled silica [240–243]. The existence of these plastic phases of confined naphthalene were independently confirmed by a combination of DSC, Raman spectroscopy, and PXRD [244].

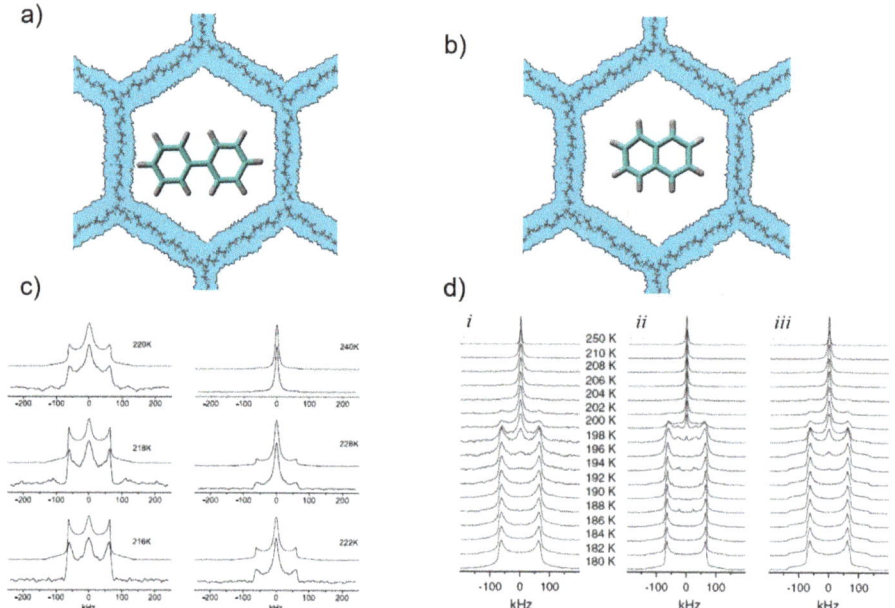

**Figure 7.** (**a**): cartoon of biphenyl [43] and (**b**) naphthalene [239] confined in MCM-41. In the case of biphenyl there are two possible rotational modes, namely, the rotation of a single phenyl ring and a molecular rotation. (**c**) experimental (lower traces) and simulated (upper traces) $^2$H solid-echo NMR spectra of the melting process of biphenyl-d$_{10}$ inside silylated MCM-41 ($d_{aver}$ = 29 Å) at different temperatures [43]. (**d**): experimental variable temperature $^2$H solid-echo NMR spectra of the melting process of naphthalene-d$_8$ inside MCM-41 ($d_{aver}$ = 33 Å) (i): experiment; (ii): simulated octahedral jump; (iii): two-phase model. All spectra are normalized to equal height (figure reproduced with permission from [176], copyright Walter de Gruyter and Company).

*3.4. Pyridine inside Mesoporous Silica*

While the nonpolar aromatic molecules described in the previous section interact only weakly with the silica surface, polar aromatic molecules such as carboxylic acids like benzoic acid [245], or nitrogen containing heterocycles such as pyridine [246–248], bipyridinyl [249], dimethylaminopyridine [250], or diethyl-2,6-di-tert.-butylaminopyridine [250], interact strongly with surface silanol groups by means of hydrogen bonds. In the case of the pyridine derivates, the ring nitrogen acts as a hydrogen bond acceptor. Since the $^{15}$N chemical shift is a very sensitive monitor of the hydrogen bond strength, these hydrogen bonds can be conveniently monitored by $^{15}$N CPMAS NMR [246,250].

In a series of seminal papers [246–248], Shenderovich and coworkers employed a combination of $^{15}$N- and $^{29}$Si-CPMAS and MAS NMR spectroscopy, line-shape analysis and structural modelling to describe the silica-surface morphology of different types of mesoporous silica on the atomic scale. While the $^{29}$Si MAS spectra revealed the ratio of Q$_3$:Q$_4$ groups and the pore wall ordering of the silica, the $^{15}$N NMR revealed the Brønsted basicity and surface morphology and surface defects of the silica. These investigations were paralleled by stray-field diffusion NMR studies, which investigated the diffusion tensor of pyridine as a function of the pore filling [251]. Later, Gurinov et al. [248] studied aluminum oxide containing SBA-15-type materials to investigate in more detail the effects of Lewis and Brønsted acidity by a combination of $^{15}$N and $^2$H NMR techniques. Lesnichin et al. [249] investigated the behavior of 2,2′-bipyridyl in confinement. They found that the molecule can only form one of the two possible hydrogen bonds to the surface and that the surface coverage grows strongly from one molecule per nm$^2$ at room temperature to 1.6 molecules per nm$^2$ at 130 K. Very recently, Shenderovich and Denisov studied the hydrogen bonding of pyridine in detail [252].

## 4. Complex Liquids in Confinement

In the previous section, the behavior of confined simple liquids was briefly reviewed. In the present section we now discuss the behavior of complex liquids, i.e., mixtures' respectively liquid solutions of two or more liquid components [253]. The investigation of these mixtures is of particular interest as they are models for many natural or technical systems, e.g., oil-water mixtures. While the phase behavior of such complex liquids is often well understood in bulk phases, there is still a very large gap in knowledge for confined systems, where the competition of liquid/liquid versus liquid/pore surface interactions creates much more complex scenarios. Understanding the effect of the confinement on the complex liquid, and analyzing the structure, dynamics and spatial distribution of the solvents on the molecular level may help in developing new applications, e.g., in chemical industry, pharmacology or oil industry or might help in developing new strategies to deal, e.g., with crude oil spills. We start with model systems, namely confined water-alcohol and water-isobutyric acid mixtures to discuss the basic behavior of these confined systems. Then we discuss recent results on confined ionic liquids and their application as solvents in catalysis in the form of supported ionic liquid phases (SILPs) respectively supported ionic liquid catalysts (SILCs). Finally, we shortly summarize recent results on confined surfactants.

### 4.1. Confined Water-Octanol Mixtures

Water-octanol mixtures are important model systems for the investigation of the phase behavior of two immiscible liquids in confinement. For their quantification the water octanol partition coefficient or $p$-value $K_{ow}$ is employed. (for details see the short review by Hermens et al. [254] and references therein). Hydrophobic liquids have a high $K_{ow}$ and hydrophilic liquids have a low $K_{ow}$. The $K_{ow}$ values are employed, e.g., in pharmacology for estimating the distribution of drugs within the body. Drugs with high $K_{ow}$ tend to accumulate in hydrophobic areas of the body such as lipid bilayers of cells and drugs with low $K_{ow}$ tend to accumulate in hydrophilic areas with high water content, e.g., the blood serum. For a detailed discussion on the application of partition coefficients see Leo at al. [255].

Kumari et al. [203] studied the phase behavior of water/octanol mixtures confined in mesoporous SBA-15 by a combination of SSNMR and MD simulations (see Figure 8). By a combination of 1D SSNMR and FSLG-NMR [204] they could analyze the strength of the magnetic dipolar interactions between the different components and thus determine the distributions of the two liquids inside the confinement. The salient idea is to search for correlations between the chemically different types of $^1$H-nuclei (e.g., aliphatic protons of the alkyl chain or hydroxyl protons of the alcohol group or water) of the confined liquid and $^{29}$Si-sites on the surface of the material. These correlations are created by the magnetic dipolar interaction between these nuclei and are indicated as cross-peaks inside the 2D-NMR spectra. Thus, they are only visible when the corresponding nuclei are in the vicinity of each other. By varying the contact time different distances are probed. A detailed analysis of the 2D-spectra is beyond the scope of the current review and can be found in [203].

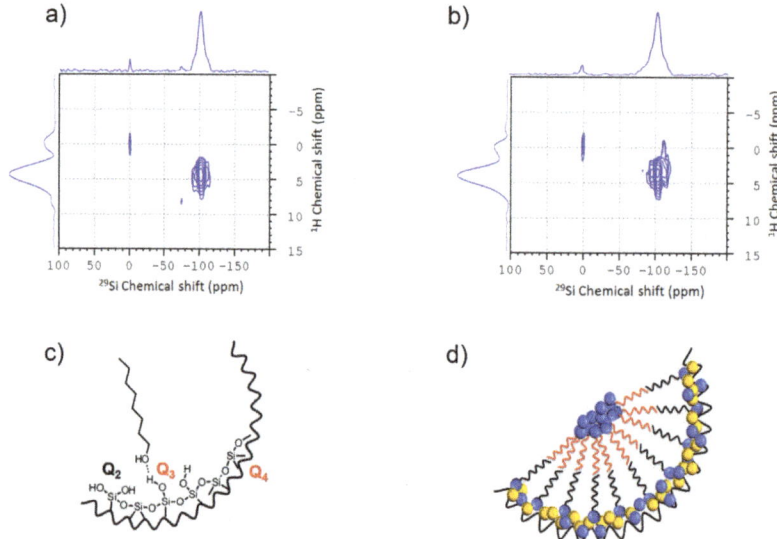

**Figure 8.** Upper panel: Room temperature $^1$H-$^{29}$Si CPMAS FSLG-HETCOR experiment measured at 8 kHz spinning of dried SBA-15 filled with a mixture of 80:20 mol% of 1-octanol and water with a contact time of (**a**) 3 ms and (**b**) 9 ms. Lower panel: (**c**) schematic models for interactions of the pore surface of SBA-15 with 1-octanol. (**d**) Graphical visualization of a feasible bilayer formation of 1-octanol inside the pore. Water molecules are concentrated near the pore wall as well as in the pore center. The intermediated area between pore wall and pore center is occupied by the aliphatic hydrophobic chains of the 1-octanol molecules (adapted from Kumari et al. [203]).

### 4.2. Water-Isobutyric Acid (IBA) Mixtures in Experiment and Simulation

Another example presents the study of binary mixtures of water and isobutyric acid (iBA, 2-methylpropanoic acid) by a combination of SSNMR spectroscopy and MD simulations. In bulk mixtures, this system has a well-known phase-diagram with a large miscibility gap as a function of temperature and mole fraction of the liquids. First NMR studies [38,39] of this system had indicated a micro-phase separation of the confined binary mixture with an anomalous temperature dependence of the self-diffusion coefficient and a bifurcation of the $T_2$-relaxation upon a critical temperature of 42 °C, proposing a structural model in the form of concentric cylindrical liquid layers below the critical temperature inside the pores. The inner cylinder was tentatively assigned to the iBA rich and the outer cylinder hull to the water rich phase.

This assignment was probed by Harrach et al. [256] by a combination of high-resolution SSNMR on frozen solutions (100 K to suppress any fluid mobility in the NMR experiments and obtain a momentary picture of the liquid distribution inside the pores) and MD simulations. By varying the contact time of $^1$H-$^{29}$Si FSLG HETCOR (see Figure 9) they mapped out different distance regimes by virtue of the strength of the magnetic dipolar interactions between protons and the silica nuclei on the surface to reveal the molecular distribution inside the pores. The latter was interpreted by MD simulations (see Figure 10), which calculated the density profile of water and iBA as a function of the distance from the pore center. An example of these calculations is shown in Figure 10. They corroborate in principle the cylindrical model but reveal that the iBA rich phase and not the water rich phase is close to the pore wall. Furthermore, the calculations indicated that the iBA molecules orient preferential like an inverted brush-like structure, i.e., radially with the carboxylic group pointing towards the pore wall and the aliphatic chains pointing radially into the direction of the pore-center.

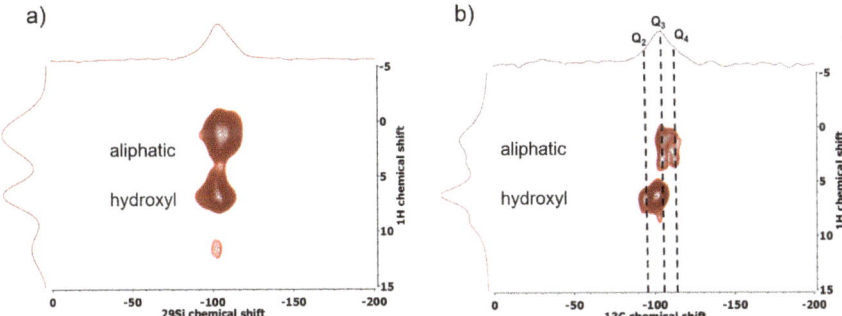

**Figure 9.** 2D $^1$H-$^{29}$Si FSLG HETCOR experiments of iBA/H$_2$O mixtures confined in SBA-15 with contact times of (**a**) 3 ms (longer distances) and (**b**) 0.5 ms (shorter distances) clearly reveal that both hydroxy- and aliphatic protons are in contact with the surface silicon nuclei (56 wt % iBA, 9.4Tesla, 100 K, 8 kHz MAS, 89 kHz FSLG homonuclear decoupling [257]). (Figure adapted from [256]).

A detailed analysis of the entropic and enthalpic parts of the free energy revealed that this unexpected phase-behavior is mainly caused by the hydrogen-bonding enthalpy and is meliorated at higher temperatures where entropic terms become stronger, leading to a more thorough miscibility. For further details see the original paper by Harrach et al. [256]

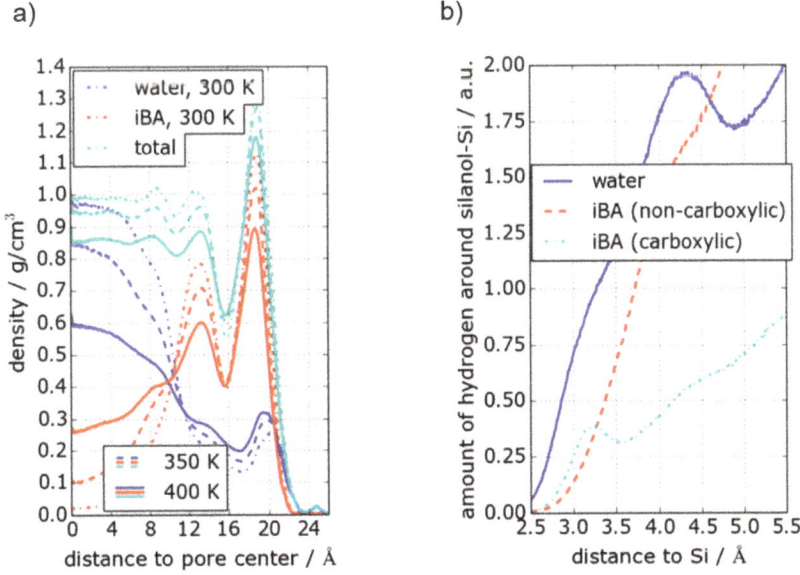

**Figure 10.** (**a**): density profiles for iBA and water calculated by molecular dynamics simulations. The center of Table 0. Å. (**b**): Density of hydrogen atoms as a function of the distance to the closest surface silanol group (figure reproduced from [256] with permission, copyright American Chemical Society).

### 4.3. Confined Water-Glycol Mixtures

Evidence for confinement-induced micro-phase separation and the confinement-enhanced tendency for crystallization were reported in $^2$H spin-lattice relaxation studies on mixtures of D$_2$O with propylene glycol (PG), propylene glycol monomethyl ether (PGME), or dipropylene glycol monomethyl ether [217,258].

For example, for a PG-D$_2$O mixture with a water concentration of 45 wt %, $^2$H spin-lattice relaxation studies revealed that crystallization is fully suppressed in the bulk but occurs in pores at T < 220 K [217] (see Figure 11a). Specifically, bimodal $^2$H spin-lattice relaxation indicated that partial freezing results in coexisting liquid and crystalline fractions inside MCM-41 pores. Recording the buildup of the magnetization in a staggered way, it was even possible to follow the crystallization process on the basis of the observation that the slow step due to the crystalline fraction grows at the expense of the fast one associated with the liquid fraction in the course of time. Similarly, $^2$H spin-lattice relaxation results for PGME-D$_2$O mixtures at 240 K indicated that freezing occurs in confinement but not in the bulk for a water concentration of 60 wt %, while such a difference was not observed at lower and higher water contents (see Figure 11b) [217]. Thus, at variance with the situation for pure liquids, confined aqueous glycol solutions with intermediate water concentrations show a higher proneness towards crystallization than their bulk counterparts. This effect was taken as evidence that, as a consequence of confinement-induced micro-phase segregation, the water concentration in some pore regions becomes sufficiently high to allow for ice formation.

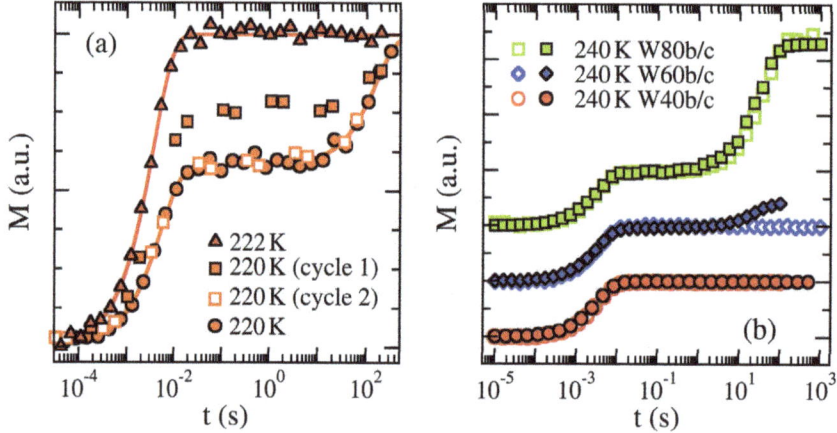

**Figure 11.** Buildup of $^2$H magnetization M(t) for water-glycol mixtures in the bulk and in MCM-41 pores (diameter 2.8 nm): (**a**) Confined PG-D$_2$O mixture (45 wt % water) together with fits to monomodal (222 K) and bimodal (220 K) spin-lattice relaxation. At the lower temperature, the relative height of the fast and slow steps differs during the first and second cycles of a staggered-range measurement performed directly after temperature equilibration (squares), while this discrepancy does not occur during a later measurement. (**b**) Bulk and confined PGME-D$_2$O mixtures at 240 K. The samples are denoted according to the weight percentages of water followed by the letters 'b' and 'c' for bulk (open symbols) and confined (solid symbols) mixtures, respectively.

### 4.4. Glass Transition of Confined Water-Alcohol Mixtures

$^2$H NMR proved also useful to ascertain the glass transition of confined aqueous solutions [217,258]. In these studies, the strong slowdown of molecular dynamics related to the increasing viscosity can be monitored by a combination of, in particular, spin-lattice relaxation and stimulated-echo experiments. Moreover, depending on the deuteration scheme of the used compounds, it is possible either to observe the dynamical behavior of a particular component selectively or to probe that of both constituents at the same time.

Figure 12 compares $^2$H NMR correlation times of a water-glycerol mixture in the bulk with that in protein and silica confinements [217]. In all samples, 25 wt % of water were mixed with selectively labelled glycerol-d$_5$. Hence, $^2$H NMR exclusively probes glycerol reorientation. The correlation times of the bulk mixture showed the characteristic non-Arrhenius temperature dependence of molecular glass-forming liquids over about 12 orders of magnitude. The agreement of the NMR data for glycerol dynamics with BDS results, which receive strong contributions from water reorientations, indicated coupled dynamics of the components. While glycerol reorientation was notably slowed down in an elastin matrix, the correlation times were unaltered when confining the liquid to MCM-41 pores [217]. This difference may suggest that glycerol interacts more strongly with elastin than with silica surfaces but it can also be caused by diverse confinement sizes in the studied samples. Specifically, the water-glycerol mixture forms only 1–2 solvation layers around the protein for the used concentration of 0.3 $g_{solvent}/g_{elastin}$, whereas interfacial and bulk behaviors can coexist in the MCM-41 pores with a diameter of 2.8 nm. Thus, the observed difference can result because slowed interfacial dynamics was probed for the elastin confinement, while bulk-like behavior in the pore center dominated the findings for the silica confinement. Consistent with the latter argument, spatially resolved analyses in molecular simulation studies of water-alcohol mixtures confined to silica pores revealed strongly retarded motion near the pore walls and bulk-like behavior in the pore center [259].

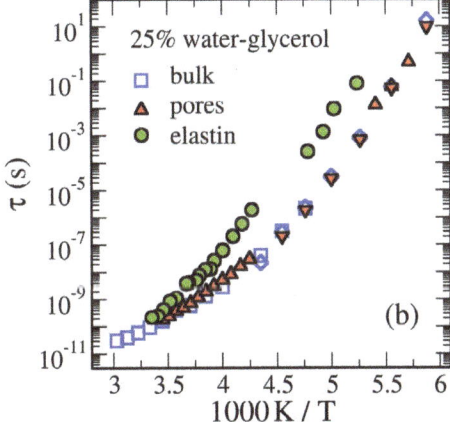

**Figure 12.** Temperature-dependent $^2$H NMR correlation times of 25 wt % water-glycerol-$d_5$ mixture in the bulk liquid, in MCM-41 pores (diameter 2.8 nm), and in an elastin matrix (0.3 $g_{solvent}/g_{elastin}$) [217]. The $^2$H NMR data (squares, circles, and up triangles) selectively characterize the rotational motion of the deuterated glycerol compound. The BDS data (diamonds, down triangles) receive contributions from both water and glycerol reorientations.

In $^2$H NMR studies on mixtures of alcohol molecules with heavy water, both components contribute to the observed signals because chemical exchange leads to perpetual redistribution of the provided deuterons [260]. Results obtained for a glass-forming mixture of water and propylene glycol in bulk and confinement are presented in Figure 13 [217]. While $^2$H spin-lattice relaxation does not yield evidence for confinement effects in the weakly supercooled regime, $^2$H rotational correlation functions from stimulated-echo experiments in the deeply supercooled regime decay slower for the mixture in silica pores than in the bulk liquid. Closer analysis showed that the common structural $\alpha$ relaxation of water and alcohol molecules is observed in the former temperature range, while faster water reorientation decouples as a secondary $\beta$ relaxation from the viscous slowdown when approaching the glass transition temperature $T_g$. Therefore, the authors concluded that the $^2$H stimulated-echo data do not yield evidence for a slowdown of the structural relaxation of the water-alcohol mixture in silica pores, but rather indicate changes of the secondary process [217].

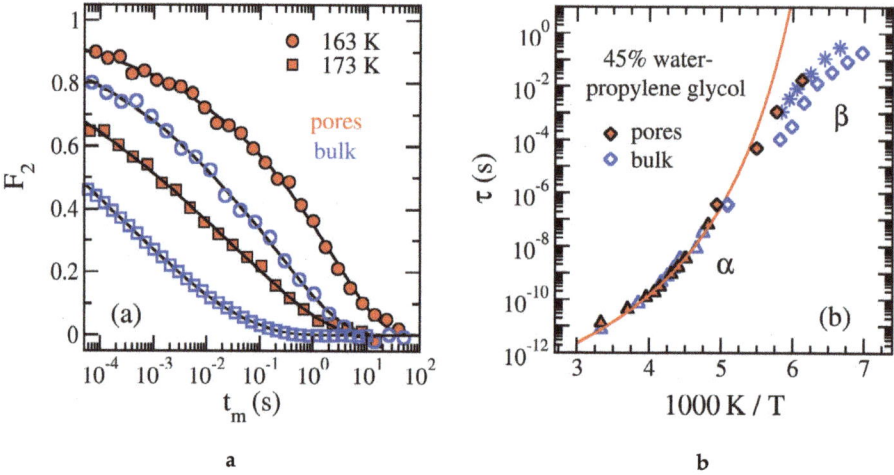

**Figure 13.** Results from $^2$H NMR studies on 45 wt % water-propylene glycol mixture in the bulk liquid and in MCM-41 pores (diameter 2.8 nm) [217]: (**a**) Rotational correlation functions $F_2(t_m)$ from stimulated-echo experiments at ~163 K and ~173 K. The lines are fits to a Kohlrausch function. (**b**) Temperature-dependent correlation times from (triangles) spin-lattice relaxation, (diamonds) line-shape analysis, and (squares) stimulated-echo experiments. The line is a Vogel-Fulcher-Tammann fit of the spin-lattice relaxation results for the confined mixture, which probe its structural α relaxation. For comparison, results for the secondary β relaxation from BDS on the bulk mixture are included as stars [261] (adapted from [217]).

## 4.5. Ionic Liquids and Surfactants in Confinement as Nonconventional Solvents

Ionic liquids (ILs) and surfactants such as poly(ethylene oxide) are versatile solvents in the field of green chemistry. Owing to favorable physical and chemical properties such as being environmentally benign and having a low vapor pressure, etc. (see, e.g., [262–265]) they are employed in a wide field of applications, ranging from basic preparative chemical synthesis to heterogeneous catalysis as a supported ionic liquid phase (SILP) catalyst [266]. In the latter application the IL is employed confined in an oxidic host material.

In the last decade, SSNMR has involved into one of the most versatile techniques to characterize the structure, dynamics, and phase behavior of ILs in general and SILP catalysts in particular, as demonstrated by the following characteristic examples.

Shylesh et al. [267] combined in situ FT-IR with $^{31}$P and $^{29}$Si MAS NMR to study the structure of sulfoxantphos (Rh-SX) in a silica supported ionic liquid film. Haumann et al. [266] studied the water-gas shift reaction employing SILP systems, consisting of the ruthenium catalyst ($[Ru(CO)_3Cl_2]_2$) and [EMIM][NTf$_2$] (1-ethyl-3-methylimidazolium bis(trifluoromethylsulfonyl)imide), respectively, [BMIM][NTf$_2$] (1-butyl-3-methylimidazolium bis(trifluoromethylsulfonyl)imide) confined in a silica gel as a function of the pore loading. Le Bideau et al. [268] investigated the dynamics of silica confined ILs by a combination of variable temperature NMR spectroscopy and relaxometry. In these experiments they observed a strong depression of the freezing point of the IL, as compared to the bulk liquid and that the confinement causes only a small slowdown of its dynamics. Rosa Castillo et al. [269] studied the phase behavior of [BMIM][PF$_6$] (1-butyl-3-methylimidazolium hexafluorophosphate) on silica and clay by multinuclear ($^1$H, $^{13}$C, $^{31}$P) SSNMR spectrometry. Waechtler et al. [270] used a combination of DSC and variable temperature $^2$H and $^{19}$F solid-state NMR to compare the phase behavior of [C$_2$Py][BTA]-d$_{10}$ (N-ethylpyridinium-bis(trifluoromethanesulfonyl)amide) in the bulk and confined in mesoporous silica gel. While two phase transitions were found for the bulk IL, one at 288–289 K indicating the onset of an intermolecular rotation and one at 295 K indicating the melting of the IL, the confined IL exhibited only a single phase transition in the lower temperature range (215–245 K).

Very recently, Hoffmann et al. [271] investigated the behavior of nonionic surfactants doped with radicals confined in SBA-15 whose surface was modified with (APTES) by a combination of DSC, SSNMR and dynamic nuclear polarization (DNP) [85,87] enhanced SSNMR spectroscopy.

## 5. Summary and Outlook

This paper reviews recent advances to the characterization of small molecules confined in microporous and mesoporous materials employing solid-state NMR techniques. It is shown that there is an exciting interplay between guest/guest and guest/host interactions, which can drastically change the physico-chemical properties of the confined systems and that solid-state NMR spectroscopy and relaxometry, combined with other techniques, such as nitrogen adsorption, differential scanning calorimetry, dielectric spectroscopy, hyperpolarization, and others, are ideal analytical tools enabling the differentiation between bound, adsorbed and free molecules inside the pores, as well as the observation of diffusion processes inside the pores of mesoporous and microporous materials. A number of examples, mainly from the groups of the two authors, were given to highlight the application of these techniques. These examples were supplemented by short references to the work of other groups in the field in order to give a broader picture of the field of imprisoned molecules.

Finally, we will end our review with giving some thoughts, where the field is moving in the next few years. Here the dramatic technical advances in sensitivity enhancement of NMR spectroscopy will enable the investigation of even more complex systems, e.g., hierarchical confinements, where the mesoporous silica-material itself is confined inside larger pores of, e.g., a polymer or paper to form a smart membrane.

**Funding:** This research was funded by the Deutsche Forschungsgemeinschaft in the framework of Forschergruppe FOR 1583, grant numbers Bu-911/18–1/2, Vo-905/8–1/2, Vo-905/9–1/2, and Vo-905/10–1/2.

**Conflicts of Interest:** The authors declare no conflicts of interest.

## References

1. Le Page, M.; Beau, R.; Duchene, J. Porous Silica Particles Containing a Crystallized Phase and Method. U.S. Patent Application No. 3,493,341, 3 February 1970.
2. Chiola, V.; Ritsko, J.E.; Vanderpool, C.D. Process for Producing Low-Bulk Density Silica. U.S. Patent Application No. 3,556,725, 19 January 1971.
3. Beck, J.; Vartuli, J.; Roth, W.; Leonowicz, M.; Kresge, C.; Schmitt, K.; Chu, C.; Olson, D.; Sheppard, E. A new family of mesoporous molecular sieves prepared with liquid crystal templates. *J. Am. Chem. Soc.* **1992**, *114*, 10834–10843. [CrossRef]
4. Zhao, D.Y.; Huo, Q.S.; Feng, J.L.; Chmelka, B.F.; Stucky, G.D. Nonionic triblock and star diblock copolymer and oligomeric surfactant syntheses of highly ordered, hydrothermally stable, mesoporous silica structures. *J. Am. Chem. Soc.* **1998**, *120*, 6024–6036. [CrossRef]
5. Zhao, D.Y.; Feng, J.L.; Huo, Q.S.; Melosh, N.; Fredrickson, G.H.; Chmelka, B.F.; Stucky, G.D. Triblock copolymer syntheses of mesoporous silica with periodic 50 to 300 angstrom pores. *Science* **1998**, *279*, 548–552. [CrossRef] [PubMed]
6. Sayari, A.; Hamoudi, S. Periodic mesoporous silica-based organic-inorganic nanocomposite materials. *Chem. Mater.* **2001**, *13*, 3151–3168. [CrossRef]
7. Linssen, T.; Cassiers, K.; Cool, P.; Vansant, E. Mesoporous templated silicates: An overview of their synthesis, catalytic activation and evaluation of the stability. *Adv. Colloid Interface Sci.* **2003**, *103*, 121–147. [CrossRef]
8. Selvam, P.; Bhatia, S.K.; Sonwane, C.G. Recent advances in processing and characterization of periodic mesoporous MCM-41 silicate molecular sieves. *Ind. Eng. Chem. Res.* **2001**, *40*, 3237–3261. [CrossRef]
9. Schreiber, A.; Ketelsen, I.; Findenegg, G.H. Melting and freezing of water in ordered mesoporous silica materials. *Phys. Chem. Chem. Phys.* **2001**, *3*, 1185–1195. [CrossRef]
10. Schüth, F.; Schmidt, W. Microporous and mesoporous materials. *Adv. Mater.* **2002**, *14*, 629–638. [CrossRef]

11. Vinu, A.; Hossain, K.Z.; Ariga, K. Recent advances in functionalization of mesoporous silica. *J. Nanosci. Nanotechno.* **2005**, *5*, 347–371. [CrossRef]
12. Wang, X.; Lin, K.S.K.; Chan, J.C.C.; Cheng, S. Direct synthesis and catalytic applications of ordered large pore aminopropyl-functionalized SBA-15 mesoporous materials. *J. Phys. Chem. B* **2005**, *109*, 1763–1769. [CrossRef]
13. Vinu, A.; Miyahara, M.; Ariga, K. Assemblies of biomaterials in mesoporous media. *J. Nanosci. Nanotechno.* **2006**, *6*, 1510–1532. [CrossRef] [PubMed]
14. Morey, M.; Davidson, A.; Stucky, G. Silica-based, cubic mesostructures: Synthesis, characterization and relevance for catalysis. *J. Porous Mater.* **1998**, *5*, 195–204. [CrossRef]
15. Chen, H.-T.; Huh, S.; Wiench, J.W.; Pruski, M.; Lin, V.S.-Y. Dialkylaminopyridine-functionalized mesoporous silica nanosphere as an efficient and highly stable heterogeneous nucleophilic catalyst. *J. Am. Chem. Soc.* **2005**, *127*, 13305–13311. [CrossRef] [PubMed]
16. Ghiaci, M.; Dokhaie, Z.; Farrokhpour, H.; Buntkowsky, G.; Breitzke, H. SBA-15 supported imidazolium ionic liquid through different linkers as sustainable catalyst for synthesis of cyclic carbonates: A kinetic study and Theoretical DFT calculations. *Ind. Eng. Chem. Res.* **2020**, *59*, 12632–12644.
17. Grünberg, A.; Breitzke, H.; Buntkowsky, G. Solid state NMR of immobilized catalysts and nanocatalysts. *Spectrosc. Prop. Inorg. Organomet. Compd.* **2012**, *43*, 289–323.
18. Mellaerts, R.; Jammaer, J.A.G.; Van Speybroeck, M.; Chen, H.; Van Humbeeck, J.; Augustijns, P.; Van den Mooter, G.; Martens, J.A. Physical state of poorly water soluble therapeutic molecules loaded into SBA-15 ordered mesoporous silica carriers: A case study with itraconazole and ibuprofen. *Langmuir* **2008**, *24*, 8651–8659. [CrossRef]
19. Ukmar, T.; Cendak, T.; Mazaj, M.; Kaucic, V.; Mali, G. Structural and dynamical properties of indomethacin molecules embedded within the mesopores of SBA-15: A solid-state NMR view. *J. Phys. Chem. C* **2012**, *116*, 2662–2671. [CrossRef]
20. Vallet-Regi, M.; Ramila, A.; Del Real, R.; Pérez-Pariente, J. A new property of MCM-41: Drug delivery system. *Chem. Mater.* **2001**, *13*, 308–311. [CrossRef]
21. Hall, S.R.; Walsh, D.; Green, D.; Oreffo, R.; Mann, S. A novel route to highly porous bioactive silica gels. *J. Mater. Chem.* **2003**, *13*, 186–190. [CrossRef]
22. de Oliveira, M.; Seeburg, D.; Weiß, J.; Wohlrab, S.; Buntkowsky, G.; Bentrup, U.; Gutmann, T. Structural characterization of vanadium environments in MCM-41 molecular sieve catalysts by solid state 51V NMR. *Catal. Sci. Technol.* **2019**, *9*, 6180–6190. [CrossRef]
23. Gedat, E.; Schreiber, A.; Albrecht, J.; Shenderovich, I.; Findenegg, G.; Limbach, H.-H.; Buntkowsky, G. 2H-solid state NMR study of benzene-d6 confined in mesoporous silica SBA-15. *J. Phys. Chem. B* **2002**, *106*, 1977. [CrossRef]
24. Medick, P.; Blochowicz, T.; Vogel, M.; Roessler, E. Comparing the dynamical heterogeneities in binary glass formers and in a glass former embedded in a zeolite - a (2) hnmr study. *J. Non Cryst. Solids* **2002**, *307*, 565–572. [CrossRef]
25. Dosseh, G.; Xia, Y.; Alba-Simionesco, C. Cyclohexane and benzene confined in MCM-41 and SBA-15: Confinement effects on freezing and melting. *J. Phys. Chem.* **2003**, *107*, 6445. [CrossRef]
26. Lusceac, S.A.; Koplin, C.; Medick, P.; Vogel, M.; Brodie-Linder, N.; LeQuellec, C.; Alba-Simionesco, C.; Roessler, E.A. Type a versus type b glass formers: NMR relaxation in bulk and confining geometry. *J. Phys. Chem. B* **2004**, *108*, 16601–16605. [CrossRef]
27. Alba-Simionesco, C.; Coasne, B.; Dosseh, G.; Dudziak, G.; Gubbins, K.; Radhakrishnan, R.; Sliwinska-Bartkowiak, M. Effects of confinement on freezing and melting. *J. Phys. Condens. Matter* **2006**, *18*, R15. [CrossRef]
28. Kiwilsza, A.; Pajzderska, A.; Gonzalez, M.A.; Mielcarek, J.; Wąsicki, J. Qens and NMR study of water dynamics in SBA-15 with a low water content. *J. Phys. Chem. C* **2015**, *119*, 16578–16586. [CrossRef]
29. Krzyżak, A.T.; Habina, I. Low field 1H NMR characterization of mesoporous silica MCM-41 and SBA-15 filled with different amount of water. *Microporous Mesoporous Mater.* **2016**, *231*, 230–239. [CrossRef]
30. Brilmayer, R.; Kübelbeck, S.; Khalil, A.; Brodrecht, M.; Kunz, U.; Kleebe, H.-J.; Buntkowsky, G.; Baier, G.; Andrieu-Brunsen, A. Influence of nanoconfinement on the pka of polyelectrolyte functionalized silica mesopores. *Adv. Mater. Interfaces* **2020**, *7*, 1901914. [CrossRef]
31. Nordberg, M.E. Properties of some vycor-brand glasses. *J. Am. Ceram. Soc.* **1944**, *27*, 299–305. [CrossRef]

32. Gelb, L.D.; Gubbins, K.E.; Radhakrishnan, R.; Sliwinska-Bartkowiak, M. Phase separation in confined systems. *Rep. Prog. Phys.* **1999**, *62*, 1573–1659. [CrossRef]
33. Ciesla, U.; Schüth, F. Ordered mesoporous materials. *Microporous Mesoporous Mater.* **1999**, *27*, 131. [CrossRef]
34. Yokoi, T.; Yoshitake, H.; Tatsumi, T. Synthesis of amino-functionalized MCM-41 via direct co-condensation and post-synthesis grafting methods using mono-, di- and tri-amino-organoalkoxysilanes. *J. Mater. Chem.* **2004**, *14*, 951–957. [CrossRef]
35. Hemminger, W.F.; Cammenga, H.K. *Methoden der Thermischen Analyse*; Springer: Berlin, Germany, 1989.
36. Grünberg, B.; Emmler, T.; Gedat, E.; Shenderovich, I.; Findenegg, G.H.; Limbach, H.H.; Buntkowsky, G. Hydrogen bonding of water confined in mesoporous silica MCM-41 and SBA-15 studied by h-1 solid-state NMR. *Chem. Eur. J.* **2004**, *10*, 5689–5696. [CrossRef] [PubMed]
37. Masierak, W.; Emmler, T.; Gedat, E.; Schreiber, A.; Findenegg, G.H.; Buntkowsky, G. Microcrystallization of benzene-d6 in mesoporous silica revealed by 2H-solid state NMR. *J. Phys. Chem. B* **2004**, *108*, 18890–18896. [CrossRef]
38. Vyalikh, A.; Emmler, T.; Gedat, E.; Shenderovich, I.; Findenegg, G.H.; Limbach, H.H.; Buntkowsky, G. Evidence of microphase separation in controlled pore glasses. *Solid State Nucl. Mag. Res.* **2005**, *28*, 117–124. [CrossRef] [PubMed]
39. Buntkowsky, G.; Breitzke, H.; Adamczyk, A.; Roelofs, F.; Emmler, T.; Gedat, E.; Grünberg, B.; Xu, Y.; Limbach, H.H.; Shenderovich, I.; et al. Structural and dynamical properties of guest molecules confined in mesoporous silica materials revealed by NMR. *Phys. Chem. Chem. Phys.* **2007**, *9*, 4843–4853. [PubMed]
40. Vyalikh, A.; Emmler, T.; Gruenberg, B.; Xu, Y.; Shenderovich, I.; Findenegg, G.H.; Limbach, H.H.; Buntkowsky, G. Hydrogen bonding of water confined in controlled-pore glass 10–75 studied by h-1-solid state NMR. *Z. Phys. Chem.* **2007**, *221*, 155–168. [CrossRef]
41. Schoen, M.; Klapp, S. Nanoconfined fluids: Soft matter between two and three dimensions. *Rev. Comput. Chem.* **2007**, *24*, 1–517.
42. Vyalikh, A.; Emmler, T.; Shenderovich, I.; Zeng, Y.; Findenegg, G.H.; Buntkowsky, G. H-2-solid state NMR and dsc study of isobutyric acid in mesoporous silica materials. *Phys. Chem. Chem. Phys.* **2007**, *9*, 2249–2257. [CrossRef]
43. Amadeu, N.d.S.; Gruenberg, B.; Frydel, J.; Werner, M.; Limbach, H.-H.; Breitzke, H.; Buntkowsky, G. Melting of low molecular weight compounds in confinement observed by h-2-solid state NMR: Biphenyl, a case study. *Z. Phys. Chem.* **2012**, *226*, 1169–1185. [CrossRef]
44. Kahse, M.; Werner, M.; Zhao, S.; Hartmann, M.; Buntkowsky, G.; Winter, R. Stability, hydration, and thermodynamic properties of rnase a confined in surface-functionalized SBA-15 mesoporous molecular sieves. *J. Phys. Chem. C* **2014**, *118*, 21523–21531. [CrossRef]
45. Sattig, M.; Reutter, S.; Fujara, F.; Werner, M.; Buntkowsky, G.; Vogel, M. NMR studies on the temperature-dependent dynamics of confined water. *Phys. Chem. Chem. Phys.* **2014**, *16*, 19229–19240. [CrossRef] [PubMed]
46. Werner, M.; Rothermel, N.; Breitzke, H.; Gutmann, T.; Buntkowsky, G. Recent advances in solid state NMR of small molecules in confinement. *Isr. J. Chem.* **2014**, *54*, 60–73. [CrossRef]
47. Werner, M.; Heil, A.; Rothermel, N.; Breitzke, H.; Groszewicz, P.B.; Thankamony, A.S.; Gutmann, T.; Buntkowsky, G. Synthesis and solid state NMR characterization of novel peptide/silica hybrid materials. *Solid State Nucl. Mag. Res.* **2015**, *72*, 73–78. [CrossRef] [PubMed]
48. Treacy, M.M.; Higgins, J.B.; von Ballmoos, R. *Collection of Simulated XRD Powder Patterns for Zeolites*; Elsevier: London, UK, 1996.
49. Marler, B.; Oberhagemann, U.; Vortmann, S.; Gies, H. Influence of the sorbate type on the XRD peak intensities of loaded MCM-41. *Microporous Mater.* **1996**, *6*, 375–383. [CrossRef]
50. Yao, M.; Baird, R.; Kunz, F.; Hoost, T. An XRD and tem investigation of the structure of alumina-supported ceria–zirconia. *J. Catal.* **1997**, *166*, 67–74. [CrossRef]
51. Brunauer, S.; Emmett, P.H.; Teller, E. Adsorption of gases in multimolecular layers. *J. Am. Chem. Soc.* **1938**, *60*, 309–319. [CrossRef]
52. Barrett, E.P.; Joyner, L.G.; Halenda, P.P. The determination of pore volume and area distributions in porous substances. I. Computations from nitrogen isotherms. *J. Am. Chem. Soc.* **1951**, *73*, 373–380.
53. Höhne, G.; Hemminger, W.F.; Flammersheim, H.-J. *Differential Scanning Calorimetry*; Springer: Berlin, Germany, 2013.

54. Freude, D.; Kärger, J. NMR techniques. *Handb. Porous Solids* **2002**, *1*, 465–504.
55. Koller, H.; Weiß, M. Solid state NMR of porous materials. In *Solid State NMR*; Springer: Berlin, Germany, 2011; pp. 189–227.
56. Haouas, M.; Martineau, C.; Taulelle, F. *Quadrupolar NMR of Nanoporous Materials*; Emagres: Orense, Spain, 2007.
57. Gutmann, T.; Groszewicz, P.B.; Buntkowsky, G. Solid-State NMR of Nanocrystals. *Ann. Rep. NMR Spectrosc.* **2019**, *97*, 1–82.
58. Faivre, C.; Bellet, D.; Dolino, G. Phase transitions of fluids confined in porous silicon: A differential calorimetry investigation. *Eur. Phys. J. B Condens. Matter Complex Syst.* **1999**, *7*, 19–36. [CrossRef]
59. Alcoutlabi, M.; McKenna, G.B. Effects of confinement on material behaviour at the nanometre size scale. *J. Phys. Condens. Matter* **2005**, *17*, R461. [CrossRef]
60. Kaerger, J.; Pfeifer; Heink, W. Principles and application of self-diffusion measurements by nuclear magnetic resonance. *Adv. Magn. Opt. Res.* **1988**, *12*, 1–89.
61. Kaerger, J.; Freude, D. In situ studies of catalytic reactions in zeolites by means of PFG and MAS NMR techniques. *Prog. Zeolite Microporous Mater. Pts a-C* **1997**, *105*, 551–558.
62. Kaerger, J.; Freude, D. Mass transfer in micro- and mesoporous materials. *Chem. Eng. Technol.* **2002**, *25*, 769–778.
63. Kärger, J.; Freude, D.; Haase, J. Diffusion in nanoporous materials: Novel insights by combining MAS and PFG NMR. *Processes* **2018**, *6*, 147. [CrossRef]
64. Kaerger, J.; Valiullin, R. Mass transfer in mesoporous materials: The benefit of microscopic diffusion measurement. *Chem. Soc. Rev.* **2013**, *42*, 4172–4197. [CrossRef]
65. Findenegg, G.H.; Jaehnert, S.; Akcakayiran, D.; Schreiber, A. Freezing and melting of water confined in silica nanopores. *ChemPhysChem* **2008**, *9*, 2651–2659. [CrossRef]
66. Geppi, M.; Borsacchi, S.; Mollica, G.; Veracini, C.A. Applications of solid-state NMR to the study of organic/inorganic multicomponent materials. *Appl. Spectr. Rev.* **2009**, *44*, 1–89. [CrossRef]
67. Vogel, M. NMR studies on simple liquids in confinement. *Eur. Phys. J.* **2010**, *189*, 47–64. [CrossRef]
68. Ishii, Y.; Tycko, R. Sensitivity enhancement in solid state 15N NMR by indirect detection with high-speed magic angle spinning. *J. Magn. Reson.* **2000**, *142*, 199–204. [CrossRef] [PubMed]
69. Mao, K.; Wiench, J.W.; Lin, V.S.Y.; Pruski, M. Indirectly detected through-bond chemical shift correlation NMR spectroscopy in solids under fast MAS: Studies of organic-inorganic hybrid materials. *J. Magn. Reson.* **2009**, *196*, 92–95. [CrossRef] [PubMed]
70. Mao, K.; Pruski, M. Directly and indirectly detected through-bond heteronuclear correlation solid-state NMR spectroscopy under fast MAS. *J. Magn. Reson.* **2009**, *201*, 165–174. [CrossRef] [PubMed]
71. Kobayashi, T.; Mao, K.; Wang, S.G.; Lin, V.S.Y.; Pruski, M. Molecular ordering of mixed surfactants in mesoporous silicas: A solid-state NMR study. *Solid State Nucl. Mag. Res.* **2011**, *39*, 65–71. [CrossRef] [PubMed]
72. Hsin, T.M.; Chen, S.; Guo, E.; Tsai, C.H.; Pruski, M.; Lin, V.S.Y. Calcium containing silicate mixed oxide-based heterogeneous catalysts for biodiesel production. *Top. Catal.* **2010**, *53*, 746–754. [CrossRef]
73. Maly, T.; Debelouchina, G.T.; Bajaj, V.S.; Hu, K.; Joo, C.; Mak-Jurkauskas, M.L.; Sirigiri, J.R.; Wel, P.C.A.v.d.; Herzfeld, J.; Temkin, R.J.; et al. Dynamic nuclear polarization at high magnetic fields. *J. Chem. Phys.* **2008**, *128*, 052211. [CrossRef]
74. Hovav, Y.; Feintuch, A.; Vega, S. Theoretical aspects of dynamic nuclear polarization in the solid state - spin temperature and thermal mixing. *Phys. Chem. Chem. Phys.* **2013**, *15*, 188–203. [CrossRef]
75. Mentink-Vigier, F.; Akbey, U.; Hovav, Y.; Vega, S.; Oschkinat, H.; Feintuch, A. Fast passage dynamic nuclear polarization on rotating solids. *J. Magn. Reson.* **2012**, *224*, 13–21. [CrossRef]
76. Hovav, Y.; Feintuch, A.; Vega, S. Theoretical aspects of dynamic nuclear polarization in the solid state - the cross effect. *J. Magn. Reson.* **2012**, *214*, 29–41. [CrossRef]
77. Overhauser, A.W. Polarization of nuclei in metals. *Phys. Rev.* **1953**, *92*, 411–415. [CrossRef]
78. Carver, T.P.; Slichter, C.P. Polarization of nuclear spins in metals. *Phys.Rev.* **1953**, *92*, 212–213. [CrossRef]
79. Carver, T.P.; Slichter, C.P. Experimental verification of the overhauser nuclear polarization effect. *Phys.Rev.* **1956**, *102*, 975–980. [CrossRef]
80. Wind, R.A.; Duijvestijn, M.J.; Vanderlugt, C.; Manenschijn, A.; Vriend, J. Applications of dynamic nuclear-polarization in c-13 NMR in solids. *Prog. Nucl. Magn. Reson. Spectrosc.* **1985**, *17*, 33–67. [CrossRef]

81. Becerra, L.R.; Gerfen, G.J.; Temkin, R.J.; Singel, D.J.; Griffin, R.G. Dynamic nuclear polarization with a cyclotron resonance maser at 5 T. *Phys. Rev. Lett.* **1993**, *71*, 3561. [CrossRef] [PubMed]
82. Thankamony, A.S.; Knoche, S.; Drochner, A.; Pandurang, J.A.; Sigurdsson, S.T.; Vogel, H.; Etzold, B.J.M.; Gutmann, T.; Buntkowsky, G. Characterization of V-Mo-W Mixed Oxide Catalyst Surface Species by 51V Solid-State Dynamic Nuclear Polarization NMR. *J. Phys. Chem. C* **2017**, *121*, 20857–20864. [CrossRef]
83. Kiesewetter, M.K.; Michaelis, V.K.; Walish, J.J.; Griffin, R.G.; Swager, T.M. High field dynamic nuclear polarization NMR with surfactant sheltered biradicals. *J. Phys. Chem. B* **2014**, *118*, 1825–1830. [CrossRef]
84. Thankamony, A.S.L.; Wittmann, J.J.; Kaushik, M.; Corzilius, B. Dynamic nuclear polarization for sensitivity enhancement in modern solid-state NMR. *Prog. Nucl. Magn. Reson. Spectrosc.* **2017**, *102*, 120–195. [CrossRef]
85. Lesage, A.; Lelli, M.; Gajan, D.; Caporini, M.A.; Vitzthum, V.; Mieville, P.; Alauzun, J.; Roussey, A.; Thieuleux, C.; Mehdi, A.; et al. Surface enhanced NMR spectroscopy by dynamic nuclear polarization. *J. Am. Chem. Soc.* **2010**, *132*, 15459–15461. [CrossRef]
86. Gajan, D.; Levine, D.; Zocher, E.; Coperet, C.; Lesage, A.; Emsley, L. Probing surface site heterogeneity through 1d and inadequate p-31 solid state NMR spectroscopy of silica supported pme3-au(i) adducts. *Chem. Sci.* **2011**, *2*, 928–931. [CrossRef]
87. Lelli, M.; Gajan, D.; Lesage, A.; Caporini, M.A.; Vitzthum, V.; Mieville, P.; Heroguel, F.; Rascon, F.; Roussey, A.; Thieuleux, C.; et al. Fast characterization of functionalized silica materials by silicon-29 surface-enhanced NMR spectroscopy using dynamic nuclear polarization. *J. Am. Chem. Soc.* **2011**, *86*–*121*. [CrossRef]
88. Conley, M.P.; Drost, R.M.; Baffert, M.; Gajan, D.; Elsevier, C.; Franks, W.T.; Oschkinat, H.; Veyre, L.; Zagdoun, A.; Rossini, A.; et al. A well-defined pd hybrid material for the z-selective semihydrogenation of alkynes characterized at the molecular level by DNP sens. *Chem. Eur. J.* **2013**, *19*, 12234–12238. [CrossRef] [PubMed]
89. Conley, M.P.; Rossini, A.J.; Comas-Vives, A.; Valla, M.; Casano, G.; Ouari, O.; Tordo, P.; Lesage, A.; Emsley, L.; Coperet, C. Silica-surface reorganization during organotin grafting evidenced by sn-119 DNP sens: A tandem reaction of gem-silanols and strained siloxane bridges. *Phys. Chem. Chem. Phys.* **2014**, *16*, 17822–17827. [CrossRef] [PubMed]
90. Valla, M.; Rossini, A.J.; Caillot, M.; Chizallet, C.; Raybaud, P.; Digne, M.; Chaumonnot, A.; Lesage, A.; Emsley, L.; van Bokhoven, J.A.; et al. Atomic description of the interface between silica and alumina in aluminosilicates through dynamic nuclear polarization surface-enhanced NMR spectroscopy and first-principles calculations. *J. Am. Chem. Soc.* **2015**, *137*, 10710–10719. [CrossRef]
91. Ong, T.C.; Liao, W.C.; Mougel, V.; Gajan, D.; Lesage, A.; Emsley, L.; Coperet, C. Atomistic description of reaction intermediates for supported metathesis catalysts enabled by DNP sens. *Angew. Chem. Int. Ed.* **2016**, *55*, 4743–4747. [CrossRef] [PubMed]
92. Delley, M.F.; Lapadula, G.; Nunez-Zarur, F.; Comas-Vives, A.; Kalendra, V.; Jeschke, G.; Baabe, D.; Walter, M.D.; Rossini, A.J.; Lesage, A.; et al. Local structures and heterogeneity of silica-supported m(iii) sites evidenced by epr, ir, NMR, and luminescence spectroscopies. *J. Am. Chem. Soc.* **2017**, *139*, 8855–8867. [CrossRef]
93. Liao, W.C.; Ong, T.C.; Gajan, D.; Bernada, F.; Sauvee, C.; Yulikov, M.; Pucino, M.; Schowner, R.; Schwarzwalder, M.; Buchmeiser, M.R.; et al. Dendritic polarizing agents for DNP sens. *Chem. Sci.* **2017**, *8*, 416–422. [CrossRef]
94. Pump, E.; Bendjeriou-Sedjerari, A.; Viger-Gravel, J.; Gajan, D.; Scotto, B.; Samantaray, M.K.; Abou-Hamad, E.; Gurinov, A.; Almaksoud, W.; Cao, Z.; et al. Predicting the DNP-sens efficiency in reactive heterogeneous catalysts from hydrophilicity. *Chem. Sci.* **2018**, *9*, 4866–4872. [CrossRef]
95. Azais, T.; Von Euw, S.; Ajili, W.; Auzoux-Bordenave, S.; Bertani, P.; Gajan, D.; Emsley, L.; Nassif, N.; Lesage, A. Structural description of surfaces and interfaces in biominerals by DNP sens. *Solid State Nucl. Mag. Res.* **2019**, *102*, 2–11. [CrossRef]
96. Eisenschmidt, T.C.; Kirss, R.U.; Deutsch, P.P.; Hommeltoft, S.I.; Eisenberg, R.; Bargon, J. Para hydrogen induced polarization in hydrogenation reactions. *J. Am. Chem. Soc.* **1987**, *109*, 8089–8091. [CrossRef]
97. Bowers, C.R.; Weitekamp, D.P. Transformation of symmetrization order to nuclear-spin magnetization by chemical reaction and nuclear magnetic resonance. *Phys. Rev. Lett.* **1986**, *57*, 2645–2648. [CrossRef]
98. Bowers, C.R.; Jones, D.H.; Kurur, N.D.; Labinger, J.A.; Pravica, M.G.; Weitekamp, D.P. Pasadena and altadena. *Adv. Magn. Res.* **1990**, *15*, 269.

99. Henning, H.; Dyballa, M.; Scheibe, M.; Klemm, E.; Hunger, M. In situ cf MAS NMR study of the pairwise incorporation of parahydrogen into olefins on rhodium-containing zeolites y. *Chem. Phys. Lett.* **2013**, *555*, 258–262. [CrossRef]
100. Hunger, M. In situ NMR spectroscopy in heterogeneous catalysis. *Catal. Today* **2004**, *97*, 3–12. [CrossRef]
101. Arzumanov, S.S.; Stepanov, A.G. Parahydrogen-induced polarization detected with continuous flow MAS NMR. *J. Phys. Chem. C* **2013**, *117*, 2888–2892. [CrossRef]
102. Heinze, M.T.; Zill, J.C.; Matysik, J.; Einicke, W.D.; Gläser, R.; Stark, A. Solid–ionic liquid interfaces: Pore filling revisited. *Phys. Chem. Chem. Phys.* **2014**, *16*, 24359–24372. [CrossRef] [PubMed]
103. Fraissard, J.; Jameson, C.; Saam, B.; Brunner, E.; Hersman, W.; Goodson, B.; Meersmann, T.; Fujiwara, H.; Wang, L.-Q.; Sozzani, P. *Hyperpolarized Xenon-129 Magnetic Resonance: Concepts, Production, Techniques and Applications*; Royal Society of Chemistry: London, UK, 2015.
104. Yang, Y.; Beele, B.; Bluemel, J. Easily immobilized di- and tetraphosphine linkers: Rigid scaffolds that prevent interactions of metal complexes with oxide supports. *J. Am. Chem. Soc.* **2008**, *130*, 3771–3773. [CrossRef]
105. Bluemel, J. Linkers and catalysts immobilized on oxide supports: New insights by solid-state NMR spectroscopy. *Coord. Chem. Rev.* **2008**, *252*, 2410–2423. [CrossRef]
106. Gutmann, T.; Grünberg, A.; Rothermel, N.; Werner, M.; Srour, M.; Abdulhussain, S.; Tan, S.; Xu, Y.; Breitzke, H.; Buntkowsky, G. Solid-state NMR concepts for the investigation of supported transition metal catalysts and nanoparticles. *Solid State Nucl. Mag. Res.* **2013**, *55*, 1–11. [CrossRef]
107. Motokura, K.; Itagaki, S.; Iwasawa, Y.; Miyaji, A.; Baba, T. Silica-supported aminopyridinium halides for catalytic transformations of epoxides to cyclic carbonates under atmospheric pressure of carbon dioxide. *Green Chem.* **2009**, *11*, 1876–1880. [CrossRef]
108. Gath, J.; Hoaston, G.L.; Vold, R.L.; Berthoud, R.; Coperet, C.; Grellier, M.; Sabo-Etienne, S.; Lesage, A.; Emsley, L. Motional heterogeneity in single-site silica-supported species revealed by deuteron NMR. *Phys. Chem. Chem. Phys.* **2009**, *11*, 6962–6971. [CrossRef]
109. Wang, Q.; Jordan, E.; Shantz, D.F. H-2 NMR studies of simple organic groups covalently attached to ordered mesoporous silica. *J. Phys. Chem. C* **2009**, *113*, 18142–18151. [CrossRef]
110. Kandel, K.; Althaus, S.M.; Peeraphatdit, C.; Kobayashi, T.; Trewyn, B.G.; Pruski, M.; Slowing, I.I. Substrate inhibition in the heterogeneous catalyzed aldol condensation: A mechanistic study of supported organocatalysts. *J. Catal.* **2012**, *291*, 63–68. [CrossRef]
111. Kandel, K.; Althaus, S.M.; Peeraphatdit, C.; Kobayashi, T.; Trewyn, B.G.; Pruski, M.; Slowing, I.I. Solvent-induced reversal of activities between two closely related heterogeneous catalysts in the aldol reaction. *ACS Catal.* **2013**, *3*, 265–271. [CrossRef]
112. Jayanthi, S.; Frydman, V.; Vega, S. Dynamic deuterium magic angle spinning NMR of a molecule grafted at the inner surface of a mesoporous material. *J. Phys. Chem. B* **2012**, *116*, 10398–10405. [CrossRef]
113. Sundaresan, J.; Werner, M.; Yeping, X.; Buntkowsky, G.; Vega, S. Restricted dynamics of a deuterated linker grafted on SBA-15 revealed by deuterium MAS NMR. *J. Phys. Chem. C* **2013**, *117*, 13114–13121.
114. Jayanthi, S.; Kababya, S.; Schmidt, A.; Vega, S. Deuterium MAS NMR and local molecular dynamic model to study adsorption–desorption kinetics of a dipeptide at the inner surfaces of SBA-15. *J. Phys. Chem. C* **2016**, *120*, 2797–2806. [CrossRef]
115. Keifer, P.A.; Baltusis, L.; Rice, D.M.; Tymiak, A.A.; Shoolery, J.N. A comparison of NMR spectra obtained for solid-phase-synthesis resins using conventional high-resolution, magic-angle-spinning, and high-resolution magic-angle-spinning probes. *J. Magn. Reson. Ser. A* **1996**, *119*, 65–75. [CrossRef]
116. Bluemel, J. Reactions of ethoxysilanes with silica: A solid-state NMR study. *J. Am. Chem. Soc.* **1995**, *117*, 2112–2113. [CrossRef]
117. Merckle, C.; Haubrich, S.; Bluemel, J. Immobilized rhodium hydrogenation catalysts. *J. Organomet. Chem.* **2001**, *627*, 44–54. [CrossRef]
118. Merckle, C.; Bluemel, J. Improved rhodium hydrogenation catalysts immobilized on silica. *Top. Catal.* **2005**, *34*, 5–15. [CrossRef]
119. Brenna, S.; Posset, T.; Furrer, J.; Blümel, J. 14n NMR and two-dimensional suspension 1H and 13C hrmas NMR spectroscopy of ionic liquids immobilized on silica. *Chem. A Eur. J.* **2006**, *12*, 2880–2888. [CrossRef] [PubMed]
120. Beele, B.; Guenther, J.; Perera, M.; Stach, M.; Oeser, T.; Bluemel, J. New linker systems for superior immobilized catalysts. *New J. Chem.* **2010**, *34*, 2729–2731. [CrossRef]

121. Posset, T.; Guenther, J.; Pope, J.; Oeser, T.; Bluemel, J. Immobilized sonogashira catalyst systems: New insights by multinuclear hrmas NMR studies. *Chem. Commun.* **2011**, *47*, 2059–2061. [CrossRef] [PubMed]
122. Cluff, K.J.; Schnellbach, M.; Hilliard, C.R.; Bluemel, J. The adsorption of chromocene and ferrocene on silica: A solid-state NMR study. *J. Organomet. Chem.* **2013**, *744*, 119–124. [CrossRef]
123. Hilliard, C.R.; Kharel, S.; Cluff, K.J.; Bhuvanesh, N.; Gladysz, J.A.; Bluemel, J. Structures and unexpected dynamic properties of phosphine oxides adsorbed on silica surfaces. *Chem. A Eur. J.* **2014**, *20*, 17292–17295. [CrossRef]
124. Kharel, S.; Cluff, K.J.; Bhuvanesh, N.; Gladysz, J.A.; Bluemel, J. Structures and dynamics of secondary and tertiary alkylphosphine oxides adsorbed on silica. *Chem. Asian J.* **2019**, *14*, 2704–2711. [CrossRef]
125. Hubbard, P.J.; Benzie, J.W.; Bakhmutov, V.I.; Blümel, J. Ferrocene adsorbed on silica and activated carbon surfaces: A solid-state NMR study of molecular dynamics and surface interactions. *Organometallics* **2020**, *39*, 1080–1091. [CrossRef]
126. Saint-Arroman, R.P.; Chabanas, M.; Baudouin, A.; Coperet, C.; Basset, J.H.; Lesage, A.; Emsley, L. Characterization of surface organometallic complexes using high resolution 2d solid-state NMR spectroscopy. Application to the full characterization of a silica supported metal carbyne: =sio-mo(=-bu-t(ch2-bu-t)(2). *J. Am. Chem. Soc.* **2001**, *123*, 3820–3821. [CrossRef]
127. Rataboul, F.; Chabanas, M.; de Mallmann, A.; Coperet, C.; Thivolle-Cazat, J.; Basset, J.M. Hydrogenolysis of cycloalkanes on a tantalum hydride complex supported on silica and insight into the deactivation pathway by the combined use of id solid-state NMR and exafs spectroscopies. *Chem. Eur. J.* **2003**, *9*, 1426–1434. [CrossRef]
128. Blanc, F.; Basset, J.M.; Coperet, C.; Sinha, A.; Tonzetich, Z.J.; Schrock, R.R.; Solans-Monfort, X.; Clot, E.; Eisenstein, O.; Lesage, A.; et al. Dynamics of silica-supported catalysts determined by combining solid-state NMR spectroscopy and dft calculations. *J. Am. Chem. Soc.* **2008**, *130*, 5886–5900. [CrossRef]
129. Kerber, R.N.; Kermagoret, A.; Callens, E.; Florian, P.; Massiot, D.; Lesage, A.; Coperet, C.; Delbecq, F.; Rozanska, X.; Sautet, P. Nature and structure of aluminum surface sites grafted on silica from a combination of high-field aluminum-27 solid-state NMR spectroscopy and first-principles calculations. *J. Am. Chem. Soc.* **2012**, *134*, 6767–6775. [CrossRef] [PubMed]
130. Conley, M.; Coperet, C.; Andersen, R. Nature of secondary interactions in molecular and silica-supported organolutetium complexes from solid-state NMR spectroscopy. *Abstr. Papers Am. Chem. Soc.* **2016**, *251*, 1155.
131. Conley, M.P.; Lapadula, G.; Sanders, K.; Gajan, D.; Lesage, A.; del Rosa, I.; Maron, L.; Lukens, W.W.; Coperet, C.; Andersen, R.A. The nature of secondary interactions at electrophilic metal sites of molecular and silica-supported organolutetium complexes from solid-state NMR spectroscopy. *J. Am. Chem. Soc.* **2016**, *138*, 3831–3843. [CrossRef] [PubMed]
132. Estes, D.P.; Gordon, C.P.; Fedorov, A.; Liao, W.C.; Ehrhorn, H.; Bittner, C.; Zier, M.L.; Bockfeld, D.; Chan, K.W.; Eisenstein, O.; et al. Molecular and silica-supported molybdenum alkyne metathesis catalysts: Influence of electronics and dynamics on activity revealed by kinetics, solid-state NMR, and chemical shift analysis. *J. Am. Chem. Soc.* **2017**, *139*, 17597–17607. [CrossRef] [PubMed]
133. Crisci, A.J.; Tucker, M.H.; Dumesic, J.A.; Scott, S.L. Bifunctional solid catalysts for the selective conversion of fructose to 5-hydroxymethylfurfural. *Top. Catal.* **2010**, *53*, 1185–1192. [CrossRef]
134. Crisci, A.J.; Tucker, M.H.; Lee, M.-Y.; Jang, S.G.; Dumesic, J.A.; Scott, S.L. Acid-functionalized SBA-15-type silica catalysts for carbohydrate dehydration. *ACS Catal.* **2011**, *1*, 719–728. [CrossRef]
135. Alamillo, R.; Crisci, A.J.; Gallo, J.M.R.; Scott, S.L.; Dumesic, J.A. A tailored microenvironment for catalytic biomass conversion in inorganic–organic nanoreactors. *Angew. Chem.* **2013**, *125*, 10539–10541. [CrossRef]
136. Sievers, C.; Noda, Y.; Qi, L.; Albuquerque, E.M.; Rioux, R.M.; Scott, S.L. Phenomena affecting catalytic reactions at solid–liquid interfaces. *ACS Catal.* **2016**, *6*, 8286–8307. [CrossRef]
137. Goldsmith, B.R.; Peters, B.; Johnson, J.K.; Gates, B.C.; Scott, S.L. Beyond ordered materials: Understanding catalytic sites on amorphous solids. *ACS Catal.* **2017**, *7*, 7543–7557. [CrossRef]
138. Moon, H.; Han, S.; Scott, S.L. Tuning molecular adsorption in SBA-15-type periodic mesoporous organosilicas by systematic variation of their surface polarity. *Chem. Sci.* **2020**, *11*, 3702–3712. [CrossRef]
139. Trebosc, J.; Wiench, J.W.; Huh, S.; Lin, V.S.Y.; Pruski, M. Studies of organically functionalized mesoporous silicas using heteronuclear solid-state correlation NMR spectroscopy under fast magic angle spinning. *J. Am. Chem. Soc.* **2005**, *127*, 7587–7593. [CrossRef] [PubMed]

140. Mao, K.; Kobayashi, T.; Wiench, J.W.; Chen, H.T.; Tsai, C.H.; Lin, V.S.Y.; Pruski, M. Conformations of silica-bound (pentafluorophenyl)propyl groups determined by solid-state NMR spectroscopy and theoretical calculations. *J. Am. Chem. Soc.* **2010**, *132*, 12452–12457. [CrossRef] [PubMed]
141. Hara, K.; Akahane, S.; Wiench, J.W.; Burgin, B.R.; Ishito, N.; Lin, V.S.Y.; Fukuoka, A.; Pruski, M. Selective and efficient silylation of mesoporous silica: A quantitative assessment of synthetic strategies by solid-state NMR. *J. Phys. Chem. C* **2012**, *116*, 7083–7090. [CrossRef]
142. Kobayashi, T.; Singappuli-Arachchige, D.; Wang, Z.R.; Slowing, I.I.; Pruski, M. Spatial distribution of organic functional groups supported on mesoporous silica nanoparticles: A study by conventional and DNP-enhanced si-29 solid-state NMR. *Phys. Chem. Chem. Phys.* **2017**, *19*, 1781–1789. [CrossRef] [PubMed]
143. Perras, F.; Kobayashi, T.; Pruski, M. Studies of silica- and alumina-supported catalysts by dynamic nuclear polarization solidstate NMR. *Abstr. Papers Am. Chem. Soc.* **2017**, *253*, 1155.
144. Kobayashi, T.; Singappuli-Arachchige, D.; Slowing, I.I.; Pruski, M. Spatial distribution of organic functional groups supported on mesoporous silica nanoparticles (2): A study by h-1 triple-quantum fast-MAS solid-state NMR. *Phys. Chem. Chem. Phys.* **2018**, *20*, 22203–22209. [CrossRef]
145. Kobayashi, T.; Pruski, M. Spatial distribution of silica-bound catalytic organic functional groups can now be revealed by conventional and DNP-enhanced solid-state NMR methods. *ACS Catal.* **2019**, *9*, 7238–7249. [CrossRef]
146. Adamczyk, A.; Xu, Y.; Walaszek, B.; Roelofs, F.; Pery, T.; Pelzer, K.; Philippot, K.; Chaudret, B.; Limbach, H.H.; Breitzke, H.; et al. Solid state and gas phase NMR studies of immobilized catalysts and catalytic active nanoparticles. *Top. Catal.* **2008**, *48*, 75–83. [CrossRef]
147. Gutmann, T.; Ratajczyk, T.; Xu, Y.P.; Breitzke, H.; Grunberg, A.; Dillenberger, S.; Bommerich, U.; Trantzschel, T.; Bernarding, J.; Buntkowsky, G. Understanding the leaching properties of heterogenized catalysts: A combined solid-state and phip NMR study. *Solid State Nucl. Mag. Res.* **2010**, *38*, 90–96. [CrossRef]
148. Grunberg, A.; Gutmann, T.; Rothermel, N.; Xu, Y.P.; Breitzke, H.; Buntkowsky, G. Immobilization and characterization of rucl2(pph3)(3) mesoporous silica SBA-3. *Z. Phys. Chem.* **2013**, *227*, 901–915. [CrossRef]
149. Srour, M.; Hadjiali, S.; Sauer, G.; Brunnengräber, K.; Breitzke, H.; Xu, Y.; Weidler, H.; Limbach, H.-H.; Gutmann, T.; Buntkowsky, G. Synthesis and solid-state NMR characterization of a novel, robust, pyridyl-based immobilized Wilkinson's type catalyst with high catalytic performance. *ChemCatChem* **2016**, *8*, 3409–3416. [CrossRef]
150. Abdulhussain, S.; Breitzke, H.; Ratajczyk, T.; Grunberg, A.; Srour, M.; Arnaut, D.; Weidler, H.; Kunz, U.; Kleebe, H.J.; Bommerich, U.; et al. Synthesis, solid-state NMR characterization, and application for hydrogenation reactions of a novel wilkinson's-type immobilized catalyst. *Chem. Eur. J.* **2014**, *20*, 1159–1166. [CrossRef] [PubMed]
151. Gutmann, T.; Alkhagani, S.; Rothermel, N.; Limbach, H.H.; Breitzke, H.; Buntkowsky, G. P-31-solid-state NMR characterization and catalytic hydrogenation tests of novel heterogenized iridium-catalysts. *Z. Phys. Chem.* **2017**, *231*, 653–669. [CrossRef]
152. Liu, J.Q.; Groszewicz, P.B.; Wen, Q.B.; Thankamony, A.S.L.; Zhang, B.; Kunz, U.; Sauer, G.; Xu, Y.P.; Gutmann, T.; Buntkowsky, G. Revealing structure reactivity relationships in heterogenized dirhodium catalysts by solid-state NMR techniques. *J. Phys. Chem. C* **2017**, *121*, 17409–17416. [CrossRef]
153. Folliet, N.; Gervais, C.; Costa, D.; Laurent, G.; Babonneau, F.; Stievano, L.; Lambert, J.-F.; Tielens, F. A molecular picture of the adsorption of glycine in mesoporous silica through NMR experiments combined with dft-d calculations. *J. Phys. Chem. C* **2013**, *117*, 4104–4114. [CrossRef]
154. Azaïs, T.; Laurent, G.; Panesar, K.; Nossov, A.; Guenneau, F.; Sanfeliu Cano, C.; Tourné-Péteilh, C.; Devoisselle, J.-M.; Babonneau, F. Implication of water molecules at the silica–ibuprofen interface in silica-based drug delivery systems obtained through incipient wetness impregnation. *J. Phys. Chem. C* **2017**, *121*, 26833–26839. [CrossRef]
155. Tielens, F.; Folliet, N.; Bondaz, L.; Etemovic, S.; Babonneau, F.; Gervais, C.; Azaïs, T. Molecular picture of the adsorption of ibuprofen and benzoic acid on hydrated amorphous silica through dft-d calculations combined with solid-state NMR experiments. *J. Phys. Chem. C* **2017**, *121*, 17339–17347. [CrossRef]
156. Klimavicius, V.; Dagys, L.; Chizhik, V.; Balevicius, V. Cp MAS kinetics study of ionic liquids confined in mesoporous silica: Convergence of non-classical and classical spin coupling models. *Appl. Magn. Res.* **2017**, *48*, 673–685. [CrossRef]
157. Riedel, E.; Janiak, C. *Anorganische Chemie*, 7th ed.; Walter De Gruyter: Berlin, Germany, 2007.

158. Zhang, C.; Lively, R.P.; Zhang, K.; Johnson, J.R.; Karvan, O.; Koros, W.J. Unexpected molecular sieving properties of zeolitic imidazolate framework-8. *J. Phys. Chem. Let.* **2012**, *3*, 2130–2134. [CrossRef]
159. Cejka, J.; Van Bekkum, H.; Corma, A.; Schüth, F. Introduction to Zeolite Science and Practice. In *Studies in Surface Science and Catalysis 168*; Elsevier: New York, NY, USA, 2007.
160. Shantz, D.F.; Fild, C.; Koller, H.; Lobo, R.F. Guest-host interactions in as-made al-zsm-12: Implications for the synthesis of zeolite catalysts. *J. Phy. Chem. B* **1999**, *103*, 10858–10865. [CrossRef]
161. Shantz, D.F.; Lobo, R.F. Guest-host interactions in zeolites as studied by NMR spectroscopy: Implications in synthesis, catalysis and separations. *Top. Catal.* **1999**, *9*, 1–11. [CrossRef]
162. Wang, B.; Côté, A.P.; Furukawa, H.; O'Keeffe, M.; Yaghi, O.M. Colossal cages in zeolitic imidazolate frameworks as selective carbon dioxide reservoirs. *Nature* **2008**, *453*, 207–211. [CrossRef] [PubMed]
163. Kärger, J.; Pfeifer, H. N.M.R. Self-diffusion studies in zeolite science and technology. *Zeolites* **1987**, *7*, 90–107. [CrossRef]
164. Kärger, J.; Vasenkov, S.; Auerbach, S.M. Diffusion in zeolites. In *Handbook of Zeolite Science and Technology*; Auerbach, S.M., Carrado, K.A., Eds.; Marcel Dekker: New York, NY, USA, 2003; pp. 458–560.
165. Nishchenko, A.M.; Kolokolov, D.I.; Gabrienko, A.A.; Stepanov, A.G. Mobility of tert-butyl alcohol in mfi framework type studied by deuterium NMR. *J. Phys. Chem. C* **2012**, *116*, 8956–8963. [CrossRef]
166. Lalowicz, Z.T.; Stoch, G.; Birczynski, A.; Punkkinen, M.; Ylinen, E.E.; Krzystyniak, M.; Gora-Marek, K.; Datka, J. Translational and rotational mobility of methanol-d(4) molecules in nax and nay zeolite cages: A deuteron NMR investigation. *Solid State Nucl. Mag. Res.* **2012**, *45*, 66–74. [CrossRef]
167. Stoch, G.; Ylinen, E.E.; Birczynski, A.; Lalowicz, Z.T.; Gora-Marek, K.; Punkkinen, M. Deuteron spin-lattice relaxation in the presence of an activation energy distribution: Application to methanols in zeolite nax. *Solid State Nucl. Mag. Res.* **2013**, *49*, 33–41. [CrossRef]
168. Arzumanov, S.S.; Kolokolov, D.I.; Freude, D.; Stepanov, A.G. Methane mobility in ag/h-zsm-5 zeolite in the presence of ethene: A view based on PFG 1H MAS NMR analysis of methane diffusivity. *J. Phys. Chem. C* **2015**, *119*, 18481–18486. [CrossRef]
169. Kärger, J.; Caro, J.; Cool, P.; Coppens, M.O.; Jones, D.; Kapteijn, F.; Rodríguez Reinoso, F.; Stöcker, M.; Theodorou, D.; Vansant, E.F.; et al. Benefit of microscopic diffusion measurement for the characterization of nanoporous materials. *Chem. Eng. Technol.* **2009**, *32*, 1494–1511. [CrossRef]
170. Karger, J. Transport phenomena in nanoporous materials. *ChemPhysChem* **2015**, *16*, 24–51. [CrossRef]
171. Naumov, S.; Valiullin, R.; Kärger, J.; Pitchumani, R.; Coppens, M.-O. Tracing pore connectivity and architecture in nanostructured silica SBA-15. *Microporous Mesoporous Mater.* **2008**, *110*, 37–40. [CrossRef]
172. Gutsze, A.; Masierak, W.; Geil, B.; Kruk, D.; Pahlke, H.; Fujara, F. On the problem of field-gradient NMR measurements of intracrystalline diffusion in small crystallites–water in naa zeolites as an example. *Solid State Nucl. Magn. Reson.* **2005**, *28*, 244–249. [CrossRef] [PubMed]
173. Pahlke, H.; Lusceac, S.A.; Geil, B.; Fujara, F. NMR study of local and long range dynamics of adsorbed water in zeolite nay(br). *Z. Phys. Chem.* **2012**, *226*, 1093–1114. [CrossRef]
174. Krylova, E.A.; Shelyapina, M.G.; Nowak, P.; Harańczyk, H.; Chislov, M.; Zvereva, I.A.; Privalov, A.F.; Becher, M.; Vogel, M.; Petranovskii, V. Mobility of water molecules in sodium- and copper-exchanged mordenites: Thermal analysis and 1 h NMR study. *Microporous Mesoporous Mater.* **2018**, *265*, 132–142. [CrossRef]
175. Brodrecht, M.; Breitzke, H.; Gutmann, T.; Buntkowsky, G. Biofunctionalization of nano channels by direct in-pore solid-phase peptide synthesis. *Chem. Eur. J.* **2018**, *24*, 17814–17822. [CrossRef]
176. Brodrecht, M.; Kumari, B.; Breitzke, H.; Gutmann, T.; Buntkowsky, G. Chemically modified silica materials as model systems for the characterization of water-surface interactions. *Z. Phys. Chem.* **2018**, *232*, 1127–1146. [CrossRef]
177. Brodrecht, M.; Kunnari, B.; Thankamony, A.S.S.L.; Breitzke, H.; Gutmann, T.; Buntkowsky, G. Structural insights into peptides bound to the surface of silica nanopores. *Chem. Eur. J.* **2019**, *25*, 5214–5221. [CrossRef]
178. Schottner, S.; Brodrecht, M.; Uhlein, E.; Dietz, C.; Breitzke, H.; Tietze, A.A.; Buntkowsky, G.; Gallei, M. Amine-containing block copolymers for the bottom-up preparation of functional porous membranes. *Macromolecules* **2019**, *52*, 2631–2641. [CrossRef]
179. Grün, M.; Unger, K.K.; Matsumoto, A.; Tsutsumi, K. Novel pathways for the preparation of mesoporous MCM-41 materials: Control of porosity and morphology. *Microporous Mesoporous Mater.* **1999**, *27*, 207–216. [CrossRef]

180. Buntkowsky, G.; Vogel, M.; Winter, R. Properties of hydrogen-bonded liquids at interfaces. *Z. Phys. Chem.* **2018**, *232*, 937–972. [CrossRef]
181. Weigler, M.; Brodrecht, M.; Breitzke, H.; Dietrich, F.; Sattig, M.; Buntkowsky, G.; Vogel, M. $^2$H NMR studies on water dynamics in functionalized mesoporous silica. *Z. Phys. Chem.* **2018**, *232*, 1041–1058. [CrossRef]
182. Richert, R. Dynamics of nanoconfined supercooled liquids. *Annu. Rev. Phys. Chem.* **2011**, *62*, 65–84. [CrossRef] [PubMed]
183. Brodrecht, M.; Klotz, E.; Lederle, C.; Breitzke, H.; Stühn, B.; Vogel, M.; Buntkowsky, G. A combined solid-state NMR, dielectric spectroscopy and calorimetric study of water in lowly hydrated MCM-41 samples. *Z. Phys. Chem.* **2018**, *232*, 1003–1016. [CrossRef]
184. Sumper, M.; Brunner, E. Silica biomineralisation in diatoms: The model organism thalassiosira pseudonana. *Chembiochem* **2008**, *9*, 1187–1194. [CrossRef] [PubMed]
185. Poulsen, N.; Sumper, M.; Kröger, N. Biosilica formation in diatoms: Characterization of native silaffin-2 and its role in silica morphogenesis. *Proc. Natl. Acad. Sci. USA* **2003**, *100*, 12075–12080. [CrossRef]
186. Brunner, E.; Groger, C.; Lutz, K.; Richthammer, P.; Spinde, K.; Sumper, M. Analytical studies of silica biomineralization: Towards an understanding of silica processing by diatoms. *Appl. Microbiol. Biotechnol.* **2009**, *84*, 607–616. [CrossRef]
187. Almqvist, N.; Delamo, Y.; Smith, B.L.; Thomson, N.H.; Bartholdson, A.; Lal, R.; Brzezinski, M.; Hansma, P.K. Micromechanical and structural properties of a pennate diatom investigated by atomic force microscopy. *J. Microsc.* **2001**, *202 Pt 3*, 518–532. [CrossRef]
188. Mann, S. *Biomineralization: Principles and Concepts in Bioinorganic Materials Chemistry*; Oxford University Press: New York, NY, USA, 2001.
189. Nassif, N.; Livage, J. From diatoms to silica-based biohybrids. *Chem. Soc. Rev.* **2011**, *40*, 849–859. [CrossRef]
190. Hartmann, M.; Kostrov, X. Immobilization of enzymes on porous silicas - benefits and challenges. *Chem. Soc. Rev.* **2013**, *42*, 6277–6289. [CrossRef]
191. Zhou, Z.; Piepenbreier, F.; Marthala, V.R.R.; Karbacher, K.; Hartmann, M. Immobilization of lipase in cage-type mesoporous organosilicas via covalent bonding and crosslinking. *Catal. Today* **2014**, *243*, 173–183. [CrossRef]
192. Matlahov, I.; Geiger, Y.; Goobes, G. Trapping rnase a on MCM41 pores: Effects on structure stability, product inhibition and overall enzymatic activity. *Phys. Chem. Chem. Phys.* **2014**, *16*, 9031–9038. [CrossRef]
193. Duer, M.J. The contribution of solid-state NMR spectroscopy to understanding biomineralization: Atomic and molecular structure of bone. *J. Magn. Reson.* **2015**, *253*, 98–110. [CrossRef] [PubMed]
194. Brodrecht, M.; Herr, K.; Bothe, S.; de Oliveira, M.; Gutmann, T.; Buntkowsky, G. Efficient building blocks for solid-phase peptide synthesis of spin labeled peptides for electron paramagnetic resonance and dynamic nuclear polarization applications. *ChemPhysChem* **2019**, *20*, 1475–1487. [CrossRef] [PubMed]
195. Trautmann, C. Micro-and nanoengineering with ion tracks. In *Ion Beams in Nanoscience and Technology*; Hellborg, R., Whitlow., H.J., Eds.; Springer: Berlin, Germany, 2009; pp. 369–387.
196. Kumari, B.; John, D.; Hoffmann, P.; Spende, A.; Toimil-Molares, M.E.; Trautmann, C.; Hess, C.; Ruff, P.; Stark, R.; Schulze, M.; et al. Surface enhanced DNP assisted solid-state NMR of functionalized sio2 coated polycarbonate membranes. *Z. Phys. Chem.* **2018**, *232*, 1173–1186. [CrossRef]
197. Febles, M.; Perez-Hernandez, N.; Perez, C.; Rodriguez, M.L.; Foces-Foces, C.; Roux, M.V.; Morales, E.Q.; Buntkowsky, G.; Limbach, H.-H.; Martin, J.D. Distinct dynamic behaviors of water molecules in hydrated pores. *J. Am. Chem. Soc.* **2006**, *128*, 10008–10009. [CrossRef] [PubMed]
198. Perez-Hernandez, N.; Trung Quan, L.; Febles, M.; Marco, C.; Limbach, H.-H.; Havenith, M.; Perez, C.; Victoria Roux, M.; Perez, R.; Martin, J.D. The mobility of water molecules through hydrated pores. *J. Phys. Chem. C* **2012**, *116*, 9616–9630. [CrossRef]
199. Hassan, J. Analysis of h-2 NMR spectra of water molecules on the surface of nano-silica material MCM-41: Deconvolution of the signal into a lorentzian and a powder pattern line shapes. *Phys. B Condens. Matter* **2012**, *407*, 179–183. [CrossRef]
200. Hassan, J.; Reardon, E.; Peemoeller, H. Correlation between deuterium NMR spectral components and MCM-41 pore surface hydration sites. *Microporous Mesoporous Mater.* **2009**, *122*, 121–127. [CrossRef]
201. Niknam, M.; Liang, J.; Walia, J.; Peemoeller, H. Chemical exchange and spectral coalescence in low-hydration MCM-41 studied by proton NMR. *Microporous Mesoporous Mater.* **2012**, *162*, 136–142. [CrossRef]

202. Weigler, M.; Winter, E.; Kresse, B.; Brodrecht, M.; Buntkowsky, G.; Vogel, M. Static field gradient NMR studies of water diffusion in mesoporous silica. *Phys. Chem. Chem. Phys.* **2020**, *22*, 13989–13998. [CrossRef]

203. Kumari, B.; Brodrecht, M.; Breitzke, H.; Werner, M.; Grunberg, B.; Limbach, H.H.; Forg, S.; Sanjon, E.P.; Drossel, B.; Gutmann, T.; et al. Mixtures of alcohols and water confined in mesoporous silica: A combined solid-state NMR and molecular dynamics simulation study. *J. Phys. Chem. C* **2018**, *122*, 19540–19550. [CrossRef]

204. Kumari, B.; Brodrecht, M.; Gutmann, T.; Breitzke, H.; Buntkowsky, G. Efficient referencing of fslg cpmas hetcor spectra using 2d h-1-h-1 MAS fslg. *Appl. Magn. Res.* **2019**, *50*, 1399–1407. [CrossRef]

205. Edzes, H.T.; Samulski, E.T. Measurement of cross-relaxation effects in proton nmr spin-lattice relaxation of water in biological-systems - hydrated collagen and muscle. *J. Magn. Reson.* **1978**, *31*, 207–229. [CrossRef]

206. Peemoeller, H. NMR spin grouping. *Bull. Magn. Reson* **1989**, *11*, 19–30.

207. Kittaka, S.; Ishimaru, S.; Kuranishi, M.; Matsuda, T.; Yamaguchi, T. Enthalpy and interfacial free energy changes of water capillary condensed in mesoporous silica, MCM-41 and SBA-15. *Phys. Chem. Chem. Phys.* **2006**, *8*, 3223–3231. [CrossRef]

208. Debenedetti, P.G. Supercooled and glassy water. *J. Phys. Condens. Matter* **2003**, *15*, 1669–1726.

209. Mishima, O.; Stanley, H.E. The relationship between liquid, supercooled and glassy water. *Nature* **1998**, *396*, 329–335. [CrossRef]

210. Cerveny, S.; Mallamace, F.; Swenson, J.; Vogel, M.; Xu, L. Confined water as model of supercooled water. *Chem. Rev.* **2016**, *116*, 7608–7625. [CrossRef]

211. Chen, S.H.; Zhang, Y.; Lagi, M.; Chong, S.H.; Baglioni, P.; Mallamace, F. Evidence of dynamic crossover phenomena in water and other glass-forming liquids: Experiments, md simulations and theory. *J. Phys. Condens. Matter* **2009**, *21*, 504102. [CrossRef]

212. Swenson, J.; Cerveny, S. Dynamics of deeply supercooled interfacial water. *J. Phys. Condens. Matter* **2015**, *27*, 033102. [CrossRef]

213. Weigler, M.; Brodrecht, M.; Buntkowsky, G.; Vogel, M. Reorientation of deeply cooled water in mesoporous silica: NMR studies of the pore-size dependence. *J. Phys. Chem. B* **2019**, *123*, 2123–2134. [CrossRef]

214. Sattig, M.; Vogel, M. Dynamic crossovers and stepwise solidification of confined water: A (2)h NMR study. *J. Phys. Chem. Lett.* **2014**, *5*, 174–178. [CrossRef] [PubMed]

215. Lederle, C.; Sattig, M.; Vogel, M. Effects of partial crystallization on the dynamics of water in mesoporous silica. *J. Phys. Chem. C* **2018**, *122*, 15427–15434. [CrossRef]

216. Yao, Y.; Fella, V.; Huang, W.; Zhang, K.A.I.; Landfester, K.; Butt, H.J.; Vogel, M.; Floudas, G. Crystallization and dynamics of water confined in model mesoporous silica particles: Two ice nuclei and two fractions of water. *Langmuir* **2019**, *35*, 5890–5901. [CrossRef] [PubMed]

217. Demuth, D.; Sattig, M.; Steinrücken, E.; Weigler, M.; Vogel, M. 2H NMR studies on the dynamics of pure and mixed hydrogen-bonded liquids in confinement. *Z. Phys. Chem.* **2018**, *232*, 1059–1087. [CrossRef]

218. Miyatou, T.; Ohashi, R.; Ida, T.; Kittaka, S.; Mizuno, M. An NMR study on the mechanisms of freezing and melting of water confined in spherically mesoporous silicas SBA-16. *Phys. Chem. Chem. Phys.* **2016**, *18*, 18555–18562. [CrossRef] [PubMed]

219. Rosenstihl, M.; Kämpf, K.; Klameth, F.; Sattig, M.; Vogel, M. Dynamics of interfacial water. *J. Non Cryst. Solids* **2015**, *407*, 449–458. [CrossRef]

220. Vogel, M. Origins of apparent fragile-to-strong transitions of protein hydration waters. *Phys. Rev. Lett.* **2008**, *101*, 225701. [CrossRef]

221. Lusceac, S.A.; Vogel, M.R.; Herbers, C.R. 2H and 13C NMR studies on the temperature-dependent water and protein dynamics in hydrated elastin, myoglobin and collagen. *Biochim. Biophys. Acta* **2010**, *1804*, 41–48. [CrossRef]

222. Lusceac, S.A.; Rosenstihl, M.; Vogel, M.; Gainaru, C.; Fillmer, A.; Böhmer, R. NMR and dielectric studies of hydrated collagen and elastin: Evidence for a delocalized secondary relaxation. *J. Non Cryst. Solids* **2011**, *357*, 655–663. [CrossRef]

223. Cerveny, S.; Schwartz, G.A.; Bergman, R.; Swenson, J. Glass transition and relaxation processes in supercooled water. *Phys. Rev. Lett.* **2004**, *93*, 245702. [CrossRef]

224. Böhmer, R.; Diezemann, G.; Hinze, G.; Rössler, E. Dynamics of supercooled liquids and glassy solids. *Prog. Nucl. Magn. Reson. Spectrosc.* **2001**, *39*, 191–267. [CrossRef]

225. Fujara, F.; Wefing, S.; Spiess, H.W. Dynamics of molecular reorientations: Analogies between quasielastic neutron scattering and deuteron NMR spin alignment. *J. Chem. Phys.* **1986**, *84*, 4579–4584. [CrossRef]
226. Vogel, M.; Medick, P.; Rössler, E.A. Secondary relaxation processes in molecular glasses studied by nuclear magnetic resonance spectroscopy. *Ann. Rep. NMR Spectrosc.* **2005**, *56*, 231–299.
227. Sjostrom, J.; Swenson, J.; Bergman, R.; Kittaka, S. Investigating hydration dependence of dynamics of confined water: Monolayer, hydration water and maxwell-wagner processes. *J. Chem. Phys.* **2008**, *128*, 154503. [CrossRef]
228. Lusceac, S.A.; Gainaru, C.; Ratzke, D.A.; Graf, M.F.; Vogel, M. Secondary water relaxation in a water/dimethyl sulfoxide mixture revealed by deuteron nuclear magnetic resonance and dielectric spectroscopy. *J. Phy. Chem. B* **2011**, *115*, 11588–11596. [CrossRef]
229. Vogel, M.; Rossler, E. Effects of various types of molecular dynamics on 1d and 2d (2)h NMR studied by random walk simulations. *J. Magn. Reson.* **2000**, *147*, 43–58. [CrossRef]
230. Böhmer, R.; Hinze, G. Reorientations in supercooled glycerol studied by two-dimensional time-domain deuteron nuclear magnetic resonance spectroscopy. *J. Chem. Phys.* **1998**, *109*, 241–248. [CrossRef]
231. Angell, C.A. Insights into phases of liquid water from study of its unusual glass-forming properties. *Science* **2008**, *319*, 582–587. [CrossRef]
232. Amann-Winkel, K.; Böhmer, R.; Fujara, F.; Gainaru, C.; Geil, B.; Loerting, T. Colloquium: Water's controversial glass transitions. *Rev. Mod. Phys.* **2016**, *88*, 011002. [CrossRef]
233. Shephard, J.J.; Salzmann, C.G. Molecular reorientation dynamics govern the glass transitions of the amorphous ices. *J. Phys. Chem. Lett.* **2016**, *7*, 2281–2285. [CrossRef]
234. Arndt, M.; Stannarius, R.; Gorbatschow, W.; Kremer, F. Dielectric investigations of the dynamic glass transition in nanopores. *Phys. Rev. E* **1996**, *54*, 5377–5390. [CrossRef] [PubMed]
235. Elamin, K.; Jansson, H.; Kittaka, S.; Swenson, J. Different behavior of water in confined solutions of high and low solute concentrations. *Phys. Chem. Chem. Phys.* **2013**, *15*, 18437. [CrossRef] [PubMed]
236. Geske, J.; Harrach, M.; Heckmann, L.; Horstmann, R.; Klameth, F.; Müller, N.; Pafong, E.; Wohlfromm, T.; Drossel, B.; Vogel, M. Molecular dynamics simulations of water, silica, and aqueous mixtures in bulk and confinement. *Z. Phys. Chem.* **2018**, *232*, 1187–1225. [CrossRef]
237. Fillmer, A.; Gainaru, C.; Böhmer, R. Broadened dielectric loss spectra and reduced dispersion strength of viscous glycerol in a connective tissue protein. *J. Non Cryst. Solids* **2010**, *356*, 743–746. [CrossRef]
238. Herbers, C.R.; Sauer, D.; Vogel, M. 2H NMR studies of glycerol dynamics in protein matrices. *J. Chem. Phys.* **2012**, *136*, 124511. [CrossRef]
239. Gruenberg, B.; Gruenberg, A.; Limbach, H.-H.; Buntkowsky, G. Melting of naphthalene confined in mesoporous silica MCM-41. *Appl. Magn. Res.* **2013**, *44*, 189–201. [CrossRef]
240. Ebener, M.; Vonfircks, G.; Günther, H. High-resolution solid-state MAS c-13 and h-1-nmr spectra of benzenoid aromatics adsorbed on alumina and silica - successful applications of 1d-pulse and 2d-pulse experiments from liquid-state nmr. *Helv. Chim. Acta* **1991**, *74*, 1296–1304. [CrossRef]
241. Ebener, M.; Francke, V.; Günther, H. Solid state 13C MAS NMR as a tool for the study of reactions between compounds adsorbed on porous materials. *Fresenius' J. Anal. Chem.* **1997**, *357*, 505–507. [CrossRef]
242. von Fircks, G.; Hausmann, H.; Francke, V.; Günther, H. High-resolution solid state 19f and 15N MAS NMR spectra of fluoroaromatics and aromatic nitrogen heterocycles physisorbed on silica and alumina. *J. Org. Chem.* **1997**, *62*, 5074–5079. [CrossRef]
243. Günther, H.; Oepen, S.; Ebener, M.; Francke, V. NMR study of adsorption techniques for organic compounds on silica surfaces. *Magn. Reson. Chem.* **1999**, *37*, 142–146. [CrossRef]
244. Lee, J.A.; Rösner, H.; Corrigan, J.F.; Huang, Y. Phase transitions of naphthalene and its derivatives confined in mesoporous silicas. *J. Phys. Chem. C* **2011**, *115*, 4738–4748. [CrossRef]
245. Azais, T.; Hartmeyer, G.; Quignard, S.; Laurent, G.; Babonneau, F. Solution state NMR techniques applied to solid state samples: Characterization of benzoic acid confined in MCM-41. *J. Phys. Chem. C* **2010**, *114*, 8884–8891. [CrossRef]
246. Shenderovich, I.G.; Buntkowsky, G.; Schreiber, A.; Gedat, E.; Sharif, S.; Albrecht, J.; Golubev, N.S.; Findenegg, G.H.; Limbach, H.H. Pyridine-n-15 - a mobile NMR sensor for surface acidity and surface defects of mesoporous silica. *J. Phys. Chem. B* **2003**, *107*, 11924–11939. [CrossRef]

247. Shenderovich, I.G.; Mauder, D.; Akcakayiran, D.; Buntkowsky, G.; Limbach, H.-H.; Findenegg, G.H. NMR provides checklist of generic properties for atomic-scale models of periodic mesoporous silicas. *J. Phys. Chem. B* **2007**, *111*, 12088–12096. [CrossRef] [PubMed]
248. Gurinov, A.A.; Rozhkova, Y.A.; Zukal, A.; Cejka, J.; Shenderovich, I.G. Mutable lewis and bronsted acidity of aluminated SBA-15 as revealed by NMR of adsorbed pyridine-n-15. *Langmuir* **2011**, *27*, 12115–12123. [CrossRef] [PubMed]
249. Lesnichin, S.B.; Kamdem, N.; Mauder, D.; Denisov, G.S.; Shenderovich, I.G. Studies of adsorption of 2,2'-bipyridyl on the surface of highly regulated silica matrix of the MCM-41 type by means of n-15 NMR spectroscopy. *Russ. J. Gen. Chem.* **2010**, *80*, 2027–2031. [CrossRef]
250. Ip, B.C.K.; Andreeva, D.V.; Buntkowsky, G.; Akcakayiran, D.; Findenegg, G.H.; Shenderovich, I.G. NMR study of proton transfer to strong bases on inner surfaces of MCM-41. *Microporous Mesoporous Mater.* **2010**, *134*, 22–28. [CrossRef]
251. Gedat, E.; Schreiber, A.; Findenegg, G.H.; Shenderovich, I.; Limbach, H.H.; Buntkowsky, G. Stray field gradient NMR reveals effects of hydrogen bonding on diffusion coefficients of pyridine in mesoporous silica. *Magn. Reson. Chem.* **2001**, *39*, 149–157. [CrossRef]
252. Shenderovich, I.; Denisov, G.S. Adduct under field—a qualitative approach to account for solvent effect on hydrogen bonding. *Molecules* **2020**, *25*, 436. [CrossRef]
253. Guo, X.-Y.; Watermann, T.; Sebastiani, D. Local microphase separation of a binary liquid under nanoscale confinement. *J. Phys. Chem. B* **2014**, *118*, 10207–10213. [CrossRef]
254. Hermens, J.L.M.; de Bruijn, J.H.M.; Brooke, D.N. The octanol–water partition coefficient: Strengths and limitations. *Environ. Toxicol. Chem.* **2013**, *32*, 732–733. [CrossRef] [PubMed]
255. Leo, A.; Hansch, C.; Elkins, D. Partition coefficients and their uses. *Chem. Rev.* **1971**, *71*, 525–616. [CrossRef]
256. Harrach, M.F.; Drossel, B.; Winschel, W.; Gutmann, T.; Buntkowsky, G. Mixtures of isobutyric acid and water confined in cylindrical silica nanopores revisited: A combined solid-state NMR and molecular dynamics simulation study. *J. Phys. Chem. C* **2015**, *119*, 28961–28969. [CrossRef]
257. van Rossum, B.J.; Förster, H.; de Groot, H.J.M. High-field and high-speed cp-MAS c-13 NMR heteronuclear dipolar-correlation spectroscopy of solids with frequency-switched lee-goldburg homonuclear decoupling. *J. Magn. Reson.* **1997**, *124*, 516–519. [CrossRef]
258. Sattig, M.; Elamin, K.; Reuhl, M.; Swenson, J.; Vogel, M. Dynamics of dipgme–water mixtures in mesoporous silica. *J. Phys. Chem. C* **2017**, *121*, 6796–6806. [CrossRef]
259. Schmitz, R.; Muller, N.; Ullmann, S.; Vogel, M. A molecular dynamics simulations study on ethylene glycol-water mixtures in mesoporous silica. *J. Chem. Phys.* **2016**, *145*, 104703. [CrossRef]
260. Sauer, D.; Schuster, B.; Rosenstihl, M.; Schneider, S.; Talluto, V.; Walther, T.; Blochowicz, T.; Stuhn, B.; Vogel, M. Dynamics of water-alcohol mixtures: Insights from nuclear magnetic resonance, broadband dielectric spectroscopy, and triplet solvation dynamics. *J. Chem. Phys.* **2014**, *140*, 114503. [CrossRef]
261. Sjostrom, J.; Mattsson, J.; Bergman, R.; Johansson, E.; Josefsson, K.; Svantesson, D.; Swenson, J. Dielectric secondary relaxation of water in aqueous binary glass-formers. *Phys. Chem. Chem. Phys.* **2010**, *12*, 10452–10456. [CrossRef]
262. Wasserscheid, P.; Keim, W. Ionic liquids - new "solutions" for transition metal catalysis. *Angew. Chem. Int. Ed.* **2000**, *39*, 3772–3789. [CrossRef]
263. Hardacre, C.; Holbrey, J.D.; Nieuwenhuyzen, M.; Youngs, T.G.A. Structure and solvation in ionic liquids. *Acc. Chem. Res.* **2007**, *40*, 1146–1155. [CrossRef]
264. Poole, C.F.; Poole, S.K. Extraction of organic compounds with room temperature ionic liquids. *J. Chromatogr. A* **2010**, *1217*, 2268–2286. [CrossRef] [PubMed]
265. Chiappe, C.; Pieraccini, D. Ionic liquids: Solvent properties and organic reactivity. *J. Phys. Org. Chem.* **2005**, *18*, 275–297. [CrossRef]
266. Haumann, M.; Schoenweiz, A.; Breitzke, H.; Buntkowsky, G.; Werner, S.; Szesni, N. Solid-state NMR investigations of supported ionic liquid phase water-gas shift catalysts: Ionic liquid film distribution vs. *Catal. Perform. Chem. Eng. Technol.* **2012**, *35*, 1421–1426.
267. Shylesh, S.; Hanna, D.; Werner, S.; Bell, A.T. Factors influencing the activity, selectivity, and stability of rh-based supported ionic liquid phase (silp) catalysts for hydroformylation of propene. *ACS Catal.* **2012**, *2*, 487–493. [CrossRef]

268. Le Bideau, J.; Gaveau, P.; Bellayer, S.; Neouze, M.A.; Vioux, A. Effect of confinement on ionic liquids dynamics in monolithic silica ionogels: H-1 NMR study. *Phys. Chem. Chem. Phys.* **2007**, *9*, 5419–5422. [CrossRef]
269. Rosa Castillo, M.; Fraile, J.M.; Mayoral, J.A. Structure and dynamics of 1-butyl-3-methylimidazolium hexafluorophosphate phases on silica and laponite clay: From liquid to solid behavior. *Langmuir* **2012**, *28*, 11364–11375. [CrossRef]
270. Waechtler, M.; Sellin, M.; Stark, A.; Akcakayiran, D.; Findenegg, G.; Gruenberg, A.; Breitzke, H.; Buntkowsky, G. H-2 and f-19 solid-state NMR studies of the ionic liquid [c2py][bta]-d(10) confined in mesoporous silica materials. *Phys. Chem. Chem. Phys.* **2010**, *12*, 11371–11379. [CrossRef]
271. Hoffmann, M.M.; Bothe, S.; Brodrecht, M.; Klimavicius, V.; Haro-Mares, N.B.; Gutmann, T.; Buntkowsky, G. Direct and indirect dynamic nuclear polarization transfer observed in mesoporous materials impregnated with nonionic surfactant solutions of polar polarizing agents. *J. Phys. Chem. C* **2020**, *124*, 5145–5156. [CrossRef]

© 2020 by the authors. Licensee MDPI, Basel, Switzerland. This article is an open access article distributed under the terms and conditions of the Creative Commons Attribution (CC BY) license (http://creativecommons.org/licenses/by/4.0/).

MDPI
St. Alban-Anlage 66
4052 Basel
Switzerland
Tel. +41 61 683 77 34
Fax +41 61 302 89 18
www.mdpi.com

*Molecules* Editorial Office
E-mail: molecules@mdpi.com
www.mdpi.com/journal/molecules